UCLA Symposia on Molecular and Cellular Biology, New Series

Series Editor, C. Fred Fox

RECENT TITLES

Volume 74
Growth Regulation of Cancer,
Marc E. Lippman, *Editor*

Volume 75
Steroid Hormone Action, Gordon Ringold, *Editor*

Volume 76
Molecular Biology of Intracellular Protein Sorting and Organelle Assembly,
Ralph A. Bradshaw, Lee McAlister-Henn, and Michael G. Douglas, *Editors*

Volume 77
Signal Transduction in Cytoplasmic Organization and Cell Motility, Peter Satir, John S. Condeelis, and Elias Lazarides, *Editors*

Volume 78
Tumor Progression and Metastasis,
Garth L. Nicolson and Isaiah J. Fidler, *Editors*

Volume 79
Altered Glycosylation in Tumor Cells,
Christopher L. Reading, Senitiroh Hakomori, and Donald M. Marcus, *Editors*

Volume 80
Protein Recognition of Immobilized Ligands,
T.W. Hutchens, *Editor*

Volume 81
Biological and Molecular Aspects of Atrial Factors, Philip Needleman, *Editor*

Volume 82
Oxy-Radicals in Molecular Biology and Pathology, Peter A. Cerutti, Irwin Fridovich, and Joe M. McCord, *Editors*

Volume 83
Mechanisms and Consequences of DNA Damage Processing, Errol C. Friedberg and Philip C. Hanawalt, *Editors*

Volume 84
Technological Advances in Vaccine Development, Laurence Lasky, *Editor*

Volume 85
B Cell Development, Owen N. Witte, Norman R. Klinman, and Maureen C. Howard, *Editors*

Volume 86
Synthetic Peptides: Approaches to Biological Problems, James P. Tam and Emil Thomas Kaiser, *Editors*

Volume 87
Gene Transfer and Gene Therapy, I. Verma, R. Mulligan, and A. Beaudet, *Editors*

Volume 88
Molecular Biology of the Eye: Genes, Vision, and Ocular Disease, Joram Piatigorsky, Toshimichi Shinohara, and Peggy S. Zelenka, *Editors*

Volume 89
Liposomes in the Therapy of Infectious Diseases and Cancer, Gabriel Lopez-Berestein and Isaiah J. Fidler, *Editors*

Volume 90
Cell Biology of Virus Entry, Replication, and Pathogenesis, Richard W. Compans, Ari Helenius, and Michael B.A. Oldstone, *Editors*

Volume 91
Bone Marrow Transplantation: Current Controversies, Robert Peter Gale and Richard E. Champlin, *Editors*

Volume 92
The Molecular Basis of Plant Development, Robert Goldberg, *Editor*

Volume 93
Cellular and Molecular Biology of Muscle Development, Laurence H. Kedes and Frank E. Stockdale, *Editors*

Volume 94
Molecular Biology of RNA, Thomas R. Cech, *Editor*

Volume 95
DNA-Protein Interactions in Transcription, Jay D. Gralla, *Editor*

Volume 96
Stress-Induced Proteins, Mary Lou Pardue, James R. Feramisco, and Susan Lindquist, *Editors*

Volume 97
Molecular Biology of Stress, Shlomo Breznitz and Oren Zinder, *Editors*

Volume 98
Metal Ion Homeostasis: Molecular Biology and Chemistry, Dean H. Hamer and Dennis R. Winge, *Editors*

Volume 99
Human Tumor Antigens and Specific Tumor Therapy, Richard S. Metzgar and Malcolm S. Mitchell, *Editors*

Please contact the publisher for information about previous titles in this series.

UCLA Symposia Board

C. Fred Fox, Ph.D., Director
Professor of Microbiology, University of California, Los Angeles

Charles J. Arntzen, Ph.D.
Director, Plant Science and Microbiology
E.I. du Pont de Nemours and Company

Floyd E. Bloom, M.D.
Director, Preclinical Neurosciences/
Endocrinology
Scripps Clinic and Research Institute

Ralph A. Bradshaw, Ph.D.
Chairman, Department of Biological
Chemistry
University of California, Irvine

Francis J. Bullock, M.D.
Vice President, Research
Schering Corporation

Ronald E. Cape, Ph.D., M.B.A.
Chairman
Cetus Corporation

Ralph E. Christoffersen, Ph.D.
Executive Director of Biotechnology
Upjohn Company

John Cole, Ph.D.
Vice President of Research
and Development
Triton Biosciences

Pedro Cuatrecasas, M.D.
Vice President of Research
Glaxo, Inc.

Mark M. Davis, Ph.D.
Department of Medical Microbiology
Stanford University

J. Eugene Fox, Ph.D.
Vice President, Research
and Development
Miles Laboratories

J. Lawrence Fox, Ph.D.
Vice President, Biotechnology Research
Abbott Laboratories

L. Patrick Gage, Ph.D.
Director of Exploratory Research
Hoffmann-La Roche, Inc.

Gideon Goldstein, M.D., Ph.D.
Vice President, Immunology
Ortho Pharmaceutical Corp.

Ernest G. Jaworski, Ph.D.
Director of Biological Sciences
Monsanto Corp.

Irving S. Johnson, Ph.D.
Vice President of Research
Lilly Research Laboratories

Paul A. Marks, M.D.
President
Sloan-Kettering Memorial Institute

David W. Martin, Jr., M.D.
Vice President of Research
Genentech, Inc.

Hugh O. McDevitt, M.D.
Professor of Medical Microbiology
Stanford University School of Medicine

Dale L. Oxender, Ph.D.
Director, Center for Molecular Genetics
University of Michigan

Mark L. Pearson, Ph.D.
Director of Molecular Biology
E.I. du Pont de Nemours and Company

George Poste, Ph.D.
Vice President and Director of Research
and Development
Smith, Kline and French Laboratories

William Rutter, Ph.D.
Director, Hormone Research Institute
University of California, San Francisco

George A. Somkuti, Ph.D.
Eastern Regional Research Center
USDA-ARS

Donald F. Stelner, M.D.
Professor of Biochemistry
University of Chicago

UCLA Symposia Board membership at the time of the meeting is indicated on the above list.

Synthetic Peptides
Approaches
to
Biological Problems

Synthetic Peptides
Approaches to Biological Problems

Proceedings of a Glaxo — UCLA Colloquium
held at Park City, Utah
January 31 - February 4, 1988

Editors

James P. Tam
Rockefeller University
New York, New York

Emil Thomas Kaiser
Department of Bioorganic Chemistry
Rockefeller University
New York, New York

Alan R. Liss, Inc. • New York

Address all Inquiries to the Publisher
Alan R. Liss, Inc., 41 East 11th Street, New York, NY 10003

Copyright © 1989 Alan R. Liss, Inc.

Printed in the United States of America

Under the conditions stated below the owner of copyright for this book hereby grants permission to users to make photocopy reproductions of any part or all of its contents for personal or internal organizational use, or for personal or internal use of specific clients. This consent is given on the condition that the copier pay the stated per-copy fee through the Copyright Clearance Center, Incorporated, 27 Congress Street, Salem, MA 01970, as listed in the most current issue of "Permissions to Photocopy" (Publisher's Fee List, distributed by CCC, Inc.), for copying beyond that permitted by sections 107 or 108 of the US Copyright Law. This consent does not extend to other kinds of copying, such as copying for general distribution, for advertising or promotional purposes, for creaing new collective works, or for resale.

Library of Congress Cataloging-in-Publication Data

Synthetic peptides.

    (UCLA symposia on molecular and cellular biology; new ser., v. 86)
    Proceedings of the UCLA Colloquium on Synthetic Peptides, Approaches to Biological Problems.
    Includes bibliographies and index.
    1. Peptides—Physiological aspects—Congresses.
2. Peptides—Immunology—Congresses. 3. Antigenic determinants—Congresses. 4. Vaccines—Congresses.
5. Peptide hormones—Congresses. I. Tam, James P.
II. Kaiser, Emil Thomas, 1938– . III. UCLA Colloquium on Synthetic Peptides: Approaches to Biological Problems (1988 : Park City, Utah) IV. Series.
QP552.P4S96 1988   612'.015756   88-32606
ISBN 0-8451-2685-7

To the memory of Emil Thomas Kaiser

Contents

Contributors xiii
Preface
    James P. Tam.. xxi

I. METHODS AND APPLICATIONS

Multiple Antigen Peptide System: A Novel Design for Synthetic Peptide Vaccine and Immunoassay
    James P. Tam 3

Cognitive Features of Continuous Antigenic Determinants
    H. Mario Geysen, Tom J. Mason, and Stuart J. Rodda 19

Synthetic Peptides: Tools for Elucidating Mechanisms of Protein Antigen Processing and Presentation
    John A. Smith 31

Use of Synthetic Peptides in Two Different Approaches to Interfere With Host Cell Penetration by *Trypanosoma cruzi*
    A. Ouaissi, J.P. Defoort, D. Afchain, H. Gras-Masse, J. Cornette, H. Caron, A. Tartar, and A. Capron 43

Glycosylation of Exogenous Peptide Acceptors by Larval Brine Shrimp Microsomes
    Michael N. Horst 51

Photochemical Approaches to the Preparation of LH-RH by Fragment Condensation
    Jean Gauthier, François Bruderlein, Norman Aubry, S. Rakhit, Serge Valois, and Yves Bousquet 63

II. PREDICTION OF PEPTIDE AND PROTEIN STRUCTURES

Insertion of a Water Molecule Into an α-Helix Backbone and Other Modes of Hydration: X-Ray Structure Analysis
    Isabella L. Karle and P. Balaram 71

Weakly Polar Interactions in Proteins
    Dagmar Ringe, Stephen K. Burley, and Gregory A. Petsko 87

Transfer of Information Through Protein/Lipid Interactions
    Charles M. Deber, G. Andrew Woolley, Raisa B. Deber, and Christopher J. Brandl....................................... 97

Conformations of Three Peptides Deduced From Experiments and Molecular Energetics
    Vincent Madison, Ziva Berkovitch-Yellin, David Fry, David Greeley, and Voldemar Toome 109

III. SYNTHETIC PEPTIDE-BASED VACCINES

The Next Generation of Foot-and-Mouth Disease Vaccines
F. Brown... 127

Biological Role of Pre-S Sequences of the Hepatitis B Virus (HBV) Envelope Protein
A. Robert Neurath, Nathan Strick, Stephen B.H. Kent, Karen Parker, Chin Sook Kim, Marc Girard, Harold E. Ralph, and Jay Valinsky......... 143

Studies on the Development of a Synthetic Peptide Vaccine Against Feline Leukemia Virus
Harish P.M. Kumar, C.K. Grant, and John H. Elder 159

Specificity at the Level of Single Amino Acids of Anti-Whole Foot-and-Mouth Disease Virus Subpopulations Present in Polyclonal Anti-Peptide Sera
Rob H. Meloen, Wouter C. Puyk, Hanneke Lankhof, Jaap G. van Bekkum, Adri Thomas, and Wim M.M. Schaaper 171

Construction of Carrier-Free Synthetic Peptide Antigens Capable of Stimulating Boostable Antibody Responses to a 75 kDa Malarial Parasite Protein
Sylvia J. Richman, Thomas Vedvick, Pawan Sharma, Janette Flint, Feroza Ardeshir, Mitchell Gross, Carol Silverman, and Robert T. Reese.... 181

IV. DETERMINATIONS OF ANTIGENIC DOMAINS

Use of Multiple Synthetic Peptide Analogs to Study Structural Features of Peptides Causing T Cell Activation
D.C. Anderson, W. van Schooten, M. Barry, A.A.M. Janson, and René R.P. de Vries... 199

The Use of Synthetic Peptides for the Analysis of the Envelope of Hepatitis B Virus (HBV)
Colin Howard, Helen Stirk, Alan Buckley, Sheila Brown, and Michael Steward 211

Analysis of Topographic Antigenic Determinants on Human Choriogonadotropin Using Synthetic Peptides
Jean-Michel Bidart, Frédéric Troalen, Claude Bohuon, and Dominique H. Bellet... 223

Analysis of the Functional and Antigenic Domains of the Herpes Simplex Virus Major Regulatory Protein α4
Jeff Hubenthal-Voss, Richard A. Houghten, and Bernard Roizman 239

V. BIOACTIVE CONFORMATIONS OF PEPTIDE HORMONES

Mimics of Secondary Structural Elements of Peptides and Proteins
W.F. Huffman, J.F. Callahan, E.E. Codd, D.S. Eggleston, C. Lemieux, K.A. Newlander, P.W. Schiller, D.T. Takata, and R.F. Walker 257

Development of a Bioactive Model of Atrial Natriuretic Factor
R.F. Nutt, S.F. Brady, T.A. Lyle, T.M. Ciccarone, C.D. Colton, W.J. Paleveda, T.M. Williams, G.M. Smith, R.J. Winquist, and D.F. Veber 267

VI. PEPTIDE HORMONES AND GROWTH FACTORS

Synthesis of Insulin-Like Growth Factor I Through Recombinant DNA Techniques and Selective Chemical Cleavage at Tryptophan
Richard DiMarchi, Harlan Long, Janet Epp, Brigitte Schoner, and Rama Belagaje 283

Elucidation of a Novel Gastrin Releasing Peptide Antagonist by Minimal Ligand Analysis
David C. Heimbrook, Mark E. Boyer, Victor M. Garsky, Nancy L. Balishin, David M. Kiefer, Allen Oliff, and Mark W. Riemen 295

Synthesis and Biological Evaluation of Growth Hormone Releasing Factor, Structural Linear and Cyclic Analogs
Edgar P. Heimer, Mushtaq Ahmad, Theodore Lambros, Timothy McGarty, Ching-Tso Wang, Thomas F. Mowles, Sarah Maines, and Arthur M. Felix 309

Synthesis and Biological Properties of a High Specific Activity Radioiodinated, Photolabile Cyclosporine
Roger Tung, Brian Dunlap, Johannes D. Aebi, William Mellon, Arnold E. Ruoho, N. Dhanasekaran, and Daniel H. Rich 321

Index 337

Contributors

Johannes D. Aebi, School of Pharmacy, University of Wisconsin-Madison, Madison, WI 53706 **[321]**

D. Afchain, Centre d'Immunologie et de Biologie Parasitaire, Institut Pasteur, 59045 Lille Cedex, France **[43]**

Mushtaq Ahmad, Department of Peptide Research, Roche Research Center, Hoffmann-La Roche Inc., Nutley, NJ 07110 **[309]**

Dave C. Anderson, Department of Pathobiology, University of Washington, Seattle, WA 98119; present address: Department of Biochemistry, NeoRx Corp., Seattle, WA 98119 **[199]**

Feroza Ardeshir, The Agouron Institute, La Jolla, CA 92037 **[181]**

Norman Aubry, Department of Chemistry, Bio-Méga Inc., Laval, Quebec, Canada H7S 2G5 **[63]**

P. Balaram, Molecular Biophysics Unit, Indian Institute of Science, Bangalore 560 012, India **[71]**

Nancy L. Balishin, Department of Cancer Research, Merck Sharp & Dohme Research Laboratories, West Point, PA 19486 **[295]**

Michael Barry, Department of Pathobiology, University of Washington, Seattle, WA 98119; present address: Department of Biochemistry, NeoRx Corp., Seattle, WA 98119 **[199]**

Rama Belagaje, Department of Molecular Biology, Lilly Resarch Laboratories, Indianapolis, IN 46285 **[283]**

Dominique H. Bellet, Département de Biologie Clinique, Institut Gustave-Roussy, 94805 Villejuif, France **[223]**

Ziva Berkovitch-Yellin, Department of Physical Chemistry, Roche Research Center, Hoffman-La Roche Inc., Nutley, NJ 07110; present address: Department of Structural Chemistry, Weizmann Institute of Science, Rehovot, 76100 Israel **[109]**

Jean-Michel Bidart, Département de Biologie Clinique, Institut Gustave-Roussy, 94805 Villejuif, France **[223]**

Claude Bohuon, Département de Biologie Clinique, Institut Gustave—Roussy, 94805 Villejuif, France **[223]**

Yves Bousquet, Department of Chemistry, Bio-Méga Inc., Laval, Quebec, Canada H7S 2G5 **[63]**

The numbers in brackets are the opening page numbers of the contributors' articles.

Contributors

Mark E. Boyer, Department of Cancer Research, Merck Sharp & Dohme Research Laboratories, West Point, PA 19486 [295]

S. F. Brady, Department of Medicinal Chemistry, Merck Sharp & Dohme Research Laboratories, West Point, PA 19486 [267]

Christopher J. Brandl, Best Department of Medical Research, University of Toronto, Toronto M5S 1A8, Ontario, Canada [97]

Fred Brown, Department of Virology, Wellcome Biotech Ltd., Beckenham, Kent BR3 3BS, England, [127]

Sheila Brown, Department of Medical Microbiology, London School of Hygiene and Tropical Medicine, London WC1E 7HT, England [211]

François Bruderlein, Department of Chemistry, Bio-Méga Inc., Laval, Quebec, Canada H7S 2G5 [63]

Alan Buckley, Department of Medical Microbiology, London School of Hygiene and Tropical Medicine, London WC1E 7HT, England [211]

Stephen K. Burley, Department of Medicine, Brigham & Women's Hospital, Boston, MA 02115 [87]

James F. Callahan, Department of Peptide Chemistry, Smith, Kline & French Laboratories, King of Prussia, PA 19406 [257]

A. Capron, Centre d'Immunologie et de Biologie Parasitaire, Institut Pasteur, 59045 Lille Cedex, France [43]

H. Caron, Centre d'Immunologie et de Biologie Parasitaire, Institut Pasteur, 59045 Lille Cedex, France [43]

T.M. Ciccarone, Department of Medicinal Chemistry, Merck Sharp & Dohme Research Laboratories, West Point, PA 19486 [267]

Ellen E. Codd, Department of Reproductive and Developmental Toxicology, Smith, Kline & French Laboratories, King of Prussia, PA 19406 [257]

C. D. Colton, Department of Medicinal Chemistry, Merck Sharp & Dohme Research Laboratories, West Point, PA 19486 [267]

J. Cornette, Centre d'Immunologie et de Biologie Parasitaire, Institut Pasteur, 59045 Lille Cedex, France [43]

Charles M. Deber, Research Institute, Hospital for Sick Children, Toronto M5G 1X8 and Department of Biochemistry, University of Toronto, Toronto M5S 1A8, Ontario, Canada [97]

Raisa B. Deber, Department of Community Health, University of Toronto, Toronto M5S 1A8, Ontario, Canada [97]

J.P. Defoort, Centre d'Immunologie et de Biologie Parasitaire, Institut Pasteur, 59045 Lille Cedex, France; present address: Service de Chimie des Biomolécules, Institute Pasteur, 59045 Lille Cedex, France [43]

René R. P. de Vries, Department of Immunohematology and Blood Bank, University Hospital of Leiden, Leiden, The Netherlands [199]

N. Dhanasekaran, Department of Pharmacology, University of Wisconsin-Madison, Madison, WI 53706 [321]

Contributors xv

Richard DiMarchi, Department of Biochemistry, Lilly Research Laboratories, Indianapolis, IN 46285 [283]

Brian Dunlap, School of Pharmacy, University of Wisconsin-Madison, Madison, WI 53706 [321]

Drake S. Eggleston, Department of Physical and Structural Chemistry, Smith, Kline & French Laboratories, King of Prussia, PA 19406 [257]

John H. Elder, Department of Molecular Biology, Scripps Clinic and Research Foundation, La Jolla, CA 92037 [159]

Janet Epp, Department of Molecular Biology, Lilly Research Laboratories, Indianapolis, IN 46285 [283]

Arthur M. Felix, Department of Peptide Research, Roche Research Center, Hoffmann-La Roche Inc., Nutley, NJ 07110 [309]

Janette Flint, The Agouron Institute, La Jolla, CA 92037 [181]

David Fry, Department of Physical Chemistry, Roche Research Center, Hoffmann-La Roche Inc., Nutley, NJ 07110 [109]

Victor M. Garsky, Department of Medicinal Chemistry, Merck Sharp & Dohme Laboratories, West Point, PA 19486 [295]

Jean Gauthier, Department of Chemistry, Bio-Méga Inc., Laval, Quebec, Canada H7S 2G5 [63]

H. Mario Geysen, Department of Molecular Immunology, Commonwealth Serum Laboratories, Parkville, Victoria 3052, Australia [19]

Marc Girard, Pasteur Vaccins, 92430 Marnes-la-Coquette, France [143]

C. K. Grant, Pacific Northwest Research Foundation, Seattle, WA 98104 [159]

H. Gras-Masse, Centre d'Immunologie et de Biologie Parasitaire, Institut Pasteur, 59045 Lille Cedex, France [43]

David Greeley, Department of Physical Chemistry, Roche Research Center, Hoffmann-La Roche Inc., Nutley, NJ 07110 [109]

Mitchell Gross, Smith, Kline & French Laboratories, King of Prussia, PA 19406 [181]

David C. Heimbrook, Department of Cancer Research, Merck Sharp & Dohme Research Laboratories, West Point, PA 19486 [295]

Edgar P. Heimer, Department of Peptide Research, Roche Research Center, Hoffmann-La Roche Inc., Nutley, NJ 07110 [309]

Michael N. Horst, Biochemistry Section, Division of Basic Science, School of Medicine, Mercer University, Macon, GA 31027 [51]

Richard A. Houghten, The Department of Molecular Biology, The Scripps Clinic and Research Foundation, San Diego, CA 92037 [239]

Colin Howard, Department of Medical Microbiology, London School of Hygiene and Tropical Medicine, London WC1E 7HT, England [211]

Contributors

Jeff Hubenthal-Voss, The Marjorie B. Kovler Viral Oncology Laboratories, The University of Chicago, Chicago, IL 60637 [239]

William F. Huffman, Department of Peptide Chemistry, Smith, Kline & French Laboratories, King of Prussia, PA 19406 [257]

A. A. M. Janson, Department of Immunohematology and Blood Bank, University Hospital of Leiden, Leiden, The Netherlands [199]

Isabella L. Karle, Laboratory for the Structure of Matter, Naval Research Laboratory, Washington, DC 20375 [71]

Stephen B.H. Kent, Division of Biology, California Institute of Technology, Pasadena, CA 91125 [143]

David M. Kiefer, Department of Cancer Research, Merck Sharp & Dohme Research Laboratories, West Point, PA 19486 [295]

Chin Sook Kim, Division of Biology, California Institute of Technology, Pasadena, CA 91125 [143]

Harish P. M. Kumar, Department of Molecular Biology, Scripps Clinic and Research Foundation, La Jolla, CA 92037 [159]

Theodore Lambros, Department of Peptide Research, Roche Research Center, Hoffmann-La Roche Inc., Nutley, NJ 07110 [309]

Hanneke Lankhof, Central Veterinary Institute, 8200 AV Lelystad, The Netherlands [171]

Carole Lemieux, Laboratory of Chemical Biology and Peptide Research, Clinical Research Institute of Montreal, Montreal, Quebec, Canada H2W IR7 [257]

Harlan Long, Department of Biochemistry, Lilly Research Laboratories, Indianapolis, IN 46285 [283]

T. A. Lyle, Department of Medicinal Chemistry, Merck Sharp & Dohme Research Laboratories, West Point, PA 19486 [267]

Vincent Madison, Department of Physical Chemistry, Roche Research Center, Hoffmann-La Roche Inc., Nutley, NJ 07110 [109]

Sarah Maines, Department of Animal Science Research, Roche Research Center, Hoffmann-La Roche Inc., Nutley, NJ 07110 [309]

Tom J. Mason, Department of Molecular Immunology, Commonwealth Serum Laboratories, Parkville, Victoria 3052, Australia [19]

Timothy McGarty, Department of Peptide Research, Roche Research Center, Hoffmann-La Roche Inc., Nutley, NJ 07110 [309]

William Mellon, School of Pharmacy, University of Wisconsin-Madison, Madison, WI 53706 [321]

Rob H. Meloen, Central Veterinary Institute, 8200 AB Lelystad, The Netherlands [171]

Thomas F. Mowles, Department of Animal Science Research, Roche Research Center, Hoffmann-La Roche Inc., Nutley, NJ 07110 [309]

Contributors   xvii

A. Robert Neurath, The Lindsley F. Kimball Research Institute of the New York Blood Center, New York, NY 10021 **[143]**

Kenneth A. Newlander, Department of Peptide Chemistry, Smith, Kline & French Laboratories, King of Prussia, PA 19406 **[257]**

R. F. Nutt, Department of Medicinal Chemistry, Merck Sharp & Dohme Research Laboratories, West Point, PA 19486 **[267]**

Allen Oliff, Department of Cancer Research, Merck Sharp & Dohme Research Laboratories, West Point, PA 19486 **[295]**

A. Ouaissi, Centre d'Immunologie et de Biologie Parasitaire, Institut Pasteur, 59045 Lille Cedex, France **[43]**

W. J. Paleveda, Department of Medicinal Chemistry, Merck Sharp & Dohme Research Laboratories, West Point, PA 19486 **[267]**

Karen Parker, Division of Biology, California Institute of Technology, Pasadena, CA 91125 **[143]**

Gregory A. Petsko, Department of Chemistry, M.I.T., Cambridge, MA 02139 **[87]**

Wouter C. Puyk, Central Veterinary Institute, 8200 AB Lelystad, The Netherlands **[171]**

S. Rakhit, Department of Chemistry, Bio-Méga Inc., Laval, Quebec, Canada H7S 2G5 **[63]**

Harold E. Ralph, The Lindsley F. Kimball Research Institute of the New York Blood Center, New York, NY 10021 **[143]**

Robert T. Reese, The Agouron Institute, La Jolla, CA 92037 **[181]**

Daniel H. Rich, School of Pharmacy, University of Wisconsin-Madison, Madison, WI 53706 **[321]**

Sylvia J. Richman, The Agouron Institute, La Jolla, CA 92037 **[181]**

Mark W. Riemen, Department of Cancer Research, Merck Sharp & Dohme Research Laboratories, West Point, PA 19486 **[295]**

Dagmar Ringe, Department of Chemistry, M.I.T., Cambridge, MA 02139 **[87]**

Stuart J. Rodda, Department of Molecular Immunology, Commonwealth Serum Laboratories, Parkville, Victoria 3052, Australia **[19]**

Bernard Roizman, The Marjorie B. Kovler Viral Oncology Laboratories, The University of Chicago, Chicago, IL 60637 **[239]**

Arnold E. Ruoho, Department of Pharmacology, University of Wisconsin-Madison, Madison WI 53706 **[321]**

Wim M. M. Schaaper, Central Veterinary Institute, 8200 AB Lelystad, The Netherlands **[171]**

Peter W. Schiller, Laboratory of Chemical Biology and Peptide Research, Clinical Research Institute of Montreal, Montreal, Quebec, Canada H2W 1R7 **[257]**

Brigitte Schoner, Department of Molecular Biology, Lilly Research Laboratories, Indianapolis, IN 46285 **[283]**

Pawan Sharma, The Agouron Institute, La Jolla, CA 92037 **[181]**

Contributors

Carol Silverman, Smith, Kline & French Laboratories, King of Prussia, PA 19406 [181]

G. M. Smith, Department of Medicinal Chemistry, Merck Sharp & Dohme Research Laboratories, West Point, PA 19486 [267]

John A. Smith, Departments of Molecular Biology and Pathology, Massachusetts General Hospital and Department of Pathology, Harvard Medical School, Boston, MA 02114 [31]

Michael Steward, Department of Medical Microbiology, London School of Hygiene and Tropical Medicine, London WC1E 7HT, England [211]

Helen Stirk, Department of Medical Microbiology, London School of Hygiene and Tropical Medicine, London WC1E 7HT, England [211]

Nathan Strick, The Lindsley F. Kimball Research Institute of the New York Blood Center, New York, NY 10021 [143]

Dennis T. Takata, Department of Peptide Chemistry, Smith, Kline & French Laboratories, King of Prussia, PA 19406 [257]

James P. Tam, The Rockefeller University, 1230 York Avenue, New York, NY 10021 [xxi,3]

A. Tartar, Centre d'Immunologie et de Biologie Parasitaire, Institut Pasteur, 59045 Lille Cedex, France; present address: Service de Chimie des Biomolécules, Institut Pasteur, 59045 Lille Cedex, France [43]

Adri Thomas, Central Veterinary Institute, 8200 AB Lelystad, The Netherlands [171]

Voldemar Toome, Department of Physical Chemistry, Roche Research Center, Hoffmann-La Roche Inc., Nutley, NJ 07110 [109]

Frédéric Troalen, Département de Biologie Clinique, Institut Gustave-Roussy, 94805 Villejuif, France [223]

Roger Tung, School of Pharmacy, University of Wisconsin-Madison, Madison, WI 53706 [321]

Jay Valinsky, The Lindsley F. Kimball Research Institute of the New York Blood Center, New York, NY 10021 [143]

Serge Valois, Department of Chemistry, Bio-Méga Inc., Laval, Quebec, Canada H7S 2G5 [63]

Jaap G. van Bekkum, Central Veterinary Institute, 8200 AB Lelystad, The Netherlands [171]

Wim van Schooten, Department of Immunohematology and Blood Bank, University Hospital of Leiden, Leiden, The Netherlands [199]

D. F. Veber, Department of Medicinal Chemistry, Merck Sharp & Dohme Research Laboratories, West Point, PA 19486 [267]

Thomas Vedvick, The Agouron Institute, La Jolla, CA 92037 [181]

Richard F. Walker, Department of Reproductive and Developmental Toxicology, Smith, Kline & French Laboratories, King of Prussia, PA 19406 [257]

Ching-Tso Wang, Department of Peptide Research, Roche Research Center, Hoffmann-La Roche Inc., Nutley, NJ 07110 **[309]**

T. M. Williams, Department of Medicinal Chemistry, Merck Sharp & Dohme Research Laboratories, West Point, PA 19486 **[267]**

R.J. Winquist, Department of Medicinal Chemistry, Merck Sharp & Dohme Research Laboratories, West Point, PA 19486 **[267]**

G. Andrew Woolley, Research Institute, Hospital for Sick Children, Toronto M5G 1XB and Department of Biochemistry, University of Toronto, Toronto M5S 1A8, Ontario, Canada **[97]**

Preface

The explosive growth in applications of synthetic peptides to the elucidation of biological problems provided the impetus to organize the UCLA Colloquium on **Synthetic Peptides: Approaches to Biological Problems**, held at Park City, Utah, January 31–February 4, 1988. Synthetic peptides, which now are widely accessible by the solid-phase synthesis method developed by Professor Bruce Merrifield 25 years ago, have become powerful tools with which to define the structure and functions of such biological activities as antigen–antibody, hormone–receptor, substrate–enzyme, and protein–DNA interactions. It is anticipated that, as the availability of amino acid sequences derived from genomic DNA increases, the understanding of the use of synthetic peptides—their syntheses, structures, and applications—will facilitate the approach to solving many biological problems.

The importance of peptides and synthetic chemistry on synthetic vaccines, mapping of immunodominant epitopes, and understanding of antigen–antibody interactions was reflected in the decision to hold this conference concurrently with the UCLA Symposium on **Technological Advances in Vaccine Development**. In his Keynote Address on Catalytic Antibodies, Dr. Richard Lerner outlined the influence of synthetic chemistry on biological problems. Approximately half of the contributions to this volume involve work related to synthetic peptides in the field of immunology.

This proceedings volume, **Synthetic Peptides: Approaches to Biological Problems**, includes 25 articles in six sections, just as they appeared in the conference program: methods and applications, prediction of peptide and protein structures, synthetic peptide-based vaccines, determinations of antigenic domains, bioactive conformations of peptide hormones, and peptide hormones and growth factors.

During the preparation of this volume, I was greatly saddened by the untimely passing of my co-organizer for this meeting, Professor Tom Kaiser, at the age of 50, on July 18, 1988. Tom will be remembered as a valued colleague, an energetic scientist, and for his brilliant work on the structural bases of biological activities of peptides and proteins.

We gratefully acknowledge Glaxo, Inc., for sponsorship of this meeting. Additional support was received from Smith Kline Beckman; Pharmacia LKB Biotechnology, Inc.; Merck Sharp & Dohme Research Laboratories; Applied Biosystems, Inc.; and O.C.S. Laboratories. In addition, we wish to thank the UCLA Symposia staff, especially Robin Yeaton and Betty Handy, for valuable and patient assistance.

James P. Tam

I. METHODS AND APPLICATIONS

MULTIPLE ANTIGEN PEPTIDE SYSTEM: A NOVEL DESIGN FOR SYNTHETIC PEPTIDE VACCINE AND IMMUNOASSAY[1]

James P. Tam

The Rockefeller University, 1230 York Ave., New York, N.Y.10021

ABSTRACT A novel approach to the synthesis of a dispersed, oligomeric, synthetic peptide system known as multiple peptide system (MAP) suitable for synthetic vaccine and immunoassay is described. A prototype of MAP contains 7 lysyl residues as an inner core matrix and 8 copies of peptide antigens as an outer surface layer this design would generally give a chemically defined macromolecule with a molecular weight exceeding 10,000 daltons. The whole MAP, core and peptide antigens, is synthesized directly and as a single unit by the solid-phase method which allows purification and characterization as an unambiguous product. Fourteen MAPs with peptide antigens, varying from 7 to 21 amino acids, were synthesized and tested in animals. All 14 MAPs were found to be immunogenic in rabbits and mice, and eleven of which elicited specific antibodies that reacted with their corresponding native proteins. MAPs containing peptides were also found to be superior as antigens for solid-phase immunoassays than their monomeric counterpart because they greatly increased coating efficiency to solid-surfaces and sensitivity for detection. More importantly, MAPs as antigens provided greatly enhanced reproducibility that was absent in immunoassays using linear, monomeric peptide antigens. Thus, the MAP system provided a general but chemically unambiguous approach for preparing synthetic peptide-based vaccine and for peptide-based antigenic substrates for solid-phase immunoassays[2].

INTRODUCTION

One of the most effective means to combat infectious diseases is through vaccines. The peptide-based vaccines provide an alternative approach to the conventional methods with the potential advantages of safety, directness, and stability. Although synthetic peptide vaccines have not been clinically approved, their feasibilities have been demonstrated in laboratories (1-4). Furthermore, peptide-based antibodies capable of reactive with their cognate sequences are useful laboratory reagents for many biological and immunological studies. However, several obstacles remain in the development of peptide-based vaccines and the foremost is the requirement of a carrier system which, at present, is both ambiguous in the structure-relationship between antigens and the carrier and undesirable in the immunological responses by host animals. In this paper, we present a novel system known as multiple antigen peptide system suitable for the preparation of peptide-based antibodies for vaccines and for laboratory uses.

The MAP system differs from the conventional approach of preparing antipeptide antibodies, which requires the conjugation of a peptide to a known protein or synthetic polymer carrier, and which usually results in a chemical undefined entity of an antigen-carrier system (1-3). The MAP system is designed specifically to avoid such a drawback. Furthermore, MAP is designed so that the size of the carrier is small when compared to the antigen peptides. A prototype design for MAP, described in this paper, would contain a small peptidyl core matrix bearing radially branching lysyl residues to which eight synthetic peptide antigens are attached. Thus, the inner core matrix is small, usually containing seven lysines in our prototype model of octabranch-

---

[1]This work was supported by PHS Grant number CA36544, awarded by the National Cancer Institute, DHHS and by Agency of International Development.

[2] Abbreviations follow the tentative rules of the IUPAC-IUB commission on Biochemical Nomenclature, published in J. Biol. Chem. (1972) 247,979-982. Others: Boc, tertbutyloxycarbonyl, BOP, benzotriazolyloxytris(dimethylamino) phosphonium hexaflurophosphate BrZ, 2-bromobenzyloxycarbonyl, Bu^t, tertbutyl Bzl, benzyl, ClZ, 2-chlorobenzyloxycarbonyl, DCC, dicyclohexylcarbodiimide, DIEA, diisopropylethylamine, DMF, N,N-dimethylformamide, DMS, dimethylsulfide Dnp, dinitrophenyl, HF, hydrogen fluoride HPLC, high-performance liquid chromatography, HOBt, 1-hydoxybenzotriazole, TGFα, transforming growth factor type α, TFA, trifluoroacetic acid, TFMSA, trifluoromethanesulfonic acid, Tos, tosyl.

ing MAPs and accounting less than 10% of the total molecular weight of the whole MAP molecules. In contrast, the outer core matrix composing of the synthetic peptide antigens are dense and accounting for more than 90% of the whole MAP system. We have tested that 14 examples of MAPS using the octabranching approach and all are immunogenic A majority of them are capable of recognizing their respective native sequences. More importantly, the whole MAP, core and peptide antigens are synthesized in a deliberate manner and in a chemically unambigous way. In addition, because of the structural multivalency, MAPS are particularly suitable as antigens for immunoassays and offer superior sensitivity and reproducibility than their monomeric counterparts.

METHODS

General Methods for the Synthesis of MAP

Either manual approach or with the aid of an automated synthesizer could be used for the synthesis of MAP and its attendant peptide antigens. Methods common to peptide chemistry are generally applicable to the synthesis of MAPs. For the synthesis of an octabranched matrix core with peptide antigen attached, it was carried out by a manual stepwise solid-phase procedure (5) on Boc-βAla-OCH$_2$-Pam resin (or Boc-Gly-Pam resin, 6) with a typical scale of 0.3 g of resin (0.03 mmol of βAla or Gly). The synthesis of the first and every subsequent level of the carrier-core was achieved using a 4 molar excess of preformed symmetrical anhydride of Boc-Lys(Boc) (0.12, 0.24, 0.48, and 0.96 mmol consecutively) in dimethylformamide (DMF, 12 ml/g resin) and followed by a second coupling via dicyclohexylcarbodiimide (DCC) alone in CH$_2$Cl$_2$ to give the octabranched core matrix containing eight functional amino groups. The protecting groups for the synthesis of the peptide antigens were tertbutyloxycarboxyl (Boc) group for the α-amino terminus and benzyl alcohol derivatives for most side chain amino acids. For all residues except Arg, Asn, Gln, and Gly, a 1-hr coupling, monitored by a quantitative ninhydrin test (7) was done with the preformed symmetrical anhydride in CH$_2$Cl$_2$. A recoupling was performed in DMF and a third coupling (if needed) in N-methylpyrrolidone (NMP) at 50°C (8). The coupling of Boc-Asn and Boc-Gln was mediated by the preformed 1-hydroxybenzotriazole ester in DMF. Boc-Gly and Boc-Arg were coupled respectively with water-soluble DCC alone to avoid the risk of formation of dipeptide and lactam. To eliminate the polycationic amino groups, which give highly charged macromolecules, the peptide chains were capped on their α-amino groups by acetylation in acetic anhydride/DMF (1 mmol) containing 0.1 mmol of N,N-dimethylaminopyridine at the completion of the

MAP. The deprotection process was initiated with the removal of the dinitrophenyl protecting group of His(Dnp) with 1 M thiophenol in DMF for 8 h (3 times and at 50° C if necessary to complete the reaction). The branched peptide-oligolysine was removed from the crosslinked polystyrene resin support with the low/high-HF method or the low-high TFMSA method (9) of cleavage to give the crude MAP (85 to 93% cleavage yield). The crude peptide and resin were then washed with cold ether-mercaptoethanol (99:1, v/v, 30 ml) to remove p-thiocresol and p-cresol and the peptide was extracted into 100 ml of 8 M urea, 0.2 M dithiothreitol in 0.1 M Tris.HCl buffer, pH 8.0. To remove all the remaining aromatic byproducts generated in the cleavage step, the peptide was dialyzed in Spectra Por 6, M.W. cutoff 1,000 tubing by equilibration in a deaerated and N_2-purged solution containing 8 M urea, 0.1 M NH_4HCO_3-$(NH_4)_2CO_3$, pH 8.0 with 0.1 M mercaptoethanol 0°C for 24 h. The dialysis was then continued in 8 M and then in 2 M urea, all in 0.1 M NH_4HCO_3-$(NH_4)_2CO_3$ buffer, pH 8.0 for 12 h and then sequentially in H_2O and 1 M HOAc to remove all the urea. The lyophilized MAP was then purified batchwise by high performance gel-permeation or ion-exchange chromatography. All of the purified materials gave satisfactory amino acid analyses.

Immunization Procedure

Rabbits (New Zealand white, two for each antigen) were immunized by subcutaneous injection (0.5 ml) of the MAP (1 mg in 1 ml phosphate buffer saline) in complete Freund's adjuvant (1:1) on day 0 and in incomplete Freund's adjuvant (1:1) on day 21 and 42 and were bled on day 49. Inbred 6- to 8-week old mice were immunized in the footpad with 80μg of MAP in complete Freund's adjuvant (1:1) on day 0 and in incomplete Freund's adjuvant at a 3 week interval (total 2 to 4 times) and bled a week after the last boosting. The antisera were used without any purification.

Assays

An enzyme-linked immunoabsorbent assay (ELISA) was used to test all antisera for their ability to react with the MAP used for immunization. Peptide antigen (0.5 μg per well) in carbonate-bicarbonate buffer (pH 9.0) was incubated at 4° C overnight in a 96-well microtiter plate. Rabbit or mouse antiserum (serially diluted in 0.01 M phosphate-buffered saline) was incubated with the microtiter plate-bound antigen for 2 h at 20°C and then washed with PBS. Goat anti-rabbit or anti-mouse IgG horseradish peroxidase conjugate was then added and incubated for an additional hour. After washing with PBS, the bound conjugate was reacted with chromogen (ortho-dianisidine dihydrochloride) at 1 mg/ml in 0.01 M phosphate buffer, pH 5.95 for 0.5 h and the absorbance of each well was determined with a micro-ELISA reader.

Solid phase immunodiometric assay (IRMA) was performed as follows: Flexible polyvinyl chloride microtiter plates (Becton Dickinson Labware) were incubated with 25μl of either syntheic peptide (monomeric form) or MAPS (oligomeric forms) dissolved in PBS. After overnight incubation at room temperature (and all subsequent operations were performed at room temperature), the wells were washed with PBS and then incubated with PBS containing 1% bovine serum albumin (PBS-BSA). After incubation of 25μl of an appropriate antiserum (diluted at 10^3 to 10^4 fold) or purified monoclonal antibody at suitable dilution for 1 hr and well washed with PBS-BSA containing 0.05% of Tween-20, 25μl of ^{125}I-labelled affinity purified goat antimouse immunoglobulins (5×10^4 cpm, Kirkegaard and Perry Lab) was placed in each well. After incubation for 1 hr, the wells were washed three times with PBS-BSA-Tween, dries and counted in a gamma counter.

RESULTS

Synthesis of MAP

The MAP comprises of essentially two structural features: (i) an inner low-molecular weight lysyl core and (ii) an outer surface layer of acetylated synthetic peptide. Both features are synthesized as a single operation on a polymeric support by the conventional stepwise solid-phase method (5). For the construction of the branching low-molecular weight lysyl core, the trifunctional Boc-Lys(Boc) was found to be suitable since the matrix branching could be effected through the N^α- and N^ε- amino groups of lysine. Thus, sequential propagation of Boc-Lys(Boc) generated 2^n amino ends that peptide antigens could be attached. In general, three generations (or levels) of Boc-Lys(Boc) giving eight reactive amino ends would be satisfactory. The octavalent lysyl core was then continued for the peptide synthesis of antigens (now eight copies) on the lysyl core to give a macromolecular MAP-antigen system. A schematic appearance of MAP containing peptide antigens is shown in Fig. 1. For the synthesis of MAP and its peptide antigen, a continuation of the Boc-benzyl protecting group strategy (Fig. 2) and a Boc-amino acid resin-bound benzyl ester or benzhydrylamide was used. Both types of resins are available through commercial sources. Boc-aminoacyl-OCH$_2$-Pam-resin (6) has an advantage of being stable to the repetitive acidolytic cleavage steps during the synthesis and which avoids premature cleavage of the peptide from the resin. Such an advantage is also being shared by the benzhydrylamine resin and has been used successfully in the preparation of the MAP system. Irrespective of the type of resin being used, it is important to use "low-loading" resins to start the synthesis. Conventional loading of a resin (i.e. the substitution of the amino acid per g. of resin) is 0.3 to 1 mmol/g and is generally too high for the preparation of MAP because the peptide content increases geometrically with each addition of the Boc-Lys(Boc). The synthesis would be difficult with the high loading resins. The preparation of a proper range of the resin loading can be achieved in several ways. The direct and reccomended method is to prepare a resin with the controlled substitution in the range of 0.1 mmol as described from the literature (5-9). An indirect method is to couple a Boc-amino acid to a resin in a limited but known quantity of 0.1 mmol/g to the commerical prepared resin that has a loading of 0.3 to 1 mmol/g and then acetylated the remaining and unreacted amino groups via acetic anhydride.

Technical Comments Concerning the Synthesis of MAP

Several comments concerning the synthesis of MAP should be emphasized. Boc-Lys(Boc) is obtained commercially as a dicyclohexylamine salt (Boc-Lys(Boc).DCHA salt). To couple the Boc-Lys(Boc) via the DCC-mediated methods to the resin support to give the octabranching lysine core, the DCHA salt has to be converted first to the Boc-Lys(Boc) carboxylic acid through acidification to pH 2, extracted into the organic phase, and then dried before use. Recently, a new coupling reagent, BOP (19), is commercially available and can be used for the coupling reaction with Boc-amino acid the DCHA salt. Thus, a conversion step of the

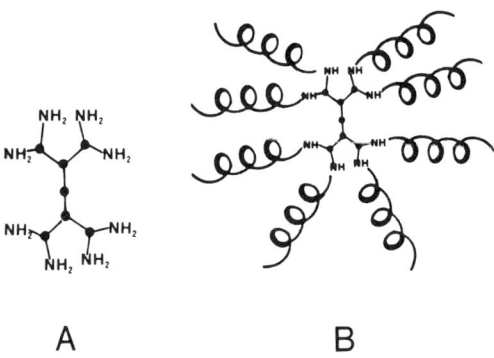

A   B

Fig. 1. Schematic representations of MAP, (A), the lysyl core (7 residues) of the octabranching MAP, (B) Map with peptide antigens.

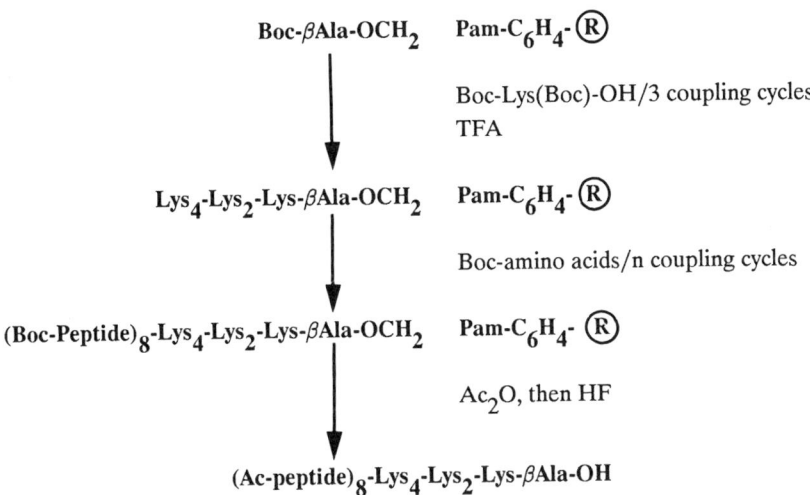

Fig. 2. A synthetic scheme for the synthesis of a MAP by the stepwise solid phase method.

Boc-Lys(Boc) DCHA is not required, and it can be used directly with the BOP reagent to give the octabranching lysine core without conversion to the free carboxylic acid.

To monitor the increase loading of Boc-Lys(Boc) on the resin support, by the quantitative ninhydrin test (7), a correction factor of 1.5 should be used since the ninhydrin test usually gives a lower value then the actual value obtained by amino acid analysis. This is largely due to the color yield difference between the N^α and the N^ε amines of the lysine.

We have found that the coupling efficiency is higher in DMF or NMP (8) than CH_2Cl_2. Usually, the active ester using either symmetrical anhydride or preactivated HOBt ester is the method of choice. Synthetic errors due to deletion peptides would be amplified and difficult to correct by conventional purification methods and the coupling efficiency should be closely monitored (7). The cleavage of MAP from resins should be conducted under the mildest and least destructive method. The low-HF or low TFMSA would be a suitable choice (9).

The MAP product after cleavage from the resin support was usually purified by extensive dislysis in a basic but strongly denaturing condition containing 8 M urea and mercaptoethanol to remove the aromatic additives of the cleavage reactions, e.g. p-cresol and thiocresol, which tended to strongly adhere to the MAP products. In addition, the base treatment would also have the beneficial effects of converting any strong acid catalyzed O-acyl rearrangment product of peptides to the serine or threoninyl containing N-acyl peptides and residual level of Met(O) to Met. The crude MAPs could be further purified by gel-permeation or ion-exchange chromatography; however, in most cases MAPs were used for immunization directly without further purifications.

MAP for Polyclonal Antibodies

Fourteen different MAPs containing peptide antigens varying from 7 to 21 amino acids using the octabranching models of MAP systems for eliciting antibody responses were tested in animals. All fourteen MAPs produced high-titered antibodies responses to the peptide antigens (Table 1). In these 14 examples of MAPs, antipeptide antibodies were elicited to cognate sequences of a diverse group of proteins including regulatory proteins (e.g. GTP α subunit, 11,17), tyrosine specific protein kinases (13,14), receptors (15), malaria surface proteins (16), growth factors (18), viral and plant proteins. Of the 14 MAP peptides 9 are hydrophilic but 5 others (AA-12, AA-15, IA-16, NP-16, FA-21) are relatively hydrophobic (see Table 1). Our results showed 11 of 14 of polyclonal antibodies using the prototype octabranching MAP model produced antisera that would recognize the cognate proteins by either immunoprecipitation of the labeled native proteins or by immunoblotting experiments with the denatured proteins. Although a direct comparison with other methods could not be made at this time, our results showed that the MAP approach would produce a high probability of producing antibodies that would recognize the cognate proteins.

In a comparative study, four different models of MAPs containing di-, tetra-, octa-, and hexadeca-branching lysyl matices were studies to determine the effect of branching on antibody responses. Our preliminary results showed that the octa-branching MAP appeared to be optimal since the hexadecabranching MAP with 16 copies of peptides showed no significant increase in antibody titer responses. Because of the ease of manipulation and purification (since octabranching MAPs with peptides of 10-15 residues often are in the range of $M_r > 10,000$ but $< 15,000$), we adopted the octabranching MAPs as a prototype design. It is also necessary to point out that it is likely that other designs of MAP with different arrangements and various degrees of branching may surpass efficacy of the present prototype. Nevertheless, the present prototype appears quite efficient for our purposes.

MAP for Monoclonal Antibodies

MAP containing a 14-residue peptide (FA-14) derived from human T-cell antigen receptor β-chain constant region gave monoclonal antibodies reacting with the intact β-chain (10). Mapping of the antigenic determinants of FA-14 using these monoclonal antibodies showed that the majority recognized the amino terminal residues of both the linear, monomeric as well as the octameric (MAP) form while no monoclonal antibodies (24 clones) to the heptalysyl core of MAP was found. Similarly, we had consistently found this was true in both polyclonal and monoclonal antibodies. These results showed that the heptalysyl core was small and would not produce any background immunological responses.

MAP in Solid-Phase Immunoassays

Synthetic peptides as antigens have been used extensively in recent years in immunoassays, particularly in the heterogenous immunoassays such as solid-phase radioimmunoassay (RIA) or enzyme-linked immunoabsorbent assay (ELISA), which require the binding of synthetic peptides on various plastic surfaces. However, synthetic peptides are not always reliable in such heterogenous immunoassays due to their low coating efficiency and poor binding to plastic surfaces. Such a difficulty was encountered in the mapping of immunodominant epitopes of the malaria CS proteins for the development of anti-peptide vaccines. In order to develop a reliable method that utilizes synthetic peptides generated by the solid-phase peptide method for solid-phase immunoassays, we tested the MAP systems as antigens to improve the reliability of binding efficiency to solid surfaces.

For comparison, we used a model sequence from the CS protein of the human malaria parasite, Plasmodium *falciparum* to test the efficacy of the monomeric synthetic peptide against the MAP approach. The antigen (NP-12), a tetrapeptide of Asn-Ala-Asn-Pro tandemly repeating three times is the immunodominant epitope of the sporozoite stage of a human malaria. One advantage of using the malaria protein sequences as a model is that polyclonal and monoclonal antibodies against the CS protein, sporozoite as well as the synthetic peptides corresponding to the immunodominant epitopes are available. The synthetic peptides, the linear monomers and their corresponding MAP (MAP-NP-12) with eight copies of the same linear peptides attached, were synthesized by the solid phase peptide synthesis method.

Table 1. Immunological Responses of the Octameric Branching MAPs as Antigens

Peptide	No. of amino acids	Protein source	Immunized animal	Half-max. response[1] log 10	Reactive to native protein[2]
1. FE-7	7	blood protein[3]	rabbit	3.0	+ (A)
2. DV-9	9	GTP protein[4]	rabbit	4.0	+ (B)
3. IG-11	11	tyrosine kinase (src)[5]	rabbit	4.6	+ (A)
4. VS-11	11	growth factor (TGFα)[6]	rabbit	4.1	+ (B)
5. AA-12	12	viral protein[7]	mouse	2.8	N.D.
6. YP-13	13	tyrosin kinase (ros)[8]	rabbit	5.5	- (A)
7. FA-14	14	T-cell receptor[9]	mouse	3.6	+ (A)
8. AA-15	15	EGF receptor[10]	rabbit	3.8	- (A)
9. AG-15	15	plant protein[11]	rabbit	3.4	+ (B)
10. IA-16	16	viral protein[12]	mouse	4.2	N.D.
11. NP-16	16	malaria protein[13]	mouse	2.2	+ (B)
12. YI-16	16	plant protein[14]	rabbit	5.2	N.D.
13. SL-17	17	EGF-receptor[15]	rabbit	3.8	+ (B)
14. FA-21	21	viral protein[16]	mouse	3.7	N.D.

1. Half-maximal response of the antiserum in the dilution vs. absorbance curve in ELISA. 2. Detection by either immunoprecipitation of the labeled protein and NaDodSO$_4$ electrophoresis of the precipitate (A) or by immunoblotting experiment (B). 3. Sequence of FE-7: FQYHSKE (single letter code for amino acid is used) and a triglycine linker with the core MAP. 4. DV-9: DSISAAKDV. 5. IG-11: IEDNEYTARQG. 6. VS-11: VVSHFNDCPDS. 7. AA-12: ALGVATSAQITA. 8. YP-13: YIQHKLQEIRHSP. 9. FA-14: FEPSEAEISHTQKA. 10. AA-15: AKPNGIFKGSTAENA. 11. AG-15: APERGAYGQQHGTG. 12. IA-16: IGTIALGVATSAQITA. 13. NP-16: NANPNANPNANPNANP. 14. YI-16: YASAYGDNEIHAKENI. 15. SL-17: SNKLTQLGTFEDHFLSL. 16. FA-21: FFGAVIGTIALGVATSAQITA.

The suitability of MAP-NP-12 as a model was tested with two different monoclonal antibodies: an antibody against the native CS protein (2A-10) and another antibody raised against the monomeric synthetic peptide conjugated to tetanus toxoid as a carrier. When MAP-NP-12 was used as a coating antigen in the solid phase IRMA,

Table 2. Comparison of NP-12 and MAP-NP-12 in IRMA

Antigen	Response (cpm x 10^{-3})				
	30	7.5	1.8	0.46	0.11
			(in µg/ml)		
MAP-NP-12	10.5	10.7	11.2	10.2	9.5
NP-12	0.5	0.4	0.4	0.4	0.4
NP-12-BSA conjugated[1]	10	9.5	9.5	7.4	3.8

1. NP-12 conjugated to bovine serum albumin

MAP-NP-12 reacted with both monoclonal antibodies and was nonreactive towards the control normal sera. The octabraching MAP containing eight copies of NP-12 peptides were then compared with the monomeric NP-12 as coating antigen in the solid phase IRMA. When the monomeric peptide of NP-12 was used, there was essentially no reactivity (background level with unrelated substrates or blank control) with the monoclonal antibody 2A11 even at a concentration of 30µg/ml. (Table 2). The lack of reactivity was contributed by the poor binding of the monomeric peptide to the plastic surface since the reactivity was restored when NP-12 was conjugated to bovine serum albumin by glutaldehyde. The lack of reactivity of the monomeric form was also observed with other monoclonal and polyclonal antibodies against the CS protein of P. falciparum (data not shown). However, the octameric MAP-NP-12 showed strong and essentially undiminished reactivity even at a concentration of 0.11 µg/ml. Thus, the MAP form of NP-12 showed strong reactivity at least 1000 fold lower concentration than the monomer.

Comparison of MAP and Monomer in ELISA

The efficacy of MAP was compared with its monomeric counterpart in another solid-phase immunoassay, the enzyme-linked immunoabsorbent assay (ELISA). We prepared five polyclonal antibodies (Pabs: Pab-73 to Pab-75 and Pab-11 in Table 3) and four monoclonal antibodies (Mabs: Mab-A13.5, Mab-B6.1, Mab-A1.5 and Mab-A9.10 in Table 3) against the native and biologically active transforming growth factor type α (18). At the same time, polyclonal antisera from two rabbits against an 11-residue peptide from the amino-terminus of TGFα in the MAP octabranching model were also prepared (anti-MAP-VS-11 sera 80 and 81). An unrelated antiserum from a 17-residue peptide, DD-17, was used as a control.

Table 3. Comparison of Reactivities of the Octameric (MAP-VS-11), Monomeric (VS-11) Peptide Antigens Related to TGFα in ELISA

Antibody[1]	Antigen (O.D. 405 nm)			
	VS-11[2] (100μg/ml)	MAP-VS-11[3] (1μg/ml)	TGFα (0.1μg/ml)	DD-17 (10μg/ml)
Anti-DD17	0.10	0.11	0.09	1.73
Pab-73	0.34	2.52	2.27	0.12
Pab-74	0.18	1.79	1.48	0.11
Pab-75	0.29	1.81	2.07	0.10
Pab-76	0.18	2.04	1.78	0.10
Pab-11	0.24	2.20	2.28	0.11
Anti-MAP-VS-11(80)	0.74	2.63	1.43	0.11
Anti-MAP-VS-11(81)	0.83	2.74	1.52	0.12
Mab-A13.5	0.11	1.37	0.37	0.09
Mab-B6.1	0.22	0.13	1.25	0.11
Mab-A1.5	0.11	0.12	1.25	0.11
Mab-A9.10	0.12	0.11	1.37	0.10

1. Anti-DD17 was an unrelated antibody derived from a 17-residue peptide, DD-17 and served as a control. Pab-73, Pab-74, Pab-75, Pab-76, and Pab-11 (polyclonal antibodies) were antisera against the native TGFα from five different rabbits.
Anti-MAP-VS-11 (80) and anti-MAP-VS-11 (81) were antisera against MAP-VS-11 (the amino terminal 11-residue peptide of TGFα, VS-11 using an octabranching MAP system) from two different rabbits. Mab-A13.5, Mab-B6.1, Mab-A1.5 and Mab-A9.10 (monclonal antibodies) were monoclonal antibodies against the native TGFα.
2. VS-11, sequence see table 1.
3. The octabranching MAP of VS-11.

Using the panel of antibodies from TGFα, the efficacies of three antigens, VS-11 (the linear monomer), MAP-VS-11 (the octabranching MAP form of VS-11) and TGFα were compared in ELISA (Table 3). At a concentration of 100µg/ml, the linear monomeric peptide, VS-11, gave a moderate color yield (0.8 O.D) against the anti-MAP-VS-11 antibodies, a weak color yield (average 0.25 O.D.) against the Pabs derived from native TGFα, and a background reading (average 0.12 O.D., except a weak response from Mab-B6.1) against the Mabs. lower than VS-11 in the same series of experiments, gave strong color yield (average >2.0 O.D.) against both the anti-MAP-VS-11 antisera and Pabs. The results using MAP-VS-11 were favorable when compared with those using TGFα as the test antigens. Furthermore, MAP-VS-11 also gave a moderate color against one of the mabs, Mab A13.5 (1.37 O.D.). This result was rather unusual since only weak to negligible response was seen with VS-11 or TGFα. One plausible explanation is that Mab-A13.5 is against a denatured but polymeric form of TGFα and MAP-VS-11 is an antigen that best resembles that denatured form.

Discussion

Two important considerations in generating antibodies against a synthetic peptide are the need to conjugating a synthetic peptide to a carrier without damaging the integrity of the synthetic peptide and the necessity to linking covalently with a macromolecular protein as a carrier. Our results in this study demonstrate that the MAP approach would provide a convenient solution to these two difficult considerations (10-12).

The MAP system with synthetic peptides attached to an octabranched heptalysyl core and synthesized *de novo* as a complete unit overcomes the ambiguity of the conventional method of chemical conjugating a synthetic peptide to a protein carrier. Furthermore, unlike the conventional method which usually provides a low molar ratio of the synthetic peptide to the protein carrier (a low density immunogen model), the MAP system gives a very high molar ratio of the synthetic peptide to the core carrier (a high density immunogen model). In the 14 examples shown in Table 1, the bulk of molecular weight accounted by the synthetic peptide is over 90% of the whole MAP. Such a high density of synthetic peptide immunogens in a macromolecular arrangement proves to be advantageous in providing faster immunological responses in test animals than the low density immunogen models of the conventional method.

A significant development of the MAP system is the demonstration that a macromolecular protein carrier is not an essential requirement to elicit antibody responses. The MAP system containing only a core branching heptalysine has a molecular weight of 915 is significantly smaller than the conventional protein carriers with M_r > 100,00 such as keyhole limpet hemocyanin or albumin. This may prove to be a valuable development towards the design of synthetic vaccines for humans since both the problems and ambiguities relating to the conjugation and protein carrier are eliminated by the MAP approach.

The obvious advantage of the MAP approach is the attainment of a macromolecule directly from the synthesis. This advantage is also exploited for immunoassays since, among the various factors that affect the efficiency and sensitivity of solid phase immunoassays using synthetic peptides as antigens, the most unpredictable and serious problem is the low coating capacity of synthetic peptides to plastic solid surfaces. It is also clear that mapping of antigenic determinants of proteins by synthetic peptides has become an useful and convenient approach to obtain detailed information concerning the antigenic sites and interactions of the antibody-protein. Our results show that the use of MAP provides a reliable and convenient method to achieve such goals. Furthermore, the advantage of polyvalency of the MAP approach over the monovalency of the synthetic peptide offers the possibility of an increase in avidity of the antibody-antigen interactions. For example the multivalency of many polysaccharides resulting from their repeating residues increases the affinity of the polysaccharides to the antibody many fold over the monomeric unit. Similarly, the reaction between anti-Dnp antibodies and the polyvalent Dnp-bacteriophage has an association constant about 10^5 greater than the reaction between the same antibodies and the univalent Dnp-lysine. The increased sensitivity of the MAP-peptides observed from these studies may reflect on the polyvalency of the MAP antigens being used.

Reference

1. Sela M, Arnon R (1980): Antiviral antibodies obtained with aqueous solution of a synthetic antigen. In Mizrahi A, Hertmann I, Klingberg MA, Kohn A (eds): "New Developments with Human and Veterinary Vaccines," New York: Alan R. Liss, pp 315-323.

2. Lerner RA (1982). Tapping the immunological repetoire to produce antibodies of predetermined specificity. Nature 299:592-596.

3. Bittle JL, Houghten RA, Alexander H, Shinnick TM, Sutcliffe JG, Lerner RA, Rowlands DJ, Brown F (1982). Protection against foot-and-mouth disease by immunization with a chemically synthesized peptide predicted from the viral nucleotide sequence. Nature(London) 298:30-33.

4. DiMarchi R, Brooke G, Gale C, Cracknell V, Doel T, Mowat N (1986). Protection against foot-and-mouth disease by a synthetic peptide. Science 232:639-641
5. Merrifield RB (1963). Solid phase peptide synthesis of a tetrapeptide. J Am Chem Soc 85:2149-2154.
6. Mitchell AR, Kent SBH, Engelhard M, Merrifield RB (1978). A New synthetic route to tert-Butyloxycarbonylaminoacyl-4-(oxymethyl)phenylacetamidomethyl-resin, and improved support for solid-phase peptide synthesis. J Org Chem 43:2845-2852.
7. Sarin VK, Kent SBH, Tam JP, Merrifield RB (1981) Quantitative monitoring of solid phase peptide synthesis by the ninhydrin reaction. Anal Biochem 117:147-157.
8. Tam JP (1985). Enhancement of coupling efficiency in solid phase peptide synthesis by elevated temperature. In Deber Cm, Kopple KD, Hruby VJ (eds): "Proc Am Pept Symp, 9th," Rockford, Ill.: Pierce Chem Co, pp 305-308.
9. Tam JP, Heath WF, Merrifield RB (1986). Mechanisms for the removal of benze protecting groups in synthetic peptides by trifluoromethanesulfonic acid-trfluoroacetic acid-dimethylsulfide. J Am Chem Soc 108:5242-5251.
10. Posnett DN, McGrath H, Tam JP (1988). A novel method for producing antipeptide antibodies. J Biol Chem 263:1719-1725.
11 Chang KJ, Pugh W, Blanchard SG, McDermed J, Tam JP (1988). Antibody specific to the /(*a subunit of the guanine nucleotide-binding regulatory protein G/do/u: Developmental appearance and immunocytochemical localization in brain. Proc Natl Acad Sci USA 85:4929-4933.
12. Tam JP (1988). Synthetic peptide vaccine design: Synthesis and properties of a high-density multiple antigen peptide system. Proc Natl Acad Sci USA 85:5409-5413.
13. Czernilofsky A, Levison A, Varmus H, Bishop JM, Tisher E, Goodman H (1980). Nucleotide sequence of an avian sarcoma virus oncogene (sic) and proposed amino acid sequence for gene product. Nature(London) 287:198-203
14. Neckameyer WS, Wang LH (1985). Nucleotide sequence of avian sarcoma virus UR2 and comparison of HS transforming gene with other members of the tyrosine protein kinase oncogene family. J Virol 53:879-884.
15. Yanagi Y, Yoshikai Y, Leggett K, Clark SP, Aleksander I, Mak TW (1984). A human T cell - specific cDNA clone encodes a protein having extensive homology to immunoglobulin chains. Nature(London) 308:145-149.

16. Dame JB, Williams JL, McCutchan TF, Weber JL, Wirtx RA, Hockmeyer WT, Maloy WL, Haynes JD, Schneider I, Roberts D, Sanders GS, Reddy EP, Diggs CL, Miller LH (1984). Structure of the gene encoding the immunodominant surface antigen on the sporozoite of the human malarian parasite *Plasmodium falciparum*. Science 225:593-599.

17. Tanabe T, Nukada T, Nishikawa Y, Sugimoto K, Suzuki H, Takahashi H, Noda M, Haga T, Ichiyama A, Kangawa K, Minamino N, Matsuo H, Numa S (1985). Primary structure of the α-subunit of transducin and its relationship to *ras* proteins. Nature 315:242-245.

18. Derynck R, Roberts AB, Winkler ME, Chen EY, Goeddel DV (1984). Human transforming growth factor-α: precursor structure and expression in *E. coli*. Cell 38:287-297.

19. Castro B, Dormoy JR, Dourtoglon B, Evin G, Selve C, Zeigler JC (1976) Peptide coupling reagents; VI. A novel cheaper preparation of benzotriazolyloxytris (dimethylamino)-hexafluorophosphate (BOP reagent). Synthesis 751-752.

COGNITIVE FEATURES OF CONTINUOUS ANTIGENIC DETERMINANTS

H. Mario Geysen, Tom J. Mason and Stuart J. Rodda

Department of Molecular Immunology, Commonwealth Serum Laboratories, 45 Poplar Road, Parkville, Victoria 3052, Australia.

ABSTRACT Antigen-antibody interactions represent a diverse set of recognition events between large molecules. The minimum repertoire of antibodies necessary to achieve specific binding depends on the cognitive features of the sites of interactions of both the antigen and the antibody. If low stringency in the requirements for successful antigen-antibody binding were a cognitive feature of epitopes, then a small minimum set of antibodies would be required in the repertoire. However, this would increase the likelihood of an induced antibody population recognizing (cross-reacting with) a self molecule. An understanding of the features of an antigen which determine recognition would be of great value to the understanding of protein-protein interactions in general, and to the design of peptides with specific biological effects. Recent advances in methods for the simultaneous synthesis of large numbers of peptides have made possible the systematic assessment of the effects of amino acid substitution on the recognition between peptide epitopes and the corresponding antibody. The ability of antibodies to bind to peptide analogues of the epitopes recognized by those antibodies was studied. This form of analysis identifies the allowed substitution pattern for each residue. A total of 103 epitopes within 63 well-defined antigenic peptides homologous with the relevant antigen sequence were identified, and the contribution of each amino acid residue to the antibody binding activity of each epitope was investigated. For each residue in the epitope, complete sets of peptide analogues containing single amino acid replacements were used to determine the alternative amino acids for which antibody binding activity was retained. The data are summarized in a replaceability matrix which indicates relationships between amino acids in terms of recognition by antibodies. In addition, the average number of residues with limited replaceability in continuous epitopes, and the frequency with which each amino acid is found in those epitopes, were determined. Finally, the potential for cross-reactivity between different antigens and a given antibody is discussed.

INTRODUCTION

Antibodies are produced in higher animals as a response by the immune system following exposure to foreign molecules (antigens). The antibodies are able to bind specifically to the antigen which stimulated their production. The minimum number of functionally different antibodies which an animal could produce depends on the size of the recognition site on an antigen and the number of different epitopes which a particular antibody can react with - the degeneracy in the specificity. Clearly, the smaller the size of the recognition site on an antigen and the greater the degeneracy in antibody specificity, the fewer would be the minimum number of functionally different antibodies. However, the likelihood that an induced antibody could react with a self-protein would be increased - a highly undesirable event.

In the past, studies of related proteins have been used to infer relationships between amino acids and their influence on protein structure and function. For instance, in a series of proteins which have retained the same function throughout evolution, conserved amino acids are identified as those which play an irreplaceable role in maintaining biological function or structural integrity. Conversely, changes of amino acids observed at the site of biological activity or at sites which determine structure have been used to postulate that such amino acids are "equivalent". Such groups of "equivalent" amino acids are frequently seen to share one or more similar physico-chemical characteristics such as charge, hydrophilicity or common functional groups in the side chain. In a similar way, comparison of the sequences of closely related antigens reveal regions of greatest sequence divergence. Examples include equivalent proteins from different isolates of the same infectious agent. It is then usually inferred that the variation in sequences has resulted from selective pressures of the immune response and therefore these are the antigenic regions of the protein. Often in using this form of analysis, some changes are ignored because the substitution is by an "equivalent" amino acid. Pairs of amino acids which are often regarded as "equivalent" and which share a physico-chemical property, include: lysine and arginine which share a positive charge; tyrosine and phenylalanine which possess an aromatic ring; serine and threonine which both have a hydroxy group; and valine and isoleucine which are both hydrophobic. However, even for the amino acids which form an "equivalent" pair there are important differences. In the same order: the size of the group carrying the charge (amine versus guanyl group); tyrosine's side chain can form a hydrogen bond whereas phenylalanine's cannot; threonine is more hydrophobic than serine; and isoleucine is larger than valine.

These earlier results have come from inferences drawn from the changes in the amino acid sequences which can be made without loss of biological function of the protein. The results presented here come from observations of the reaction of antibodies with a large number of similar epitopes. Thus, in these studies the "equivalence" of amino acids has been experimentally determined. Furthermore, the frequency with which pairs of amino acids can be interchanged without loss of antibody binding will indicate the residues which contribute to the overall degeneracy in antibody specificity.

METHODS AND MATERIALS

The general methods used in this study have been described previously (1). In brief, antibodies were reacted with every overlapping peptide homologous with the sequence of the inducing antigen. The length of the peptides varied from 5 to 8 residues. Peptides which reacted strongly with the antibody were then used as the "parent" sequence of a replacement set of peptides. This is the set of peptides which consists of all peptides whose sequence is the same as the parent except at one residue - at that residue one of the 20 commonly occurring amino acids has been substituted. Reactions between a particular replacement peptide and the antibody were taken to be positive if the ELISA response was at least 20% of that of the parent peptide. The varied amino acid of a positively reacting peptide was deemed to be a "replacement" for that particular residue in the parent peptide. In this way a replaceability pattern was built up for each antibody/parent peptide pair.

RESULTS and DISCUSSION

Replaceability Patterns

Altogether 103 replaceability patterns (each representing an individual recognition event of antigen by an antibody), were analyzed. This database comprises an assessment of the replaceability of 640 amino acids in peptides ranging from 5 to 8 residues in length. Whilst this database is limited to sequential epitopes, i.e., those where the amino acid residues involved in binding to the antibody are a part of a continuous stretch of the primary sequence of the respective antigen, relationships determined are likely to apply to topographic epitopes as well.

The replaceability of each residue in the database, defined as the number of allowed alternative common amino acids for that residue (0-19), is summarized in Fig. 1 as a frequency distribution. As expected, residues within an epitope tended to show either a very limited replaceability, or they were generally replaceable (1). This is consistent with the concept that epitopes comprise essential residues (those with limited replaceability), or non-essential residues (those with a general replaceability) when assessed in terms of making a contribution to antibody binding. Frequency distributions for each of the 20 common amino acids are shown in Fig. 1.

Residues were divided into two groups based on the number of allowed replacement amino acids, with the dividing line chosen as the nadir of the frequency distribution. Thus, essential residues comprised residues for which 0 to 9 replacements were allowed, and non-essential residues comprised residues for which 10 to 19 replacements were allowed. For 171 of the 435 essential residues, replacement by any amino acid at all prevented antibody binding.

Data about the identity of acceptable replacements for essential residues are tabulated in the form of a replaceability matrix in Table 1. It is clear from this table that when replacements are allowed, they usually bear a physico-chemical relationship to the parent residue. For amino acids traditionally considered conservative substitutions for each

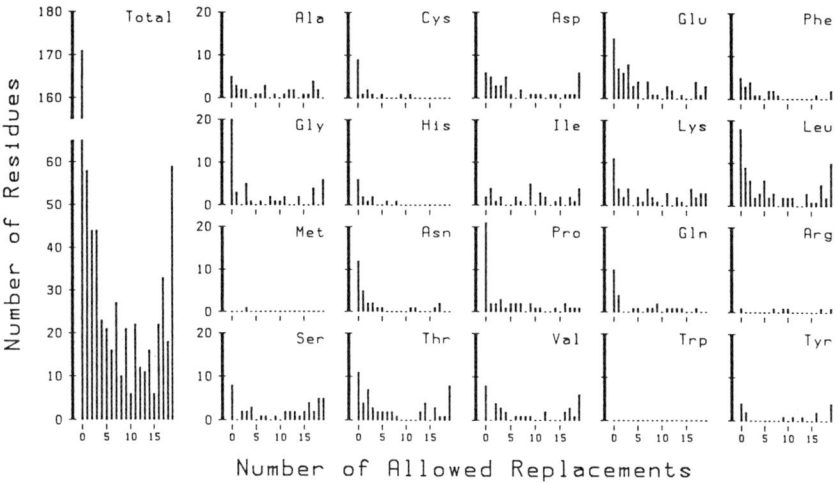

FIGURE 1. Frequency of occurrence of residues with different degrees of replaceability. The left hand figure shows all 640 residues of the database and the other figures shows the replaceability for each amino acid. The abscissa gives the number of alternative amino acids acceptable as replacements.

other, replaceabilities shown to be significantly higher than random range from a low of 29% to a high of 65% acceptability (methionine for leucine and valine for isoleucine respectively). As expected however, obvious clashes such as change of charge (lysine to glutamic [0%], aspartic [10%], glutamic to lysine [6%], arginine [4%]), aromatic hydrophobic to aliphatic hydrophobic (phenylalanine to isoleucine [16%], leucine [37%] and valine [16%], small residue to large (glycine to alanine [24%], valine [3%], isoleucine [6%], leucine [6%], are infrequently allowed, although not all of these values reached significance level.

Contrary to the concept of junior and senior antigens (2), the replacement of a large residue with a small one is also disallowed. If we assume that compensatory adjustments of side-chains do not occur, this can be looked on as creation of a void in the cluster of residues of an epitope in contact with its antibody. For example, replacement of an essential residue by glycine, would create a void between the interacting molecules the size of the replaced side-chain. The energy penalty associated with a void has been estimated

TABLE 1
REPLACEABILITY MATRIX FOR ESSENTIAL RESIDUES IN EPITOPES

Parent amino acid	No.	Replacement amino acid																			Av Rep
		A	C	D	E	F	G	H	I	K	L	M	N	P	Q	R	S	T	V	W	Y	
A	18	100	6	17	11		50		11	6	11	33	22	17	17		50	22	28			3.00
C	15	7	100	7			7	13		7		20	13	7	20	27	7	7		7		1.47
D	26	12	38	100	50			23			4	12	42	4	12		27	15		19		2.62
E	48	21	15	42	100	4	13	8	6	6	8	21	13	13	33	4	15	10	6	8	6	2.50
F	19	11	16	16	11	(2)		5	16	5	37	11	11	11	11	5	32	26	16	16	21	2.74
G	33	24	3	6	100	100	6	12	6	12	6	15	3	23	9	18	24	3	3	3	3	1.61
H	13				8		23	100		15		8	15	23	23	23	8	15	8	8	8	1.85
I	17	47	29	18	29	35	(0)	(0)	100	24	59	18	12	6	29	24	24	24	65	6	(0)	4.47
K	31	16	(0)	10	(0)	10	10	26	13	100	10	32	23		26	39	29	23	13	6	(0)	2.87
L	51	12	10	8	(2)	22	(2)	8	49	18	100	29	6	4	12	10	4	4	22	6	10	2.35
M	1									100		100		100	100							3.00
N	23		4	4	4	4	4	9		9	4	13	100		13	17	9	4		4		1.04
P	37	24	8	3	5	8	14	11	11	16	3	16	11	100	16	5	14	11	14	8	(0)	1.97
Q	20	20	15	5	15	5	20	10	5	5	5	25	25	10	100	10	15	10	10	10	10	2.30
R	3	67				33		67	33	67	33	67		33	33	100			33	33	33	5.33
S	18	33	11		17	11	6	39	6	11	6	22		6		28	100	22	6	17	6	2.44
T	34	12	6	6	15	6	3	9	15	12	9	24	12	6	12	3	44	100	21	15	6	2.32
V	21	24	5			14		19	52	10	33	5	14	14		10	5	29	100	14	6	2.62
W	0																					
Y	7	14	14			29			14	14	14	14				14			14	14	100	1.57

The replaceability of all essential residues in the database is expressed as Percent replaceability = (No. of time the replacement amino acid was allowed)x100/(No. of occurrences of the parent amino acid, column 2). Shown underlined are those percent replaceabilities which are significantly ($P<0.05$) different from that expected by chance. The significance was calculated in a 2x2 Chi-square test against the null hypothesis that all alternative amino acids are equally likely. Those frequencies which are significantly ($P<0.05$) lower than expected are shown in parentheses. Zeros have been omitted except for where they are statistically significant. Av Rep is the average replaceability expressed as the number (not percentage) of times a replacement was acceptable for that parent residue, averaged over the nineteen replacements possible for each parent residue.

at about 70 cal per Å² surface area (3). Thus the change alanine to glycine, observed frequency 50%, is associated with an estimated energy loss of less than 1 Kcal/mole, whereas valine to glycine (0%) and leucine to glycine (2%) are associated with estimated energy losses of 2.7 and 3.6 Kcal/mole respectively. The energy loss of the latter two examples is a significant proportion of the energy of binding between antigens and antibodies, which is typically in the range 8-16 Kcal/mole. It is therefore not surprising that these changes are infrequently allowed. These calculations also support the idea that the packing density at the interface is high and that dispersion forces account for a significant proportion of the interactions maintaining the complex of antigen and antibody.

Frequency data for residues comprising protein epitopes

The potential diversity of the immune repertoire in terms of the number of different antibody molecules which can be generated has been suggested to be of the order of 10^7. However, this in itself does not imply that each possible antibody is functionally distinct. Nor, is it axiomatic that the inherent likelihood of inducing any particular antibody is the same as that for any other antibody. Furthermore, it is well recognized that in response to a protein antigen, antibody production is not equally distributed over all the possible epitopes, but is "focussed" on a sub-set of these (4). Epitopes in this latter category are usually designated as the immunodominant determinants of the particular protein antigen.

It is useful to imagine induction of each particular antibody to be associated with a probability factor, which is a function of several variables. Some of these variables have been discussed in other publications for example, local mobility of the antigen at the site of the epitope (5), protrusion of the epitope (6), charge effects (7), etc. A further possibility exists, namely that the residue composition of individual epitopes influences the likelihood of the induction of a corresponding antibody. This could arise either from a bias for individual amino acids inherent in the conserved structure/framework motif of all antibodies, or alternatively from the necessity to include residues able to contribute to the energetics of binding between the antibody and the antigen.

In order to test for an observed bias for individual amino

acids, the frequency with which each residue is found in known epitopes can be compared with the expected frequency based on the overall amino acid composition of proteins. The ratio of the observed frequency and the expected frequency is the a measure of the propensity (or tendency) for a given residue to be a part of an epitope.

Propensity factors

In two earlier studies, propensity factors were calculated based on a set of identified epitopes. In one (8), hexapeptides homologous with the primary sequence of 12 proteins were assessed for an ability to bind antibodies in polyclonal sera raised to each of the respective antigens. In total, 310 antigenic hexapeptides were identified and their residue composition analyzed to provide a table of propensity factors for each of the 20 common amino acids. In the other study (9), epitopes taken from published data for 20 proteins, were analyzed in a similar manner. It was further suggested that these factors would provide a basis for the prediction of antigenic regions in a protein and be helpful in a rational approach to the synthesis of peptides corresponding to continuous (sequential) epitopes which may elicit antibodies reactive with the intact protein.

In both of these earlier studies, some of the residues included for analysis are unlikely to be directly involved in binding antibody because a typical epitope of 5-7 residues contains one or more non-essential residues (see below). The consequence of this is to "dilute" the significance of the propensity factors by the proportion of "essential" residues to total residues. A further restriction on the usefulness of propensity factors for predicting antigenic regions in a protein is that no account is taken of any need to provide a "cocktail" of residues rather than a continuous stretch of the same or similar amino acids with the highest individual factors. For example, proteins rich in lysine (propensity factor = 1.61) such as histones, and even polylysine, contain regions predicted to have relatively high antigenicity values, whereas in fact polylysine is considered non-immunogenic and histones are among the least antigenic proteins known.

A consequence of classifying residues as either essential or non-essential is that it provides a rationale for defining the length of epitopes. We define the epitope as consisting all residues between and including the outer-most essential residues. This definition removes the "surplus" residues

TABLE 2
PROPENSITY FACTORS[a]

Amino acids	Essential residues	Non-essential residues
Tryptophan	0.0	0.0
Arginine	0.24*	0.54
Methionine	0.26	0.0
Alanine	0.39*	1.05
Cysteine	0.43	0.32
Serine	0.59	1.55
Tyrosine	0.67	1.03
Glycine	0.71	0.64
Isoleucine	0.94	1.11
Valine	0.97	1.11
Aspartic acid	1.07	0.77
Leucine	1.15	1.66
Lysine	1.20	1.63
Threonine	1.22	1.52
Histidine	1.26	0.0
Phenylalanine	1.34	0.37
Asparagine	1.41	0.96
Glutamine	1.57	0.67
Proline	1.77	0.96
Glutamic acid	1.98*	0.92

[a]Propensity factors for occurrence of amino acids in epitopes. Factors significantly different from 1 are indicated by *.

included in longer antigenic peptides from any consideration of the composition of that epitope.

From the results obtained in this work, 63 sequential epitopes from 17 protein antigens were analyzed, and propensity factors calculated for both categories of residues (table 2). Comparison of the propensity factors calculated for essential residues with other parameters such as hydrophilicity values, hydropathicity values or values related to surface exposure of a protein, fail to show a correlation.

TABLE 3.
EPITOPE LENGTH[a]

Epitope length	Mean number of essential residues		
	Polyclonal	Monoclonal	Total
3	2.7	0.0	2.7
4	3.7	3.3	3.6
5	4.3	4.8	4.4
6	4.6	5.2	4.8
7	5.1	6.8	5.7
8	0.0	5.0	5.0
Overall	4.0	5.0	4.2

[a]Data for all 103 epitopes in the database. 81 epitopes with a mean length of 4.9 residues were defined by polyclonal sera and 22 epitopes with a mean length of 5.5 residues were defined by monoclonal sera.

This suggests that the bias for residues facilitating antigen recognition is inherent in the overall structure of the antigen combining site of antibodies and is not a consequence of antigen structure. The poor immunogenicity observed for histones may partially be explained by the particularly low value for arginine as these proteins are rich in this amino acid. Furthermore it is consistent with the finding that closely related proteins can often be distinguished serologically by their reactivity with antisera to an epitope, in which the non-reacting protein includes an arginine and the reacting protein another residue. Some examples are 1. human IgG, Oz locus where Oz+ equates to a lysine at position 193 and Oz- to arginine, and 2. Thy 1.1 and Thy 1.2 where the latter has an arginine instead of a glutamine at position 89. The high propensity factors for the hydrophobic residues proline, phenylalanine and leucine suggests that these play a dominant role in the binding energetics by way of "hydrophobic" (entropic) as well as van der Waals interactions.

The database was analyzed for the frequency of epitopes by epitope length. Table 3 summarizes the results, and suggests that the average or typical sequential epitope comprises a total of 5.0 (sd=1.3) residues of which an average of 4.2 (sd=1.2) are residues essential to binding antibody. When epitopes defined by monoclonal antibodies are analyzed, the average number of residues is 5.5, with 5.0 being essential. This difference is explained by the, as expected, diversity of antibodies to a given epitope present in a polyclonal serum. These different clonal antibody populations may "recognize" the same linear sequence of amino acids in alternative ways, i.e., they have different replaceability patterns. This would have the effect of degrading the observable specificity for individual residues within an epitope by increasing the number of acceptable replacements, accounting for the difference in the perceived lengths for epitopes defined by polyclonal or monoclonal antibodies.

ACKNOWLEDGEMENTS

The authors are particularly indebted to the Chief Executive of the Commonwealth Serum Laboratories, Dr N. McCarthy, for his support, advice, and encouragement, and to Gordon Tribbick, Richard Lauricella, Steve Laurie, Heather Gould, and the rest of the technical staff for their enthusiastic and skilled assistance. Our special thanks to Jan Bartley for her help in, and her patience during the writing of this paper.

REFERENCES

1. Geysen HM, Rodda SJ, Mason TJ, Tribbick G, and Schoofs PG (1987). Strategies for epitope analysis using peptide synthesis. J Immunol Meth 102:259.
2. Fazekas de St Groth S (1975). The phylogeny of influenza. In Mahy BWJ and Barry RD (eds): "Negative Strand Viruses," London: Academic Press, p741.
3. Kuntz ID, and Kauzmann W (1975). Hydration of proteins and polypeptides. In Anfinsen CB, Edsall JT, and Richards FM (eds): "Advances in Protein Chemistrys," New York: Academic Press ,p 239.
4. Lerner RA (1983). Synthetic Vaccines. Sci Amer 248: No 2, 48.
5. Westhof E, Altschuh D. Moras D, Bloomer AC,

Mondragon, Klug A and Van Regenmortel MHV (1984). Correlation between segmental mobility and the location of antigenic determinants in proteins. Nature (London) 311: 123.
6. Thornton JM, Edwards MS, Taylor WR and Barlow DJ (1986). Loacation of continuous antigenic determinants in the protruding regions of proteins. EMBO J 5:409.
7. Geysen HM, Tainer JA, Rodda SJ, Mason TJ, Alexander H, Getzoff ED and Lerner RA (1987). Chemistry of antibody binding to a protein. Science 235:1184
8. Geysen HM, Mason TJ, Rodda SJ, Meloen RH and Barteling SJ (1985). Amino acid composition of antigenic determinants:Implication for antigen processing by the immune system of animals. In Lerner RA, Chanock RM and Brown F (eds): "Vaccines 85," New York: Cold Spring Harbor Laboratory, p 133.
9. Welling GW, Wicher JW, Van der Zee R and Welling-Webster S (1985). Prediction of sequential antigenic regions of proteins. FEBS Lett 188:215.

SYNTHETIC PEPTIDES: TOOLS FOR ELUCIDATING MECHANISMS OF PROTEIN ANTIGEN PROCESSING AND PRESENTATION[1]

John A. Smith

Departments of Molecular Biology and Pathology, Massachusetts General Hospital and Department of Pathology, Harvard Medical School, Boston, MA 02114

ABSTRACT Synthetic peptides are indispensable reagents for elucidating (i) the biochemical reactions involved in protein antigen processing and (ii) the molecular interactions between protein fragments, resulting from such processing, and peptide-binding receptors (so-called, class I and II histocompatibility antigens) on antigen-presenting cells, as well as between those bound (referred to as "presented") peptides and receptors present on cytotoxic (T_C) and helper (T_H) T-cells. This article summarizes how synthetic peptides are currently being used for examining the molecular details of protein antigen processing and presentation and describes how synthetic peptides might be used in the future.

INTRODUCTION

There are five signal advances in our understanding of protein antigen presentation that involve the use of synthetic peptides. First, synthetic peptides, presumably mimicking processed protein fragments and containing between 7 and 20 amino acid residues have been substituted for intact protein antigens in order to locate T-cell

[1] This work was supported by Hoechst Aktiengesellschaft (Frankfurt am Main, West Germany).

epitopes in a protein and to determine the types of class I and II[2] histocompatibility antigens to

TABLE 1
PEPTIDES AS T-CELL EPITOPES AND THEIR CLASS II MHC-RESTRICTION

Peptide	Restriction	Synthetic Peptide	Ref.
p/h Cytc(45-58)	A^b/A^k	+	2
HEL(74-86)	A^b/A^k	+	3
HEL(81-96)	A^b	+	3
Herpes gpD(1-23)	A^b	+	4
Nase(91-110)	A^b	+	5
b Cytc(13-25)	A^d/E^d	−	6
Inf HA(111-120)	A^d/E^d	+	7
Inf HA(130-142)	A^d	+	7,8
λ_R(12-26)	A^d/E^k	+	9
sw Mb(106-118)	A^d	+	8,10
Ova(323-339)	A^d	+	11
Nase(61-80)	A^d	+	5
HEL(46-61)	A^k	+	12
Inf HA(48-68)	A^k	+	13
Mal CS(326-343)	A^k	+	14
m Cytc(89-103)	E^b/E^k	+	15
HSV (8-23)	E^d/E^k	+	8
sw Mb(110-121)	E^d	+	16
sw Mb(136-146)	E^d	+	17
sw Mb(69-78)	E^k	+	16
Nase(51-70)	E^k	+	5
Nase(81-100)	E^k	+	5

Abbreviations: b, bovine; B, B-chain; CS, circumsporite protein; Cytc, cytochrome c; gp, glycoprotein; h, horse; HA, hemagglutinin; HEL, hen egg white lysozyme; HSV, human stomatitis virus; Inf, influenza virus; λ_R; m, moth; Mal, malaria; Mb, myoglobin; Nase, staphylococcal nuclease; p, pigeon; sw, sperm whale.

[2] Class II histocompatibility antigens are also known as Ia molecules.

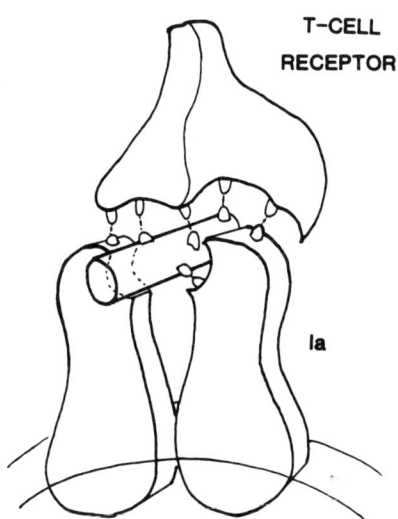

FIGURE 1. Schematic representation of the Ia-peptide antigen-T cell receptor complex. The specific contact residues between peptide and Ia would be different for Ia molecules of different haplotypes (e.g., I-A^b *versus* I-A^k), although the overall geometry of the complex should be the same. The peptide antigen (i.e., the cylinder) binds in the polymorphic, Ia-binding site. The T-cell receptor interacts with both the peptide antigen and Ia molecule.

which these peptides will bind and in turn be presented to the receptors of T_C- and T_H-cells, respectively (1-17) (Fig. 1). Further, by synthesizing peptide analogues, it is possible to determine which residues account for the binding to the histocompatibility antigens and which residues account for the epitope recognized by the T-cell receptors (18,19). Second, it has been shown in binding assays that labelled peptides bind to isolated Ia molecules. Unanue and coworkers (20) determined by classical equilibrium dialysis methods that a fluorescent labelled hen egg lysozyme (HEL) peptide (46-61) binds, albeit weakly ($K_D = 10^{-6}$), to isolated I-A^k restricted class II molecules (and not with I-A^d). Using a gel filtration assay, Grey and coworkers (21) determined a similar value for the K_D of radiolabelled chicken ovalbumin (Ova) peptide (323-339) bound to isolated I-A^d molecules, and they observed that the peptide would not bind to I-A^k, I-E^k, and I-E^d molecules. They also studied the kinetics of the peptide-Ia molecule interaction by measuring the $k_d = 2.8 \times 10^{-6}$ sec^{-1} and $k_a = 1.2$ M^{-1} sec^{-1}. By fluorescence energy transfer experiments utilizing a fluorescein-labelled peptide, reconstituted planar membranes containing affinity purified Ia molecules, and protein specific T_H cells, McConnell and coworkers (22) have proven the existence of a trimolecular complex consisting of (i) an I-A^d molecule, (ii) a synthetic peptide (Ova (323-339) and (iii) I-A^d restricted, Ova-specific T-cell hybridoma cells. Third, Gefter and coworkers (23) used synthetic peptides from the lambda repressor cI protein (λ_R), staphylococcal nuclease (Nase), and Ova to demonstrate by peptide-binding assays that there is one peptide binding site per Ia molecule and that these peptides could bind to the same Ia and compete competitively with one another for binding. In additional studies, conducted by Guillet *et al.* (24) and Buus *et al.* (8), the generality of this conclusion was proven. Fourth, Townsend *et al.* (25) showed that synthetic peptides could be presented to T_c-cells by class I histocompatibility antigens, as has been demonstrated for Ia molecules. Fifth, Wiley, Strominger, and co-workers (26) determined by x-ray crystallography the structure of a human class I histocompatibility antigen (i.e., HLA-A2) and demonstrated that it contained a single peptide binding site per molecule. They also showed that the "hypervariable" residues, identified by comparing protein sequences from many class I histocompatibility antigens determined by other investigators, are located primarily in the peptide binding site (27), and thereby they accounted for the promiscuous peptide binding observed by Gefter, Grey, Smith, and coworkers for the Ia molecule (8,24,25). Although the structure of an Ia molecule is not yet published, it is likely that it will resemble the class I histocompatibility antigen, because both are two chain, major histocompatibility complex (MHC) restricted, peptide-binding receptors.

Antigen Processing and Presentation

RESULTS AND DISCUSSION

Location of T-Cell Antigenic Sites in Proteins

A. Comparison of protein species variants with clustered point mutations

Protein	Species	T-cell response
------R--------	Bovine	+
------K--------	Porcine	±
------L--------	Ovine	−

Advantage:
1. Inexpensive (if multiple species variants of a model proteins are available and have been rigorously characterized; may take decades to do this)

Disadvantages:
1. Few protein models (e.g., myoglobin, cytochrome c, lysozyme)
2. Mutations not distributed uniformly and number of substitutions at a given residue may be small
3. Localization of T-cell antigenic site is imprecise, determines only critical residues

B. Enzymatic or chemical fragmentation of a protein antigen and isolation of peptides recognized by T-cells

Protein Antigen
↓ Enzymatic or chemical cleavage

Protein fragments
↓ Reversed-phase HPLC separation of peptides

Assay for T-cell recogition

Amino acid analysis and sequence analysis of peptide(s)

Advantage:
1. Cheap and relatively precise

Disadvantage:
1. Limited number of specific proteases to control cleavage
2. Cleavage may inadvertently destroy epitope
3. Access to expensive analysis equipment

C. Nested Set of Overlapping Peptides

```
1--------20  21--------40  41--------60  61--------80
    11----------30  31---------50  51--------70
```

Advantage:
1. "Complete" and rapid mapping of restricted T-cell epitopes

Disadvantage:
1. Requires peptide synthesis expertise
2. Expensive (depends on scale)
3. Additional overlapping peptides required to completely map the linear T-cell antigenic sites (see below)

```
6--------25  26--------45  46--------65  66--------85
    16----------35  36--------55  56--------75
```

D. Predictive algorithms

DeLisi & Berzofsky (Ref. 28): Amphipathic α-helices
Rothbard & Taylor (Ref. 29): Specific motifs of sequence

FIGURE 2. Schemes for locating T-cell antigenic sites in proteins

There are four schemes for locating T-cell antigenic sites in proteins, and the advantage and disadvantages of each approach are summarized in FIG. 2. The preferred approach is to synthesize nested sets of overlapping peptides (5; Z. Lui and J.A. Smith, unpublished). Although it is known that T-cells and B-cells recognize different antigenic sites in proteins, there are still no predictive algorithms that can determine the location of either type of antigenic sites with certainty, although two algorithms have been used to predict the location of T-cell antigenic sites (28, 29).

Synthesis of Peptide Mimicking Processed Protein Fragments Presented by Histocompatibility Molecules to Receptors on T-Cells

In contrast to antibody recognition, T-cell recognition cannot distinguish between "native" and "denatured" proteins (31), because in order to invoke T-cell recognition, protein antigens must be chemically and physically modified (processed) within acidic intracellular compartments (endosomes) of antigen-presenting cells, themselves bearing histocompatibility molecules. These molecules in turn function as receptors for antigenic peptides generated by these poorly understood, processing events (reviewed in 32).

Synthetic peptides, containing between 7 and 17 residues, can be substituted for cellularly processed protein fragments and are the mainstay for determining the molecular details of the interactions between antigenic peptide and binding site of either a histocompatibility antigen or a T-cell receptor (FIG. 1). The use of such peptides is currently shedding light on the mechanisms of antigen presentation and in the future will help clarify the mechanisms involved in antigen-processing.

Different histocompatibility molecules interact with different antigenic peptides from multisite protein antigens, although at the present time it is unclear why some peptides interact preferentially with certain Ia molecules cells and others do not (TABLE 1). However, it is now clear that peptides from a large number of different protein antigens that are restricted by the same Ia will bind to and mutually compete for binding to the Ia molecule (23,24). The chemical basis for how a potentially large number of peptides, derived from a variety of foreign proteins, are able to form unique complexes with a very limited number of histocompatibility molecules appears to be the propitious localization of the hypervariable residues of these molecules in a solitary binding site.

A. Overlapping, truncated peptides

Peptide	Residue #	Ia binding	T-cell response
81-100	20	+	+
71-85	15	ND	±
81-95	15	±	±
86-95	10	−	−
86-100	15	+	+
91-105	15	±	−

B. Equivalent substitutions

Peptide	Ia binding	T-cell response
GRGLAYIYADGKMVN	+	+
-A-------------	±	−
-----A---------	−	−
--------A------	+	+
---------A-----	−	−
-----------A---	+	−

(Ref. 19)

C. Hypervariable substitutions

Peptide	Ia binding	T-cell response
GRGLAYIYADGKMVN	+	+
-K-------------	−	−
-M-------------	−	−
-A-------------	−	−
---------G-----	+	−
---------L-----	+	±
---------V-----	±	±

(Ref. 18)

D. Conservative substitutions

Peptide	Ia binding	T-cell response
GRGLAYIYADGKMVN	+	+
-----F---------		
------F--------	ND	−
---------N-----	±	−
------------L--	+	+

FIGURE 3. Strategies for synthesizing peptides in order determination of the structural characteristics of a T-cell antigen required for its interactions with histocompatibility antigens and recognition by T-cells.

As shown in FIG. 3, there four strategies for preparing synthetic peptides in order to characterize the residues that interact with the histocompatibility molecule and that are recognized by the receptors of T_C or T_H cells. The overlapping, trucated peptide strategy is best suited for determining the minimum size of an antigenic peptide and should be used before beginning to synthetize numerous analogues by the other strategies. The choice between equivalent substitution (19) or hypervariable substitution (18) is an individual preference. The choice of conservative substitution may reduce the number of syntheses required but interpretation of experimental results may be equivocal.

Use of Synthetic Peptides for Studying Protein Antigen Processing and Presentation

PROCESSING

A. Determination of the minimum and maximum length of a processed protein fragment.

B. Determination of molecular mechanisms utilized during antigen processing utilizing radio- and photoaffinity-labelled peptides (Ref. 8).

PRESENTATION

A. Mapping of a protein's T cell antigenic sites, using nested sets of overlapping peptides (see Fig. 2C).

B. Fine specificity analysis of histocompatibility antigen interactions and T-cell recognition (Ref. 18 & 19).

C. Determination of the range of peptide sequences that may be accommodated by class I and II histocompatibility antigens (class I and II) (Ref. 18 & 24).

  1. Is there a a preferred conformation for peptides binding to histocompatibility antigens (Ref. 28)?

  2. Are there really definitive sequence patterns required for peptide binding by a histocompatibility antigen (Ref. 29)?

  3. How specific are these interactions?

  4. Will identical peptide antigens be presented by both class I or II histocompatibility antigens or are these antigenic sites mutually exclusive?

D. Preparation of specific immunostimulatory peptides ("tailor-made" for eliciting the best immune response against a pathogenic organism in an outbred individual).

E. Preparation of hybrid peptide replacing normally immunogenic residues with residues for which there is immunological tolerance. Such a peptide should bind to and block the binding site of the histocompatibility antigen binding site but should not be recognized by T-cells (Ref. 24)

F. Preparation of immunogenic peptides containing both a sequence which binds to a histocompatibility antigen and is recognized by T-cells and another sequence against which a immune response is not normally observed (see below). The tolerated sequence may be attached either N- or C-terminally to the carrier peptide. This approach forms the basis for suppression of immunological tolerance and for vaccination against tolerized antigens (Ref. 30).

	Peptide		
	A	B	A+B or B+A
Ia binding	+	+	+
T-cell response against B	NA	–	+

FIGURE 4. Present and future uses for synthetic peptides to elucidate the mechanisms of protein antigen processing and presentation.

FIG. 4 summarizes how peptides are being and might used to study processing and presentation. However, the potential uses for peptides and analogues are boundless and are limited only by the imagination and synthetic capability of a researcher.

Clearly, rational design of synthetic vaccines and immunosuppressive agents depends on a complete understanding of the nature of antigenic sites. Such knowledge will come from a close collaboration between cellular immunologists, cell biologists, and peptide chemists.

ACKNOWLEDGMENTS

The author wishes to thank Professors M.L Gefter and H.M. Grey, as well as Drs. J-G. Guillet, M-Z. Lai, T.J. Briner, and D. Perkins for numerous helpful discussions.

REFERENCES

1. Atassi MZ, Young CR (1985). Discovery and implications for the immunogenicity of free small synthetic peptides: powerful tools for manipulating the immune system and for production of antibodies and T cells of preselected submolecular specificities. CRC Crit Rev Immunol 5:387.
2. Suzuki G, Schwartz RH (1986). The pigeon cytochrome c-specific T cell response of low responder mice. I. Identification of antigenic determinants on fragment 1 to 65. J Immunol 136:230.
3. Shastri N, Oki A, Miller A, Sercarz EE (1985). Distinct recognition phenoypes exist for T cell clones specific for small peptide regions of proteins. J Exp Med 162:332.
4. Heber-Katz E, Hollosi M, Dietzschold B, Hudecz F, Fasman GD (1985). The T cell response to the glycoprotein D of the herpes simplex virus: The significance of antigen conformation. J Immunol 135:1385.
5. Finnegan A, Smith MA, Smith JA, Berzofsky J, Sachs DH, Hodes RJ (1986). The T cell repertoire for recognition of a phylogenetically distant protein antigen:

peptide specificity and MHC restriction of staphylococcal nuclease-specific T cell clones. J Exp Med 164:897.
6. Corradin G, Juillerat MA, Vita C, Engers HD (1983). Fine specificity of a Balb/c T cell clone directed against beef apo cytochrome c. Mol Immunol 20:763.
7. Hackett CJ, Hurwitz JL, Dietzschold B, Gerhard W (1985). A synthetic decapeptide of influenza virus hemagglutinin elicits helper T cells with the same fine recognition specificities as occur in response to whole virus. J Immunol 135:1391.
8. Buus S, Sette A, Colon SM, Miles C, Grey HM (1987). The relation between major histocompatibility complex (MHC) restriction and the capacity of Ia to bind immunogenic peptides. Science 235:1353.
9. Lai M-Z, Ross DT, Guillet J-G, Briner TJ, Gefter ML, Smith JA (1987). T lymphocyte response to bacteriophage λ repressor cI protein. Recognition of the same peptide presented by Ia molecules of different haplotypes. J Immunol 139:3973
10. Berkower I, Kawamura H, Matis LA, Berzofsky JA (1985). T cell clones to two major T cell epitopes of myoglobin: effect of I-A/I-E restriction on epitope dominance. J Immunol 135:2628.
11. Shimonkevitz R, Colon S, Kappler JW, Marrack P, Grey HM (1984). Antigen recognition by H-2-restricted T cells. II. A tryptic ovalbumin peptide that substitutes for processed antigen. J Immunol 133:2067.
12. Babbitt BP, Matsueda G, Haber E, Unanue ER, Allen PM (1986). Antigenic competition at the level of peptide-Ia binding. Proc Natl Acad Sci USA 83:4509.
13. Mills KHG, Skehel JJ, Thomas DB (1986). Extensive diversity in the recognition of influenza virus hemagglutinin by murine T helper clones. J Exp Med 163:1477.
14. Good MF, Maloy WL, Lunde MN, Margalit H, Cornette JL, Smith GL, Moss B, Miller LH, Berzofsky JA (1987). Construction of synthetic immunogen: use of new T-helper epitope on malaria circumsporozoite protein. Science 235:1059.

15. Hedrick SM, Matis LA, Hecht TT, Samelson LE, Longo DL, Heber-Katz E, Schwartz RH (1982). The fine specificity of antigen and Ia determinant recognition by T cell hybridoma clones specific for pigeon cytochrome c. Cell 30:141.
16. Livingstone AM, Fathman CG (1987). The structure of T-cell epitopes. Ann Rev Immunol 5:477.
17. Berkower I, Buckenmeyer GK, Berzofsky JA (1986). Molecular mapping of a histocompatibility-restricted immunodominant T cell epitope with synthetic and natural peptides: implications for T cell antigenic structure. J Immunol 136:2498.
18. Sette A, Buus S, Colon S, Smith JA, Miles C, Grey HM (1987). Structural characteristics of an antigen required for its interaction with Ia and recognition by T cells. Nature 328:395.
19. Allen PM, Matsueda GR, Evans RJ, Dunbar JB, Jr, Marshall GR, Unanue ER (1987). Identification of the T-cell and Ia contact residues of a T-cell antigenic epitope. Nature 327:713.
20. Babbitt BP, Allen PM, Matsueda G, Haber E, Unanue ER (1985). Binding of immunogenic peptides to Ia histocompatibility molecules. Nature 317:359.
21. Buus S, Sette A, Colon SM, Jenis DM, Grey HM (1986). Isolation and characterization of antigen-Ia complexes involved in T cell recognition. Cell 47:1071.
22. Watts TH, Gaub HE, McConnell HM (1986). T cell-mediated association of peptide antigen and major histocompatibility complex protein detected by energy tranfer in an evanescent wavefield. Nature 320:176.
23. Guillet J-G, Lai M-Z, Briner TJ, Smith JA, Gefter ML (1986). Interaction of peptide antigens and class II major histocompatibility complex antigens. Nature 324:260.
24. Guillet J-G, Lai M-Z, Briner TJ, Buus S, Sette A, Grey HM, Smith JA, Gefter ML (1987). Immunological self, nonself discrimination, Science 235:865.
25. Townsend ARM, Rothbard J, Gotch FM, Bahadur G, Wraith D, McMichael AJ (1986). The

epitopes of influenza nucleoprotein recognized by cytotoxic T lymphocytes can be defined with short synthetic peptides. Cell 44:959.
26. Bjorkman PJ, Saper MA, Samraoui B, Bennett WS, Strominger JL, Wiley DC (1987). Structure of the human class I histocompatibility antigen, HLA-A2. Nature 329:506.
27. Bjorkman PJ, Saper MA, Samraoui B, Bennett WS, Strominger JL, Wiley DC (1987). The foreign antigen binding site and T cell recognition regions of class I histocompatibility antigens. Nature 329:512-518.
28. DeLisi C, Berzofsky JA (1986). T-cell antigenic sites tend to be amphipathic structures. Proc Natl Acad Sci USA 82:7048.
29. Rothbard JB, Taylor WR (1988). A sequence pattern common to T cell epitopes. EMBO J 7:93.
30. Francis MJ, Hastings GZ, Syred AD, McGinn B, Brown F, Rowlands DJ (1987). Non-responsiveness to a foot-and-mouth disease virus peptide overcome by addition of foreign helper T-cell determinants. Nature 330:168.
31. Gell PGH, Benacerraf B (1959). Studies on hypersensitivity. II. Delayed hypersensitivity to denatured proteins in guinea pigs. Immunology 2:64.
32. Unanue ER, Allen PM (1987). The basis for the immunoregulatory role of macrophages and other accessory cells. Science 236:551.

USE OF SYNTHETIC PEPTIDES IN TWO DIFFERENT APPROACHES TO INTERFERE WITH HOST CELL PENETRATION BY *TRYPANOSOMA CRUZI*

A. Ouaissi, J. P. Defoort[1], D. Afchain, H. Gras-Masse[1], J. Cornette, H. Caron, A. Tartar[1], A. Capron.

Centre d'Immunologie et de Biologie Parasitaire, Institut Pasteur, rue Calmette, 59045 LILLE CEDEX, FRANCE.

ABSTRACT The trypomastigote stage of *Trypanosoma cruzi* is able to invade a variety of vertebrate cell types. Studies of molecular interactions involved in the trypomastigote interiorization suggest that host-cell fibronectin, and a trypomastigote-specific surface antigen might mediate the attachment of the pathogen to the host cell. Synthetic peptides have been used in two different strategies aimed at interfering with the invading phase : 1- peptides containing the RGD sequence corresponding to the cell attachment site inhibited the binding of fibronectin to the parasite surface and the infection of 3T3 fibroblasts by the pathogen. 2- a peptide containing the RGD sequence, and a peptide copying a repeated sequence present in the trypomastigote-surface antigen were used to immunize mice. They induced a certain degree of protection against *Trypanosoma cruzi* infection.

[1]Present address: Service de Chimie des Biomolécules, Institut Pasteur, rue Calmette, 59045 LILLE CEDEX, FRANCE.

INTRODUCTION

Attachment of *Trypanosoma cruzi*, the causative agent of Chagas'disease, to mammalian cells is believed to be of fundamental importance in the host cell invasion by the pathogen.

In an early attempt to identify molecules that might be involved in this interaction, we have shown that fibronectin, a major component of the extracellular matrix, binds to a specific receptor of molecular weight 85 and 68 kDa on the *Trypanosoma cruzi* trypomastigote (Tehuantepec strain) (1), and enhances their association with host cells.

Recently, a genomic DNA fragment encoding antigenic determinants present in a 85 kDa *T.cruzi* trypomastigote surface antigen (Peru strain) has been cloned (2). This 85 kDa, presenting a nonapeptide unit tandemly repeated five times, is known to be necessary for efficient interiorization of trypomastigotes in mammalian cells, and might thus be identical to the parasite fibronectin receptor.

To interfere with the invading phase in two differents strategies, as shown in the figure 1, we synthesized peptides containing the cell attachment site of fibronectin (3) **RGD**(AVTG**RGD**SPC=Fibronectin-Peptide) and an octadecapeptide representing two tandem repeats (DKKESGDSEDKKESEDSE=Repeat-Peptide).

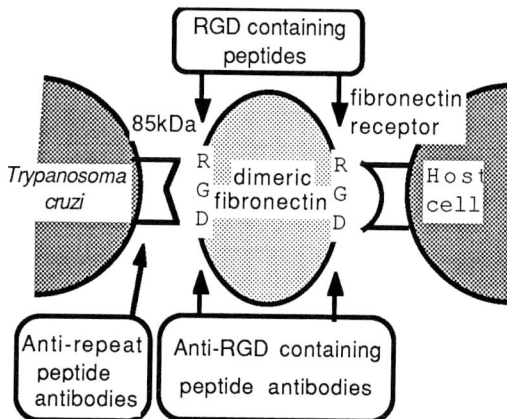

FIGURE 1

MATERIALS AND METHODS

The Y strain of *T.cruzi* used in this study is being maintained in our laboratory by serial passages of blood-induced infections in BALB/c mice (4).

The peptides were synthesized by solid-phase method (5). They were coupled to tetanus toxoid through an additional N-terminal cysteyl residue (added to the octadecapeptide) using the heterobifunctional reagent maleimidocaproyl-N-hydrosuccinimide ester.

The inhibition of binding of ^{125}I-labeled fibronectin to trypomastigotes with Fibronectin-Peptide and analogs and the inhibition of infectivity of 3T3 fibroblasts by trypomastigotes with Fibronectin-Peptide and analogues were performed as described (6).

Immunization were performed as follows: two groups of BALB/c mice (n=10, n=11) were immunized subcutaneously with 25 µg of either Repeat-Peptide or Fibronectin-Peptide conjugated to tetanus toxoid in 0,1 ml of Freund's Complete Adjuvant (FCA). Three identical injections were given to each mouse in Freund's Incomplete Adjuvant (FIA) at two weeks intervals. A third group (n=7) received similar schedules of injections of tetanus toxoid and FCA or FIA, the last group was untreated and used for the control inoculum..

The protection induced by immunization was challenged by inoculation of lethal dose (10^6) of blood stream trypomastigotes intraperitoneally (two weeks after the last injection).

Blood smears from the tails of all mice were examined as described previously (6). Briefly, the number of parasites in 30 microspic field (10 x ocular, 40 x objective) was counted on a thin smear of tail blood. The mice were checked for survivors regulary.

Sera were essayed by ELISA methodology (two weeks after the last injection) in which Repeat-Peptide and Fibronectin-Peptide were adsorbed to polystyrene microtiters wells.

RESULTS

Inhibitory Effects on the Binding of ^{125}I-labeled Fibronectin to *T. cruzi* Using Fibronectin-Peptide and Analogues.

When trypomastigotes were incubated with Fibronectin-Peptide or peptides containing the RGD sequence, a concentration-dependent inhibition of the binding of ^{125}I-labeled fibronectin to the parasite was observed (Fig 2). Unrelated peptides had no significant effect on the binding of iodinated fibronectin to the parasite surface.

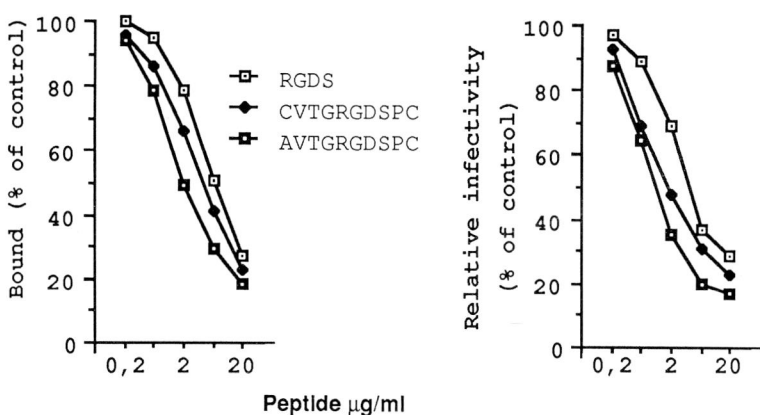

FIGURE 2.   FIGURE 3.

Inhibitory Effects on the Infectivity of 3T3 Fibroblasts by *T. cruzi* Using Fibronectin-Peptide and Analogues.

We observed a significant decrease of the infectivity of 3T3 fibroblasts when the trypomastigotes were incubated with the Fibronectin-Peptide or other RGD containing analogues (Fig 3). Unrelated peptides had no significant effect on the infectivity of 3T3 fibroblasts.

Effects of Antipeptide Antibodies.

The ELISA tests showed that sera of mice immunized with the Repeat-Peptide or with the Fibronectin-Peptide developed antibodies that reacted with the corresponding immobilized peptides. None of the Tetanus Toxoid-immunized animals showed positive reactions against these peptides.

Sera from Repeat-Peptide-immunized mice immunoprecipitated three radiolabeled trypomastigote surface antigens of molecular weight 68, 85 and 160 KDa (Fig 4)

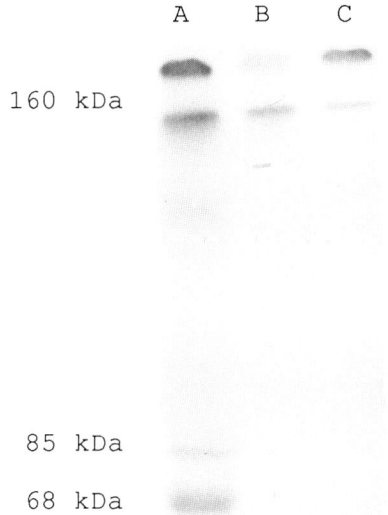

A: sera of mice immunized with Repeat-Peptide
B: sera of mice immunized with Fibronectin-Peptide
C: sera of non immunized mice

FIGURE 4.

In vivo Protection Induced by Immunization With the Peptides.

Figure 5 shows the survival of control and immunized mice following challenge infection. Both control groups showed high mortality between days 10 and 13 post infection. Immunization with synthetic peptides reduced the mortality rate and increased the time of survival.

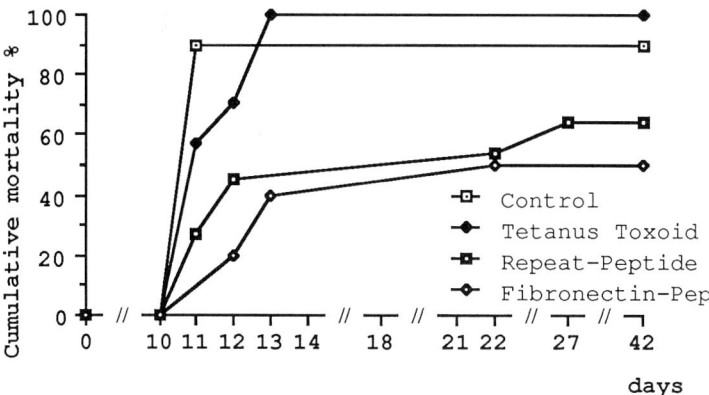

FIGURE 5.

TABLE 1.

Treatments	Days Postinfection	Parasitemia
Repeat Peptide		13,5 ± 13,2
Fibronectin peptide	11	16,5 ± 10,9
Tetanus Toxoid		18,3 ± 8,3
Control		22,0 ± 0,0
Repeat Peptide		3,7 ± 1,0
Fibronectin peptide	14	5,5 ± 5,6
Tetanus Toxoid		−
Control		5,0 ± 0,0
Repeat Peptide		1,5 ± 1,8
Fibronectin peptide	18	2,1 ± 2,9
Control		2,0 ± 0,0
Repeat Peptide		0,8 ± 0,7
Fibronectin peptide	21	1,5 ± 1,7
Control		1,0 ± 0,0

A low level of parasitemia was detected after 14th day of infection (Table 1). Except one mouse who survived the infection and did not develop high level of parasitemia, the untreated mice in the control group died before day 11 post-infection.

DISCUSSION

Analysis of the data in Fig. 1 shows that the fibronectin sequence recognized by the parasite receptor is the same as the sequence recognized by the cellular fibronectin receptor. The dimeric fibronectin acts as a bridge between the host cell surface and *T. cruzi*. When increasing concentrations of RGD containing peptides were added, saturation of fibronectin receptor on target cell or/and parasite receptor induced a significant decrease in the infectivity.

When mice were immunized with the Fibronectin-Peptide, a certain degree of protection has been observed. The partial protection obtained supposed that the constant presence of antibodies to fibronectin cell attachment domain in the circulation might bind to the fibronectin domain containing the RGD sequence. Thus, fibronectin is no more able to act as a bridge. However, due to the high concentration of circulating fibronectin, another mechanism involving for instance antiidiotypes can not be excluded. In any case, due to autoimmune reactions, this approach is not suitable for vaccination.

In case of *T. cruzi*, the need for predetermined specificity of the neutralizing immune response is enhanced by the potential risk of autoimmune response elicited by the parasite native immunogens. The parasite receptor could be a major target for the development of a protective immunity. The repeated structure present in a 85 kDa glycoprotein probably implied in this mechanism offers an alternate possibility to interfere with the interaction without targeting the host fibronectin.

CONCLUSION

Synthetic peptides offer a new tool to study the biochemical basis of parasite-cell interaction processes at the submolecular level. These studies can help in the design of strategies for interacting with these processes, such as synthetic vaccines. Moreover, the predetermined specificity of the immunity elicited by a synthetic immunogen allows the control of the risk of an autoimmune response by a parasite total antigen.

REFERENCES

1. Ouaissi M. A, Cornette J, Capron A (1986). Identification and isolation of *Trypanosoma cruzi* cell surface protein with properties expected of a fibronectin receptor. Molecular and Biochemical Parasitology, 19, 201-211.
2. Peterson D. S, Wrightsman R. A, Manning J. E (1986). Cloning of a major surface antigen gene of *Trypanosoma cruzi* and identification of a nonapeptide repeat. Nature, 322, 566-568.
3. Pierschbacher M. D, Ruoslahti E (1984). Cell attachment activity of fibronectin can be duplicated by small synthetic fragments of the molecule. Nature, 309, 30-33.
4. Rodriguez A. M, Afchain D, Santoro F, Bazin F, Capron A (1983). Parasitological and immunological aspects of *Trypanosoma cruzi* infection in Nude Rats. Parasitenkd, 69, 141-147.
5. Merrifield R. B (1963). Solid phase peptide synthesis.I. The synthesis of a tetrapeptide. Journal of American Chemistry Society, 85, 2149.
6. Ouaissi M. A, Cornette J, Afchain D, Capron A, Gras-Masse H, Tartar A (1986). *Trypanosoma cruzi* infection inhibited by peptides modeled from a fibronectin cell attachment domain. Science, 234, 603-607.

GLYCOSYLATION OF EXOGENOUS PEPTIDE ACCEPTORS BY LARVAL BRINE SHRIMP MICROSOMES[1]

Michael N. Horst

Division of Basic Science
School of Medicine
Mercer University
Macon, GA 31027

ABSTRACT Glycosylation of synthetic peptides by oligosaccharyl transferase was examined using crude microsomes prepared from larval brine shrimp, Artemia. The enzyme glycosylates several synthetic peptides including ^{3}H Asn-Leu-Thr-NH$_2$ and ^{3}H-Asn-Tyr-Thr-NH$_2$. Maximal enzymatic activity was observed with 10 mM MgCl$_2$ while 50 mM EDTA blocked activity. Formation of product was linear with time of incubation and added microsomal protein. When reaction mixtures were analyzed by gel or paper chromatography, several radiolabeled products were observed. Products could be purified by WGA-agarose chromatography; they were sensitive to treatment with chitinase but were resistant to endo H.

INTRODUCTION

The biosynthesis of the crustacean cuticle or exoskeleton involves the formation of a chitoprotein which serves as a primer for polymerization of polysaccharide chains by chitin synthetase. Several of the enzymes in this pathway appear to utilize an isoprenoid lipid, dolichol,

---

[1]This work was supported by NIH grant GM-30952.

in the assembly of an oligosaccharide which is transferred to protein in a cotranslational manner. The initial biosynthetic step involves the phosphorylation of dolichol to yield dolichol phosphate (1). The next reaction is the glycosylation of dolichol phosphate to yield dolichol-pyrophosphate-N-acetyl-D-glucosamine (2). After transfer of additional GlcNAc residues, the dolichol linked oligosaccharide reaches a size of 3 to 8 GlcNAc residues (3). The oligosaccharide is then transfered en bloc to a protein acceptor via an oligosaccharyl transferase, forming a chitoprotein primer. Extension of the primer oligosaccharides occurs via chitin synthetase; the properties of this enzyme from larval Artemia have been described (4).

This paper describes some properties of the oligosaccharyl transferase from larval Artemia which transfers chitin oligosaccharides from lipid to protein. Although similar enzymes have been described in mammalian systems (5,6), the crustacean enzyme appears to be unique in that only chitin oligosaccharides are transferred to protein acceptors; in mammals, high mannose oligosaccharides (GlcNAc 2, Man 9, Glc 3) are normally transferred (6).

MATERIALS AND METHODS

Chemicals. ^{3}H Acetic anhydride (30 Ci/mmol) was obtained from ICN. UDP-6-^{3}H N-acetyl-D-glucosamine (20.4 Ci/mmol) was purchased from NEN. All reagents for peptide synthesis were from Applied Biosystems. WGA-Agarose was from E-Y Labs. Endo H was purchased from Miles.

Preparation of peptides. Asn-Leu-Thr was kindly provided by Dr. W.J. Lennarz. Additional amounts of this peptide plus two others, Asn-Tyr-Thr and Dansyl-Ala-Ile-Glu-Asn-Ala-Thr-Leu were synthesized by the Merrifield technique using an Applied Biosystems Model 430A peptide synthesizer in the University of Florida Protein Chemistry Core Facility, Gainesville, FL.

Preparation of ^{3}H-peptide acceptors. Peptides were acetylated using ^{3}H acetic anhydride by the method of Welply et al.(6). After acetylation, the sample was treated with hydroxylamine and chromatographed on Sephadex G-10. All peptides were blocked at their carboxyl termini with methylamine via carbodiimide condensation, re-chromatographed on Sephadex G-10, lyophilized and stored at -20°C.

Oligosaccharyl transferase assay. Crude microsomes were prepared from larval brine shrimp (2-3) and resuspended in 20 mM HEPES buffer, pH 7.1, containing 0.2 mM PMSF, 10 mM $MgCl_2$ and 5 mM $MnCl_2$. After addition of peptide acceptor (200,000 cpm), samples (0.1 to 0.3 ml) were incubated (37°C/1 h) and then placed on ice. Samples were precipitated with TCA and the resultant supernatants were analyzed by paper chromatography as described (6).

Incubation of the heptapeptide acceptor was carried out as described above except that increasing amounts of unlabeled peptide and UDP-^{3}H-GlcNAc (200,000 cpm) were added to the incubations. After 1 h, samples were adjusted to 50% methanol and centrifuged; the supernatant was dried, redissolved in water and applied to an Amberlite XAD-7 column (1 x 5 cm). After washing the column with water, bound material was eluted with 100% methanol and counted.

Enzymatic and chromatographic procedures. Samples were digested with chitinase from Streptomyces griseus in 20 mM potassium phosphate buffer, pH 6.3, as described (7). Endoglycosidase H digestions were carried out in 20 mM acetate buffer, pH 5.9, using 50 milliunits enzyme per sample.

Samples (2 ml) were chromatographed on a BioGel P-4 column (1.5 x 95 cm) equilibrated with 25 mM acetic acid. Fractions (2 ml) were collected and analyzed for radioactivity by liquid scintillation counting. Analysis of samples by paper chromatography was carried out as described bu Welply et al. (6). Afterward, strips were dried, cut into 1 cm sections and radioactivity was measured in a liquid scintillation spectrometer.

Affinity Chromatography on WGA-agarose. Samples of reaction products were dissolved in 10 mM ammonium formate buffer, pH 7.4, and mixed with 5 ml wheatgerm agglutinin (WGA)-agarose at 4°C for 4 to 15 h. The matrix was then packed into a column and washed with the same buffer; the bound material was eluted with 0.2 M N-acetyl-D-glucosamine in the buffer.

RESULTS

Artemia microsomes glycosylate ^{3}H peptide acceptors. When larval brine shrimp microsomes were incubated with ^{3}H Ac-Asn-Tyr-Thr-NH$_2$, two new radioactive components were detected by paper chromatography (Fig 1B). The control peptide exhibited two radioactive components (Fig 1A:6-9;10-12); only the first component (6-9) appears to be a substrate for the oligosaccharyl transferase. The slower mobility of the two new radiolabeled species suggests that they have been glycosylated to varying degrees. The product obtained after incubation of larval brine shrimp microsomes with ^{3}H Ac-Asn-Leu-Thr-NH$_2$ is shown in Fig 1D. The untreated starting material (Fig 1C) contains one radiolabeled component (12-16) whereas after incubation, the sample contains two new peaks, one at the origin and a large amount of a second component (7-10). These results indicate that up to 70% of the added peptide may be glycosylated by the brine shrimp microsomes.

Properties of the glycosylation reaction. When Artemia microsomes were incubated with UDP-^{3}H N-acetylglucosamine and a dansylated heptapeptide containing the sequence -Asn-Ala-Thr-, a linear relationship between added peptide and radioactive product was seen from 40 to 400 µg added peptide (data not shown). These results support the notion that added peptides are glycosylated by Artemia microsomes, since the

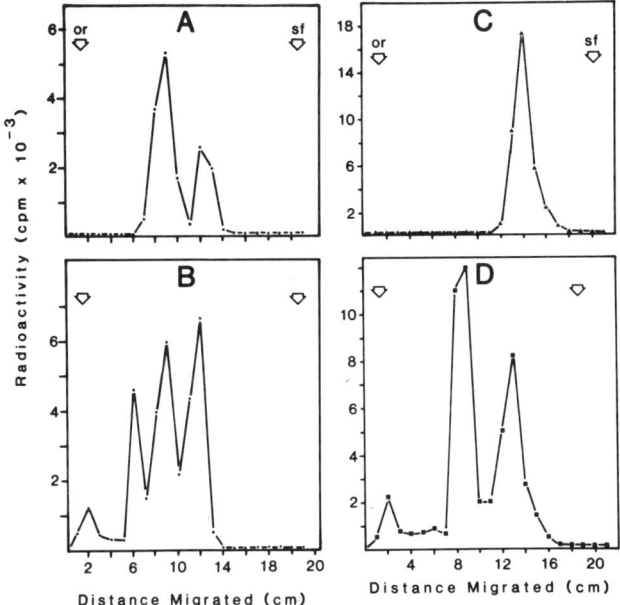

FIGURE 1. Paper chromatography of peptides and products after incubation with <u>Artemia</u> microsomes. A: ^{3}H Ac-Asn-Tyr-Thr-NH$_2$; B: Asn-Tyr-Thr after incubation. C: ^{3}H Ac-Asn-Leu-Thr-NH$_2$; D: Asn-Leu-Thr after incubation. O, origin; SF, solvent front.

radiolabeled product was purified by XAD-7 chromatography after incubation. Controls indicated that there was little or no glycosylation of endogenous peptides which were bound to the XAD-7 matrix.

When ^{3}H Ac-Asn-Leu-Thr-NH$_2$ was incubated with increasing amounts of membrane protein, a linear increase in radiolabeled product was observed up to 40 mg protein per sample (data not shown). In a separate experiment, activity was also shown to be proportional to time of incubation up to 60 min (data not shown).

When incubations with ^{3}H peptides were carried out with various cations, maximal acti-

TABLE 1
EFFECT OF DIVALENT CATIONS ON OLIGOSACCHARYL TRANSFERASE ACTIVITY[a]

Addition	Radioactivity	
	Total CPM	Percent
None	1,000	4
$MgCl_2$ (10 mM)	24,900	100
$MnCl_2$ (10 mM)	18,800	76
$CaCl_2$ (10 mM)	13,700	55
$ZnCl_2$ (50 mM)	180	1
EDTA (50 mM)	2,000	8

[a]Microsomes were incubated with ^{3}H Ac-Asn-Leu-Thr-NH_2 plus various divalent cations or EDTA (see Methods). After chromatography, radioactivity appearing in the 7-10 cm zone (see Fig 1D) was measured.

vity was observed with 10 mM $MgCl_2$ (Table 1). Little activity was observed in the absence of added cation or in the presence of a chelating agent such as 50 mM EDTA.

Characterization of the reaction product. When ^{3}H Asn-Tyr-Thr-NH_2 was incubated with brine shrimp microsomes and the product was analyzed by gel permeation chromatography, two new radioactive peaks were observed (Fig 2, fractions 32-40 and 41-51). In comparison to the starting material, the elution position of the products indicates the addition of 1 to 3 GlcNAc residues. Upon further incubation, larger products are observed (fractions 20-22), indicating additional glycosylation (data not shown).

When fractions 41-51 (Fig 2) were pooled and digested with chitinase, a smaller radiolabeled component was obtained which migrated like the original peptide (Fig 3A). On the other hand, digestion of the glycopeptide product with endoglycosidase H yielded no change in the elution position (Fig 3B). Thus, the glycopeptide does not appear to contain a high

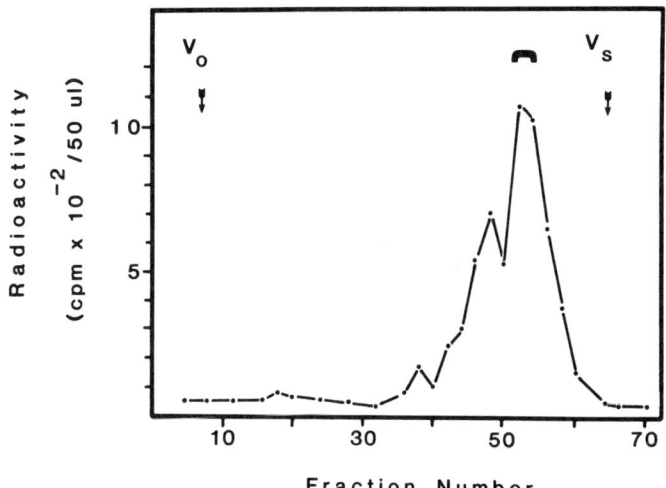

FIGURE 2. BioGel P-4 chromatography of ^{3}H-Ac-Asn-Tyr-Thr-NH$_2$ reaction mixture after incubation and TCA precipitation. Bar indicates position of starting material.

mannose oligosaccharide. These results indicate that the oligosaccharide attached to synthetic peptides in the <u>Artemia</u> system contains GlcNAc residues exclusively. In a related experiment, greater than 70% of the ^{3}H GlcNAc incorporated into the synthetic heptapeptide from UDP-^{3}H GlcNAc was removed by digestion with chitinase.

<u>Purification of radiolabeled product by affinity chromatography.</u> The ^{3}H-Ac-Asn-Tyr-Thr-NH$_2$ reaction product was purified after incubation by affinity chromatography on WGA-agarose. The bound material eluted by 0.2 M GlcNAc was analyzed by gel chromatography (Fig 4). The size of the product indicates that from two to three GlcNAc residues have been added to the peptide acceptor.

Labeling of the Oligosaccharide-Lipid Pool and Transfer to Peptide Acceptors. When ^{3}H peptide was incubated with microsomes plus ^{14}C UDP-GlcNAc, a double labeled product was detected by gel permeation chromatography (data not shown). Since this material could contain unreacted UDP-^{14}C GlcNAc, the peak was pooled and analyzed by paper chromatography. A double labeled peak distinct from nucleotide-sugar was observed; these results suggest that both ^{3}H and ^{14}C radiolabels reside in the same component.

Oligosaccharyl Transferase Activity During Larval Development. When the specific activity of oligosaccharyl transferase was assayed in membranes prepared from larval brine shrimp at various stages of development, a biphasic pattern was observed with maximal activity at

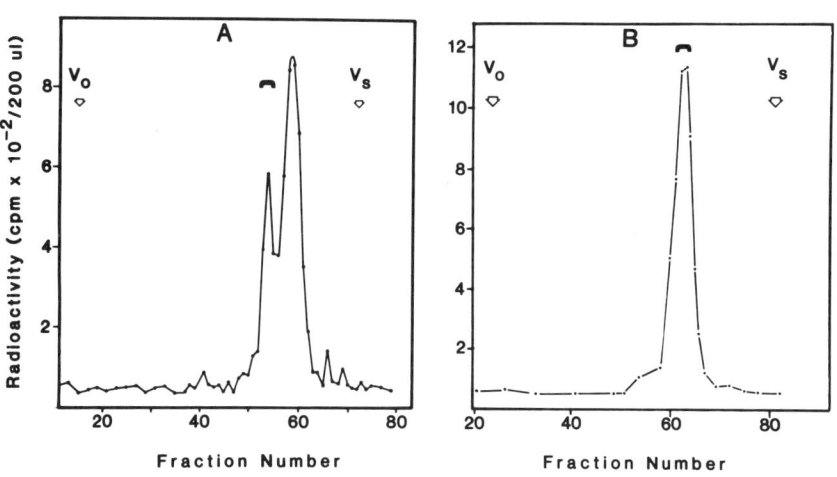

FIGURE 3. BioGel P-4 chromatography of peptide reaction product (Fig 2, #41-51) after digestion with A: chitinase or B: endoglycosidase H for 15 h (see Methods). Bracket indicates position of starting material.

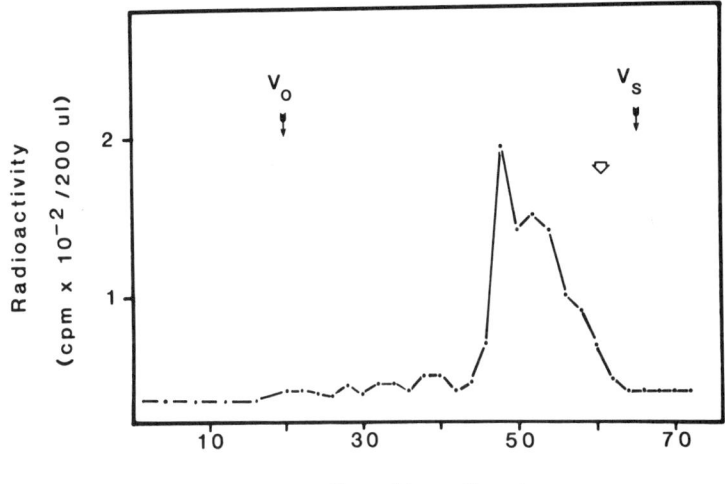

FIGURE 4. BioGel P-4 chromatography of reaction product purified by affinity chromatography on WGA-agarose. Arrow indicates elution position of the starting material.

24 and 72 h of growth (data not shown). Similar results have been obtained with the dolichol kinase of Artemia (1). The biphasic behavior of the enzyme corresponds to those times in early larval development of the brine shrimp when chitin synthesis and cuticle formation occur.

DISCUSSION

Glycosylation of exogenous peptide acceptors containing the sequence -Asn-Axx-Thr- by an enzyme from larval brine shrimp microsomes has been detected. The enzyme appears to require Mg^{+2} ions and transfers a chitin oligosaccharide from a lipid intermediate to either an endogenous acceptor or an exogenous peptide. The structure of such lipid-linked chitin oligosaccharides from Artemia has been described previously (3); oligosaccharides range from 2 to 8 GlcNAc residues.

In this study, two glycopeptides were isolated by gel permeation chromatography on BioGel P-4 after incubation of tripeptide acceptors with brine shrimp microsomes. The size of the glycopeptides indicates that the added oligosaccharide is quite small, eg. 2 to 4 GlcNAc residues. Similar observations have been made in mammalian systems by Chen and Lennarz (8) who detected the transfer of chitobiose to glycoprotein acceptors in hen oviduct preparations.

The assay procedure described in this paper utilizes endogenous oligosaccharide lipids as glycosyl donors. Although UDP-GlcNAc is not required for activity, the oligosaccharide pool can be labeled with UDP-^{14}C GlcNAc. No increased transferase activity was detected when incubations were carried out in 0.5 mM UDP-GlcNAc. Preliminary experiments were conducted to attempt the purification of the oligosaccharide lipids from _Artemia_ followed by addition back to fresh microsomes, with the intent of chasing the oligosaccharide into a protein pool. All efforts in this area failed, despite attempts using detergents or liposomes as carriers for the added oligosaccharide-lipids. The present assay procedure obviates the need for add back experiments by using a synthetic ^{3}H peptide.

The specific activity of the oligosaccharyl transferase and dolichol kinase from larval _Artemia_ rises as the time for molting approaches. Thus, the two enzymes appear to be developmentally regulated in response to the molt cycle of the organism.

ACKNOWLEDGEMENTS

I thank Drs. Ben M. Dunn and W.J. Lennarz for providing the peptides. Ms. S. Martin, Ely Klar, and Mr. J.W. Hightower provided technical assistance. Ms. Ginger Sanders typed the manuscript.

REFERENCES

(1) Horst, M.N. Dolichol phosphorylation occurs via a CTP-dependent reaction in larvae of Artemia salina. Submitted to J Exp Zool.
(2) Horst, M.N. Isolation of a crustacean N-acetyl-D-glucosamine-1-phosphate transferase and its activation by phospholipids. Submitted to J Cellular Bioch.
(3) Horst M N (1981). The biosynthesis of crustacean chitin. Isolation and characterization of polyprenol-linked intermediates from brine shrimp microsomes. Arch. Biochem. Biophys. 223:254-263.
(4) Horst M N (1980) The biosynthesis of crustacean chitin by a microsomal enzyme from larval brine shrimp. J Biol Chem 256:1412-1419.
(5) Ronin C, Granier C, Van Reitschoten J, and Bouchilloux S (1978). Enzymatic transfer of oligosaccharide from oligosaccharide-lipids to an Asn-Ala-Thr containing heptapeptide. Biochem Biophys Resh Comm 81:772-778.
(6) Welply J, Shenbagamurthi P, Lennarz WJ and Naider F (1983). Substrate recognition by oligosaccharyltransferase. Studies on glycosylation of modified Asn-X-Thr/Ser tripeptides.J Biol Chem 258:11856-11863.
(7) Molano J, Duran A and Cabib E (1977). A rapid and sensitive assay for chitinase using tritiated chitin. Anal Biochem 83:648-656.
(8) Chen W and Lennarz WJ (1977). Metabolism of lipid-linked N-acetylglucosamine intermediates. J Biol Chem 252:3473-3479.

PHOTOCHEMICAL APPROACHES TO THE PREPARATION
OF LH-RH BY FRAGMENT CONDENSATION

Jean Gauthier, François Bruderlein, Norman Aubry,
S. Rakhit, Serge Valois and Yves Bousquet

Department of Chemistry, Bio-Méga Inc.,
Laval, Québec, Canada, H7S 2G5

INTRODUCTION

Recently in our laboratories we have developed photo-chemical techniques suitable to prepare a variety of protected peptides on a large scale. In search for a more efficient process to prepare human LH-RH (I), we have studied several strategies based on the solid phase approach. All fragments generated in the protected form were utilized to find an efficient condensation procedure to generate LH-RH in good yield and purity.

$$\langle Glu^1-His^2-Trp^3-Ser^4-Tyr^5-Gly^6-Leu^7-Arg^8-Pro^9-Gly^{10}-NH_2$$
LH-RH(I)

RESULTS AND DISCUSSION

A. COUPLING OF FRAGMENT IN SOLUTION

A stepwise solid phase synthesis of segment 1-9 using a photochemical support was achieved using short washing cycles. A detailed solid phase procedure was described recently by Castro et al (1). The nonapeptide thus produced and generated photochemically in very good yield has the following side-chain protections: Z for $\langle$Glu, Tos for His and Arg, N-formyl group for Trp, Bzl(2,6-diCl) for Tyr and Bzl for Ser (SCHEME 1).
The activation of the BOC-amino acid (symmetrical anhydrides in the case of His and $\langle$Glu) was carried out with BOP in methylene chloride. All couplings were achieved in the presence of DIEA within thirty (30) minutes.

(1+9) SEGMENT CONDENSATION APPROACH

BOC-Pro-O-PHOTORESIN

1) 30% TFA/CH_2Cl_2 BOC-Arg-(Tos)-OH/BOP/DIEA
2) 30% TFA/CH_2Cl_2 BOC-Leu-OH/BOP/DIEA
3) 30% TFA/CH_2Cl_2 BOC-Gly-OH/BOP/DIEA
4) 30% TFA/CH_2Cl_2 BOC-Tyr(2,6-diCl-Bzl)OH/BOP/DIEA
5) 30% TFA/CH_2Cl_2 BOC-Ser(Bzl)-OH/BOP/DIEA
6) 30% TFA/CH_2Cl_2 BOC-Trp(formyl)-OH/BOP/DIEA
7) 30% TFA/CH_2Cl_2 BOC-His(Tos)-OH/DCC
8) 30% TFA/CH_2Cl_2 <Glu(Z)-OH/DCC

<Glu^1-His^2-Trp^3-Ser^4-Tyr^5-Gly^6-Leu^7-Arg^8-Pro^9-O-PHOTORESIN
    Z   Tos   CHO   Bzl   Bzl(2,6-diCl)   Tos

9) 350 nm ↓ hv

<Glu-His-Trp-Ser-Tyr-Gly-Leu-Arg-Pro-OH
   | | | | | |
   Z   Tos CHO Bzl Bzl(2,6-diCl) Tos

10) H-Gly-NH_2, BOP/NMM
11) HOBT/DMF ↓

<Glu-His-Trp-Ser-Tyr-Gly-Leu-Arg-Pro-Gly-NH_2
   | | | | |
   Z   CHO Bzl Bzl(2,6-diCl) Tos

12) HF   ↓ $(CH_2SH)_2$/anisole

<Glu-His-Trp-Ser-Tyr-Gly-Leu-Arg-Pro-Gly-NH_2

SCHEME 1: Solid Phase Synthesis of Nonapeptide on Photoresin followed by Photolysis and Condensation of Protected Peptide with Glycinamide.

Further condensation of the nonapeptide with glycinamide in DMF with BOP (benzotriazolyloxytris (dimethylamino) phosphonium hexafluorophosphate) in the presence of NMM afforded the decapeptide in very good yield. The tosyl group of His was partly cleaved under these conditions (FIG. 1) and further treated with HOBT in DMF for complete removal.

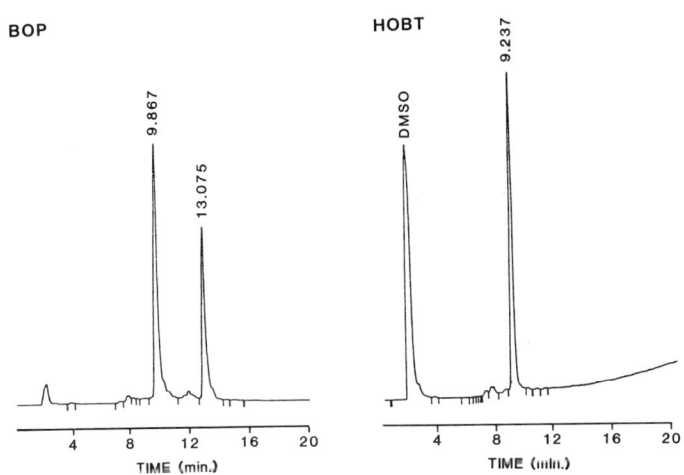

FIG. 1: Effect of HOBT on the Selective Deprotection of Histidine

Final deprotection with hydrogen fluoride in the presence of anisole and ethanedithiol (1 hr. at 0°) afforded the crude LH-RH (FIG. 2). Purification by reverse phase HPLC gave the pure trifluoroacetate salt.

B. COUPLING OF FRAGMENTS ON SOLID SUPPORT

Two other protected segments have been prepared in good yield and purity in the manner described earlier. The condensation of the resin-peptides prepared from BHA resin and the two protected segments was achieved using BOP/NMM in DMF as coupling conditions as shown in the following:

1) <Glu-His-Trp-Ser-OH & Tyr-Gly-Leu-Arg-Pro-Gly-R
          Z   Tos CHO Bzl   Bzl(2,6-diCl) Tos

2) <Glu-His-Trp-Ser-Tyr-OH & Gly-Leu-Arg-Pro-Gly-R
          | | | | |
          Z Tos CHO Bzl   Bzl(2,6-diCl)   Tos

Segment condensations of the (4+6) and (5+5) types were less attractive for preperative purposes. The HPLC patterns of the post HF material showed contaminants near the main respective peaks. The decapeptide obtained by the (4+6) scenario had a major contaminant (25%) which was separated by preparative HPLC. The amino acid analysis and mass spectrum of this material are comparable to LH-RH data and suggestive for racemization at serine.

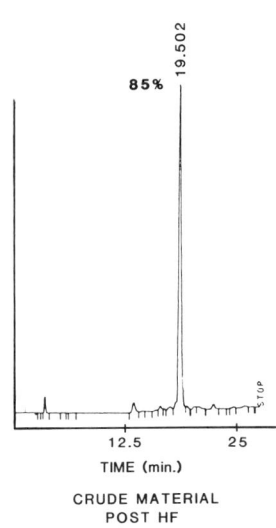

FIG. 2: HPLC Chromatogram of Crude LH-RH.

IDENTIFICATION

Amino acid analysis was carried out on all peptides after chemical hydrolysis (only tryptophane had lower values).
The protected segments were cleaved with hydrogen fluoride, purified and characterized: Aminoacid analysis, mass spectrometry and HPLC analysis.
HPLC co-injection with reference LH-RH showed that the two products co-migrate.

D-Pro^9-(LH-RH) was prepared by solid phase synthesis from BHA resin as a reference material. The degree of racemization of proline after condensation was evaluated by HPLC analysis. The HPLC pattern of a reference mixture and LH-RH obtained by condensation (FIG. 3) showed minor contaminants (<0.7%).

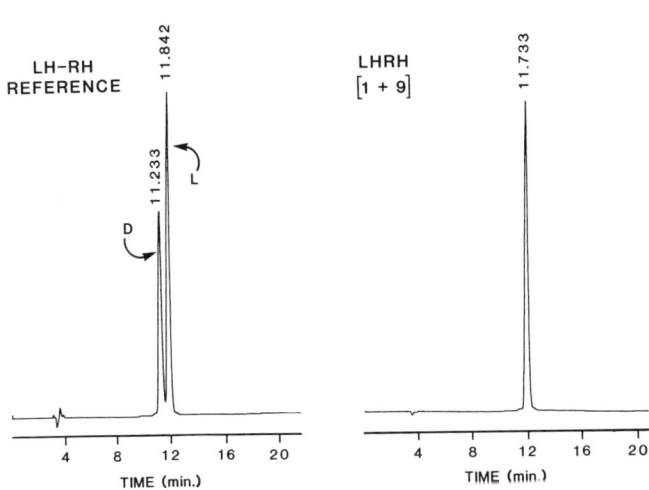

FIG. 3: HPLC Chromatograms of LH-RH
  Column: C_{18}Vydac (30 X 0.39 cm, 10 mm particle size)
  Eluent: A: 0.1% TFA in acetonitrile
          B: 0.1% TFA in water
  Flow-rate: 1 ml / min., Detection: 230 nm

REFERENCE

1. D. Le Nguyen, Heitz A, and Castro J (1987). J. Chem. Soc., Perkins Trans, 1;1915

II. PREDICTION OF PEPTIDE AND PROTEIN STRUCTURES

INSERTION OF A WATER MOLECULE INTO AN α-HELIX BACKBONE AND OTHER MODES OF HYDRATION: X-RAY STRUCTURE ANALYSIS[1]

Isabella L. Karle and P. Balaram

Laboratory for the Structure of Matter, Naval Research Laboratory, Washington, D. C. 20375-5000 and Molecular Biophysics Unit, Indian Institute of Science, Bangalore 560 012, India

ABSTRACT Single crystal x-ray diffraction structure analyses have identified the manner in which water molecules are essential to creating mini-polar areas on apolar helices. Amphiphilic helical structures have a sequence of residues such that the polar side chains occur on one side of the helix (with affinity for water and ions) and the apolar side chains occur on the other side of the helix (with hydrophobic characteristics). Completely apolar peptides, such as analogs of membrane-active peptides, can acquire amphiphilic character by at least two different means, for example: (I) Curving of the helix by the presence of Pro residues; and (II) Insertion of a water molecule into the helical backbone. In each case, the distortion of the helical backbone by curvature or by stretching exposes carbonyl groups to the outside environment with the consequence of attracting several water molecules by hydrogen bonding. As a result, one side of the helical molecules, even though there are only apolar residues, acquires a polar character. Examples will be taken from the crystal structures of a number of 10 to 16 residue peptides.

[1] This work was supported in part by National Institutes of Health Grant GM30902 and in part by a grant from the Department of Science and Technology, India.

INTRODUCTION

The characterization and synthesis of amphiphilic helical peptides has been presented by Kaiser and Kezdy (1,2). In helical amphiphilic peptides, residues with lipophilic sidechains occur on one side of the helical backbone, whereas residues with hydrophilic side-chains occur on the other side of the helix. Many peptides that act at lipid-water interfaces assume a unique amphiphilic secondary structure which is compatible with the anisotropy of the interface.

The peptides described in this paper, on the other hand, have been synthesized solely with apolar residues Trp, Ala, Val, Leu, Ile, Phe, Pro and Aib. (The Aib residue, α-aminoisobutyric acid, is a common component of naturally occurring peptides that are involved in transmembrane pore formation.) The peptides have been designed to be apolar analogs of naturally occurring ionophores such as zervamicin IIA (3). Initially, the intent was to study the role of specific side chains in helix formation and helix aggregation (4). The serendipitous discovery of water molecules that are hydrogen-bonded to C=O and NH moieties in the helical backbones and that form mini-polar areas on the exterior of these peptides has expanded the scope of our investigations. The molecular aggregate in the crystal of one of the model apolar peptides suggests a mode of ion transport through membranes by largely hydrophobic peptides that have been hydrated.

METHODS

X-ray diffraction data have been collected with CuKα radiation from single crystals of the peptides grown from various solvents such as methanol/water, dioxane, dimethyl sulfoxide/water, etc. For data collection, a single crystal of most of the peptides was sealed in a very thin-walled glass capillary along with a drop of the mother liquor, since the crystals lost their integrity upon drying. Each of the peptides scattered very well so that thousands of independent reflections were measured for each (8000 data for the 16-residue peptide). None of the crystals contained any atoms heavier than oxygen, therefore direct phase determination methods were used to solve the initial helical structures (5), and rotation/translation functions (6,7) based on a fragment of the initial structure were used to derive the structures of succeeding helical peptides. Least-square refinement was performed on all the coordinates of C,

N and O atoms and their anisotropic thermal parameters. Hydrogen atoms bonded to C atoms were placed in idealized positions, whereas the H atoms bonded to N atoms were found in difference maps and their coordinates were refined. The agreement factor between the observed intensities of the reflections and the calculated structure factors based on the refined positions of the atoms and their thermal parameters (R values) ranged from 5% to 7.5%. The standard deviations for the bond lengths are near ± 0.01 Å, for the bond angles they are near $\pm 0.8°$ and for the torsion angles they are near $\pm 0.9°$

CONFORMATIONAL STABILITY

The question of variability in the conformation of a particular peptide in solvents of different polarity or in crystals grown from solvents of different polarity can be examined by deriving the structures of different polymorphs of the same peptide. Polymorphs are different crystal forms of a particular substance with different cell dimensions and usually different space groups, packing, and molecular environment. For peptides having 10 or more residues, examination of polymorphs shows a surprising degree of constancy of conformation. Antamanide, for example, a cyclic decapeptide, crystallized from solvents ranging from CH_3CN/H_2O to n-hexane/ethyl acetate (dried) to form five different polymorphs, having three to sixteen water sites per peptide molecule, has very nearly the same folding for both the backbone and the side chains in each of the quite different crystal forms (8). The conformational stability is maintained by numerous internal hydrogen bonds and by internal hydrophobic attractions between rings in Pro and Phe residues. External attractions between molecules are weak compared to the internal attractions.

Linear peptides have similar conformational stability. The 1-10 residue apolar analog of zervamicin IIA, Boc-Trp-Ile-Ala-Aib-Ile-Val-Aib-Leu-Aib-Pro-OMe, crystallized from $DMSO/H_2O$ (9), dioxane (10), and isopropanol (11), forms three different polymorphs having space groups P1, $P2_1$, and P1, respectively. The second polymorph is anhydrous, while the first has two water molecules of crystallization in the head-to-tail region and the third has no water but a co-crystallized isopropanol molecule. In all three polymorphs, the backbones have almost identical torsion angles. There are some differences in the conformations of the side chains in the Ile residues. The molecule in the first polymorph (9) is shown in Fig. 1.

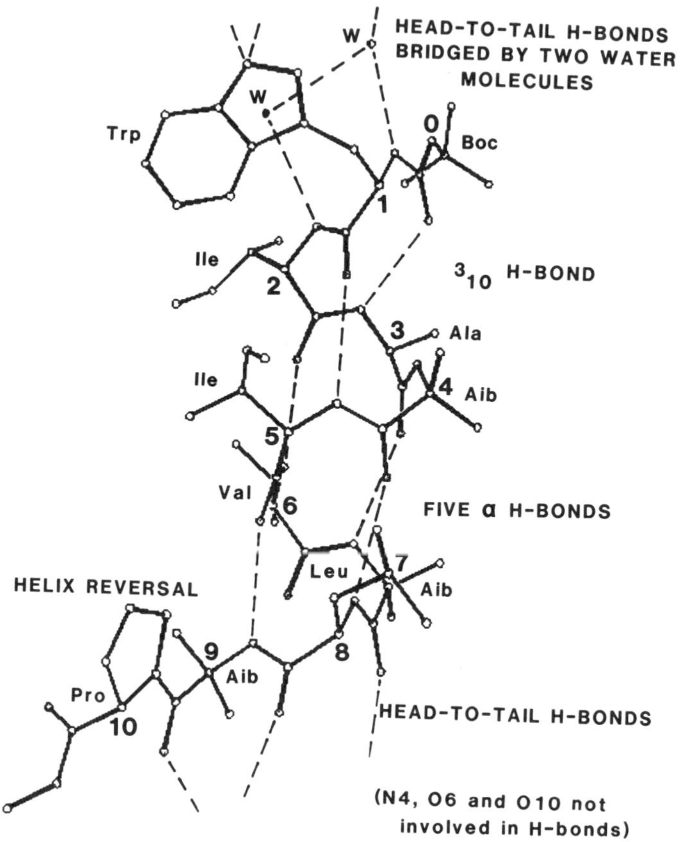

FIGURE 1. Conformation of the apolar decapeptide and the two molecules of water (W) in the head-to-tail hydrogen-bonding region (9), drawn by computer using the experimentally determined coordinates. The C^{α} atoms are labeled 1-10. Hydrogen bonds are indicated by dashed lines.

Furthermore, the 16-residue apolar analog of zervamicin IIA (vide infra) has the same 1-10 sequence as the 10-residue analog above. The conformation of residues 1-9 is the same as shown in Fig. 1.

HELIX TYPES

The 16-residue apolar peptide mentioned above has the sequence Boc-Trp-Ile-Ala-Aib-Ile-Val-Aib-Leu-Aib-Pro-Ala-Aib-Pro-Aib-Pro-Phe-OMe. The peptide backbone makes a continuous spiral that begins as a 3_{10}-helix at the N-terminus, changes to an α-helix for two turns, and ends in a spiral

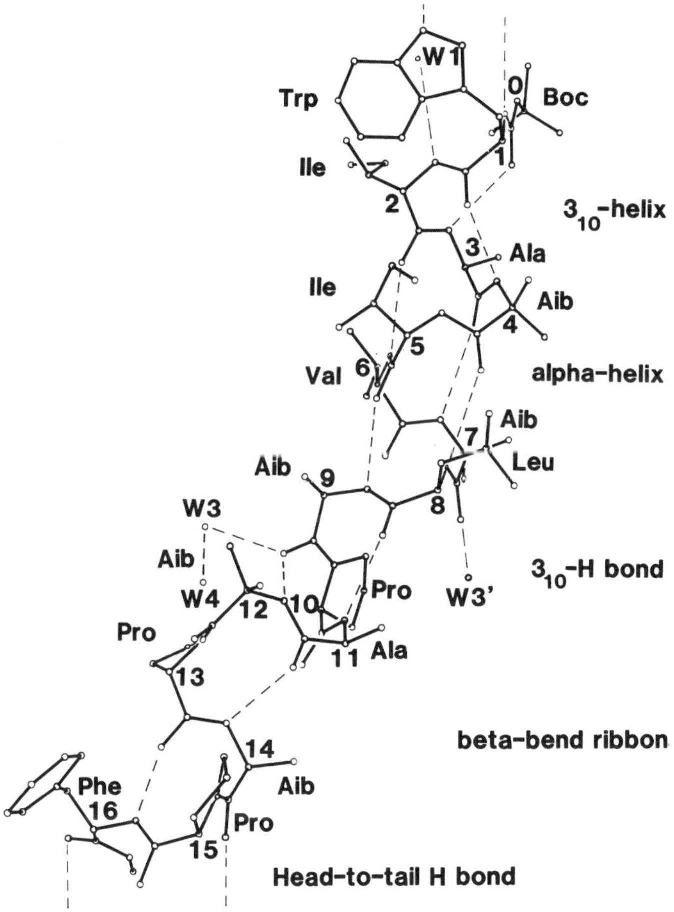

FIGURE 2. Conformation of the 16-residue peptide with three different types of helices. The C^{α} atoms are labeled 1-16. Water molecules are labeled W. Hydrogen bonds are indicated by dashed lines (12).

of three β-bends in a ribbon, Fig. 2 (12). A schematic
representation of the β-bend is shown in Fig. 3, although
the spiral twist in the ribbon is not shown. Each of the
β-bends has a proline residue at one of the corners, although
not at the same corner in the three bends. The entire structure can be described as a mixed $3_{10}/\alpha$-helix with the three
proline residues being incorporated into the helix with a
loss of three hydrogen bonds. Distortions from idealized
conformational angles reflect mainly the positioning of the
proline residues. It is notable that the helix continues
even in the presence of three proline residues.

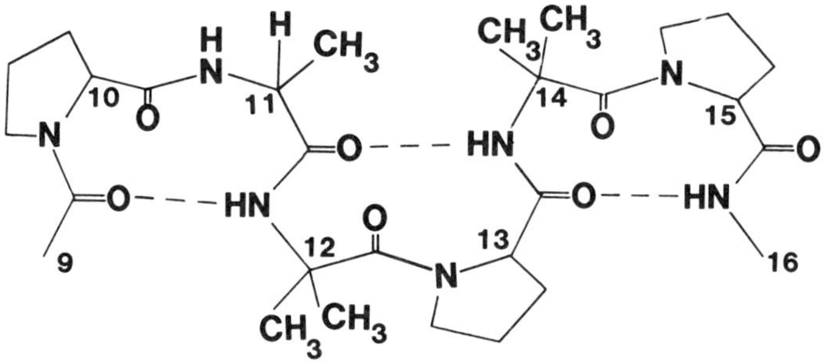

FIGURE 3. Schematic representation of β-bend ribbon.
Note that the proline residues occur either at the i+1 or
i+2 positions in the β-bends (12).

PARALLEL PACKING

In the three polymorphic crystals of the 10-residue
analog above, as well as in the crystal of the 16-residue
analog above, the helical molecules form columns by head-to-tail hydrogen bonding, both directly between the peptide
molecules and mediated by water or other solvent molecules.
The unusual feature in all four crystals is the parallel
packing of the helical columns. In a survey of protein
structures (13), neighboring α-helices have been found to
be antiparallel or tilted up to ~90°, but not approaching
parallel packing.

Three of the four peptide structures discussed above
crystallize in space group P1 with one peptide molecule per
cell. Since there are no symmetry elements in space group
P1, the molecules must repeat only by translation along the
three axes of the cell. Thus the aggregation of the helical
columns must necessarily be parallel. The parallel packing

of the 16-residue helical peptide is shown in Fig. 4.

One of the polymorphs of the 10-residue peptide crystallizes in space group $P2_1$ (10). This space group has one twofold screw axis, thus it could be possible that adjacent helices be related in an antiparallel mode. However, the cell has a very small cross-section, with a = 9.42 Å and c = 10.55 Å, just enough area for the cross-section of one helical molecule with its side chains. The b axis is long, 36.39 Å, and accommodates two peptide molecules, head-to-tail along the helix axis, that are related by the two-fold screw. The result is that in the directions perpendicular to the helix axis, all the packing is parallel.

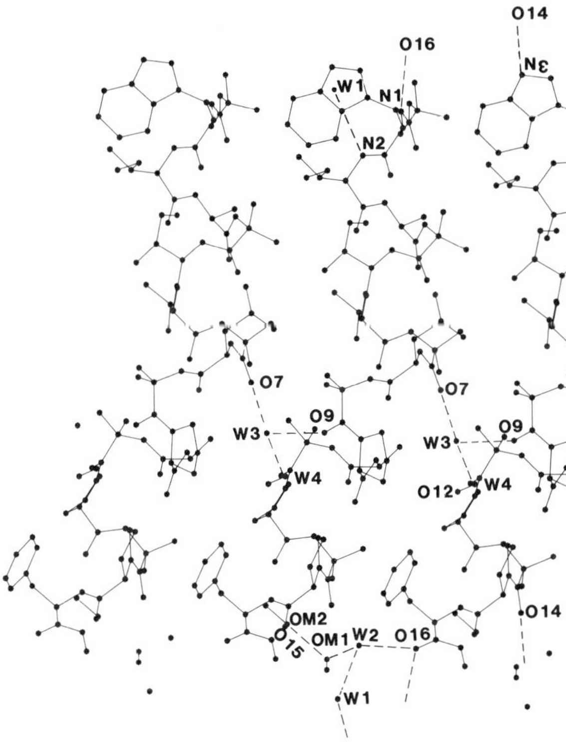

FIGURE 4. Parallel packing of helical molecules of the 16-residue peptide in three adjacent P1 cells (12). In the vertical direction, there is head-to-tail hydrogen bonding to adjacent cells. In the direction perpendicular to the paper, the helical columns also repeat in a parallel fashion.

HYDRATION OF HELIX BACKBONES

Apolar peptides, such as membrane-active peptides and their analogs, can acquire polar character by at least two different means, for example: Bending of the helix by the presence of Pro residues; and insertion of a water molecule directly into the helix backbone.

Bending of the Helix Backbone

The helix of the 16-residue peptide shown in Figs. 2 and 4 is bent by ~33° near Pro-10. The bend is caused by replacing the amide hydrogen atom by the bulky pyrrolidine ring. Obviously there is a loss of an intrahelical hydrogen bond at every proline residue. The bend in the helix causes carbonyl oxygens O(7), O(9), O(10) and O(12) to be exposed to the exterior environment. Water molecules are attracted to the protruding C=O groups and form hydrogen bonds with them (except for O(10)). In this manner, mini-polar areas are formed on a helix composed of apolar residues. The mini-polar area is in the vicinity of small apolar residues, that is, Ala and Aib. Residues with large chains, Ile, Leu and Val, are located nearer to the N terminus.

Neighboring helices are linked laterally by means of hydrogen bonds between exposed carbonyl oxygens and water molecules serving as bridges, see, e.g. O(7)···W(3)···O(9) and O(9)···W(3)···W(4)···O(12) in Fig. 4. It may be possible that such mini-polar areas assist membrane active peptides in aggregating for pore formation.

Structural similarities exist between melittin (14), a 26-residue membrane surface-active peptide, alamethicin (15), a 20-residue trans-membrane pore-forming peptide, and the present 16-residue apolar peptide. Each of the three peptides form a bent helix with the bend occurring at a proline residue near the middle of the sequence. Each peptide has predominantly, if not completely, hydrophobic residues. The larger, more hydrophobic side chains occur mostly on the concave side of the bend of the helix. Bound water molecules near the exposed carbonyl at the bend of the helix have not been reported yet for melittin or alamethicin. However, melittin and alamethicin occur in crystals with two and three molecules per asymmetric unit, respectively, making them much larger structural problems than the 16-residue peptide described here. So far, the least-squares refinement for melittin has been reported to a resolution of 2.5 Å and R = 28% and that for alamethicin to a resolution of 1.5 Å and R = 15.5%. This compares to a resolution of 0.9 Å and R = 7.3% for the 16-residue peptide. Hence it is not unusual that bound water molecules have not been characterized yet in the larger structures.

Insertion of Water into Helix Backbone

The recent crystal structure determinations of two apolar peptides, Boc-Aib-Ala-Leu-Aib-Ala-Leu-Aib-Ala-Leu-Aib-OMe (16) and Boc-Ala-Leu-Aib-Ala-Leu-Aib-OMe (17) have

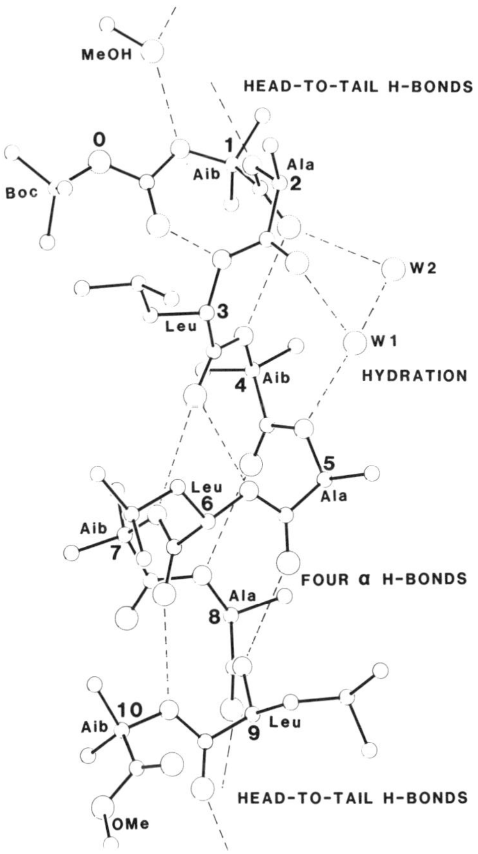

FIGURE 5. Conformation of Boc-Aib-(Ala-Leu-Aib)$_3$-OMe. Included are the methanol molecule that bridges the head-to-tail hydrogen bonding, and the two water molecules (W) that bind to backbone atoms. Water W(1) is inserted into the helical backbone.

shown an unusual occurrence. In both peptides, a water molecule is inserted into the helical backbone between the carbonyl of the first Ala residue and the amide group of the second Ala residue. The conformation of the decapeptide is shown in Fig. 5. The helix of the backbone near the C-terminus assumes an α- form with torsional angles φ and ψ near -60° and -45°, respectively, whereas the helix near the N-terminus is distorted by an entry of a water molecule between N of Ala(5) and O of Ala(2), spreading N(5) and O(2) to a distance of 5.05 Å. The water molecule forms hydrogen bonds with backbone atoms, N(2)···W(1) and W(2···O(2) with distances 2.93 Å and 2.86 Å, respectively. The distortion of the helix exposes carbonyls in residues (1) and (4) to the external environment with the consequence that additional water molecules are attracted. Hydrogen bonds are formed between W(2)···O(1), W(1)···W(2), and W(2)···O(4) (the latter with a symmetry related molecule).

Amphiphilic mimic. The two water molecules create a mini-solvation region in very close contact with the peptide backbone. The representation of the peptide by a helical

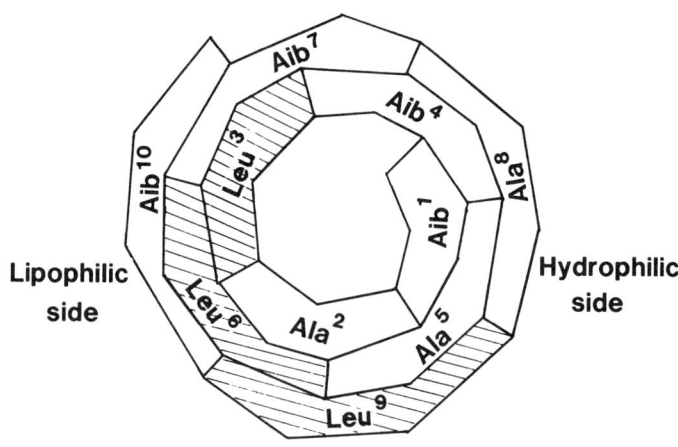

FIGURE 6. Two-dimensional axial projection (2) of the helical conformation of Boc-Aib-(Ala-Leu-Aib)$_3$-OMe. The Leu residues are shaded.

wheel (2,18) in Fig. 6 shows that the large hydrophobic side-chains are grouped on the left side, while the residues whose backbones form hydrogen bonds with the water molecules, that is, Aib-1, Ala-2, Aib-4 and Ala-5, have small apolar side-chains and are grouped on the right side. The right side has become polar by virtue of exposure of carbonyl and amide moieties to the external environment and the presence of inserted and bound water molecules. As a consequence, this apolar peptide mimics an amphiphilic helix.

Amphiphilic packing mode. Along the direction of the helix axis (vertical as viewed in Fig. 7), there is head-to-tail hydrogen bonding between successive molecules that are repeated simply by translation. Owing to the two-fold screw

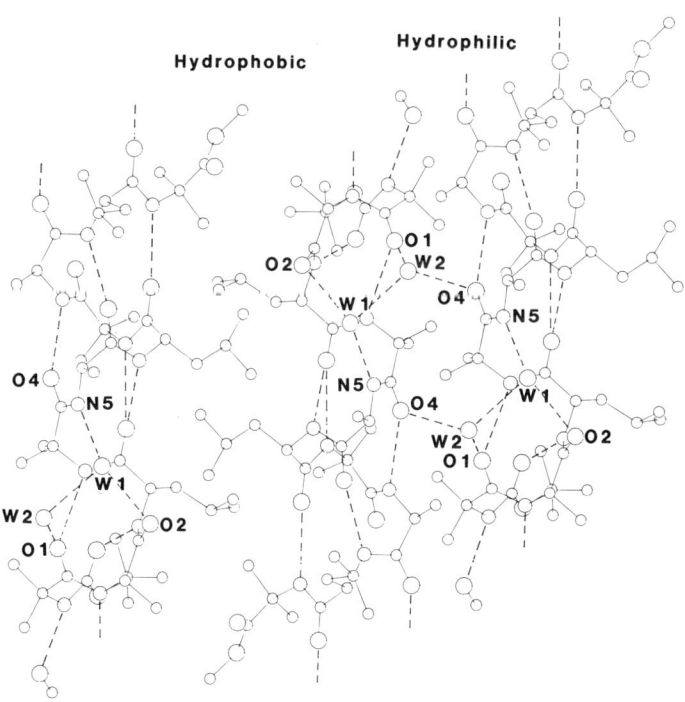

FIGURE 7. Molecular packing in the crystal of Boc-Aib-(Ala-Leu-Aib)$_3$-OMe showing the hydrophobic and hydrophilic contacts between the three antiparallel helices shown. The first and third molecules are related by translation of one cell length (16).

symmetry operation in space group P2$_1$ (perpendicular to the paper as viewed in Fig. 7) neighboring helices run in antiparallel directions. Between the left and center molecules all the intermolecular contacts are hydrophobic, mainly characterized by the interdigitation of leucyl side-chains from neighboring helices. Between the center and right molecules there are polar contacts in the form of hydrogen bonds, mediated by the presence of four molecules of water per neighboring pair of peptide molecules. The occurrence of hydrophobic contacts on one side of the helix and polar contacts on the other side give rise to a helix with amphiphilic character despite the presence of only nonpolar residues.

Polar channel formation. A view of the packing of molecules (looking down the helices; perpendicular to the view in Fig. 7) is shown schematically in Fig. 8. The shaded areas on either side indicate the hydrophobic contacts. The

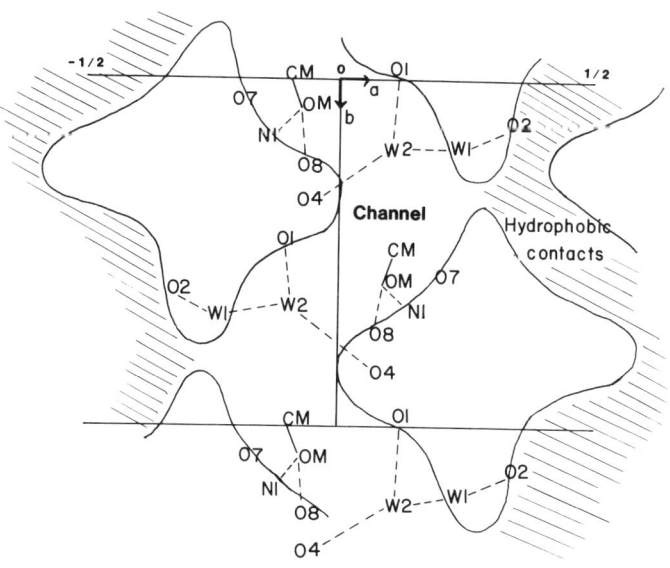

FIGURE 8. A correctly proportioned, schematic diagram of the packing of the helices in the crystal of Boc-Aib-(Ala-Leu-Aib)$_3$-OMe. Only the outline of the helical molecules is shown along with the O and N atoms that are on the surface. Positions of water molecules (W) and methanol (CM-OM) are shown (16).

middle of the diagram shows the polar region between helices that contains the water molecules and a number of carbonyl oxygens that extend away from the helix backbone and line the polar region. The direction of these carbonyls toward the surface of the peptide is mainly caused by the bending of the helix backbone as discussed above. Furthermore, the carbonyl oxygens are not shielded by the side chains. The last component of the polar region is a methanol solvent molecule (CM-OM in Fig. 8) that mediates head-to-tail hydrogen bonding in the columns of helical molecules. The methanol molecules cocrystallized from the methanol/water solution from which crystals were grown. From a different solvent mixture, the crystals presumably would have a different cocrystallized molecule, perhaps even water, in this location.

The polar region has been labeled a channel in Fig. 8. This nomenclature is a misnomer since "channel" implies a finite cross-section. In this structure, the polar region extends the length of the crystal in the direction of the helix axes and meanders along the width of the crystal in

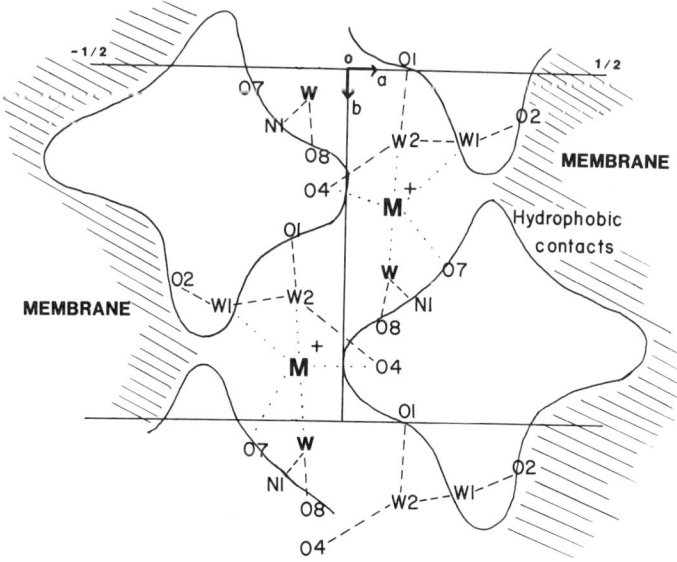

FIGURE 9. A hypothetical multiple channel for ion transport based on the crystal structure results shown in Fig. 8.

the b axis direction. Perhaps the nomenclature "polar crevasse" would be more descriptive.

<u>Speculation for ion transport.</u> The crystal packing of the helical molecules shown in Fig. 8 can serve as a model for a hypothetical pore that serves to transport ions. Let us assume that the hydrophobic contacts are interfaces with a membrane as shown in Fig. 9. Of course, the helical peptide should be long enough to span the thickness of a membrane. Furthermore, let us replace the methanol molecules (CM-OM) in Fig. 8 with water molecules (bold-face W) in Fig. 9. With these changes, it appears to be possible to pass a small cation, Na^+, for example, along the helix direction, in the locations marked with M^+. Transient $M^+\cdots O$ ligands would form to carbonyl oxygens or water molecules at various levels as indicated by the dotted lines. An intriguing feature of this model is that the helical molecules do not aggregate to form isolated pores, but rather, a polar crevasse with multiple pore areas that would accommodate simultaneous ion transport for a number of cations upon voltage stimulation.

REFERENCES

1. Kaiser ET, Kezdy FJ (1983). Secondary structures of proteins and peptides in amphiphilic environments (A review). Proc Natl Acad Sci USA 80:1137.
2. Kaiser ET, Kezdy FT (1984). Amphiphilic secondary structure: design of peptide hormones. Science 223:249.
3. Rinehart KL Jr, Pandey RC, Moore ML, Tarbox SR, Snelling CR, Cook JR, Milberg RH (1979). Mass spectroscopic studies of peptide structures. In Gross E, Meienhofer J (eds): "Peptides and Biological Function, Proc Sixth Amer Peptide Symp," Rockford, IL: Pierce Chemical Co, p. 59.
4. Mathew MK, Balaram P (1983). A helix dipole model for alamethicin and related transmembrane channels. FEBS Lett 157:1.
5. Karle J, Karle IL (1966). The symbolic addition procedure for phase determination for centrosymmetric and noncentrosymmetric crystals. Acta Cryst 21:849.
6. Beurskens PT, Bosman WP, Doesburg HM, van den Hark T, Prick PAJ, Noordik JH, Beurskens G, Gould RO, Parthasarathi V (1983). DIRDIF: computational and statistical aspects of direct methods applied to difference structure factors. In Srinivasan R, Sarma RH (eds): "Conformation in Biology," New York: Adenine, p. 389.

7. Egert E, Sheldrick GM (1985). Search for a fragment of known geometry by integrated Patterson and direct methods Acta Cryst A41:262.
8. Karle, IL, Wieland T (1987). Fourth polymorph of [Phe^4 Val^6]antamanide (pentahydrate). Int J Peptide Protein Res 29:596.
9. Karle, IL, Sukumar M, Balaram P (1986). Parallel packing of α-helices in crystals of the zervamicin IIA analog Boc-Trp-Ile-Ala-Aib-Ile-Val-Aib-Leu-Aib-Pro-OMe· 2H$_2$O. Proc Natl Acad Sci USA 83:9284.
10. Karle IL, Flippen-Anderson JL, Sukumar M, Balaram P (1988). Monoclinic polymorph of Boc-Trp-Ile-Ala-Aib-Ile-Val-Aib-Leu-Aib-Pro-OMe (anhydrous): Parallel packing of 3_{10}-/α-helices and a transition helix type. Int J Peptide Protein Res, In press.
11. To be published.
12. Karle IL, Flippen-Anderson J, Sukumar M, Balaram P (1987). Conformation of a 16-residue zervamicin IIA analog peptide containing three different structural features: 3_{10}-helix, α-helix, and β-bend ribbon. Proc Natl Acad Sci USA 84: 5087.
13. Richardson JS (1981). The anatomy and taxonomy of protein structure. In Anfinson CB, Anson ML, Edsall JT, Richards FM (eds): "Advances in Protein Chemistry," New York: Academic Press, p. 169.
14. Terwilliger TC, Weissman L, Eisenberg D (1982). The structure of melittin in the form I crystals and its implications for melittin's lytic and surface activities. Biophys J 37:353.
15. Fox RO Jr, Richards FM (1982). A voltage-gated ion channel model inferred from the crystal structure of alamethicin at 1.5 Å resolution. Nature (London) 300: 325.
16. Karle IL, Flippen-Anderson, J, Uma K, Balaram P (1988). Aqueous channels within apolar peptide aggregates. Solvated helix of Boc-Aib-Ala-Leu-Aib-Ala-Leu-Aib-Ala-Leu-Aib·2H$_2$O·CH$_3$OH in crystals. Proc Natl Acad Sci USA 85: In press.
17. Karle IL, Balaram P. Ms in preparation.
18. Schiffer M, Edmundson AB (1967). Use of helical wheels to represent the structures of proteins and to identify segments with helical potential. Biophysical J 7:121.

Weakly Polar Interactions in Proteins

Dagmar Ringe[*], Stephen K. Burley[#], and Gregory A. Petsko[*]

[*] Department of Chemistry, M.I.T.
77 Massachusetts Avenue, Cambridge, MA 02139

[#] Department of Medicine, Brigham & Women's Hospital, 75 Francis Street, Boston, MA 02115

ABSTRACT The three-dimensional structure of a protein is determined by its amino acid sequence and the forces which drive a disordered polypeptide chain into its final, folded conformation, and once there, maintain this compact structure. There are two broad classes of noncovalent interactions that are traditionally thought to contribute to the forces stabilizing protein structure: electrostatic interactions and hydrophobic interactions. Recent geometric analyses of structural data bases of proteins suggest that the packing of atoms in a protein is, at least in part, also determined by the interactions of polar atoms in a protein.

Proteins are complex molecules whose structures and complexes with ligands and other proteins are stabilized by a variety of noncovalent electrostatic interactions. The results of recent crystallographic and energetic analyses of mutants and wild-type proteins allow one to estimate the importance of such interactions. Although it has been known for several decades that the three-dimensional structure of a protein is determined by its amino acid sequence, the forces which allow a disordered polypeptide chain to fold into its final conformation, and once there, maintain this compact structure, are only understood in very general terms. Traditionally, the forces which contribute to the stabilization of protein structure have been broadly classed into two categories of noncovalent interactions. Those forces which arise from electrostatic interactions (of which ion pairs, hydrogen bonds, and van der Waal's interactions are the best characterized) fit into one, and the hydrophobic interaction (which is not really a force since it results primarily from entropic changes) fits into the other. However, recent geometric analyses of structural data bases of proteins and small organic molecules suggest that the packing of atoms in a protein is, in part, also determined by interactions between asymmetric distributions of the electrons surrounding the nuclei of atoms (Thomas et al., 1982; Reid et al., 1985; Singh and Thornton, 1985; Gould et al., 1986). Such interactions are enthalpically favourable and have been termed weakly polar because they are essentially electrostatic in origin, but may involve amino acid side chains which are traditionally considered to be nonpolar. Thus, stabilizing forces are categorized here according to the type of electrostatic interaction observed and its distance dependence. The categories are: ion pairs, hydrogen bonds, charge-dipole, dipole-dipole, London dispersion forces, and quadrupolar interactions.

Ion Pairs. At the appropriate pH, the carboxy and amino acid termini of proteins possess unit electronic charges. Similarly, the side chains of aspartic acid, glutamic acid, lysine, arginine, and histidine can be ionized depending on the effective pH. Under physiological conditions, these ionizable groups are at least partially charged, and are mainly found on the surface of a protein, where they can interact with other charged groups or become solvated by water molecules in the hydration shell surrounding the protein. When charged groups are found in the interior of a protein where they are inaccessible to solvent, oppositely charged ionizable groups are usually found within 3.5 Å of them, and the two charges form an ion pair or a salt-bridge (if they are close enough to form an H-bond). If the interior ionizable group is not balanced by a neighbouring charge of opposite sign, it is usually found near an electrically neutral group, forming a hydrogen bond (charge-dipole interaction).

Charge-Charge Interactions. These have been extensively described by Rashin and Honig (1984), and Barlow and Thornton (1983), who documented their important role in the stabilization of protein structure. The potential energy of interaction between two charges separated by a distance r in a medium of homogeneous dielectric constant is given by Coulomb's law and is proportional to $1/r$. It must be remembered that this relationship is not valid in the vicinity of boundaries between media of differing dielectric properties, for which the Poisson-Boltzmann equation must be solved explicitly. The stabilization energy of this interaction has been both calculated and observed experimentally and lies in the range of 3-4 kcal/mol.

An example of this interaction occurs in the specificity site of the enzyme aspartate amino transferase. In the wild type enzyme, the γ-carboxylate of the substrate L-aspartic acid is held in place by a charge-charge interaction with the side chain of arginine 292. When this residue is changed (by site-directed mutagenesis) to an aspartic acid, the γ-carboxylate can no longer make a favourable charge-charge interaction and its binding is stabilized no better than that of an uncharged substrate side chain. The difference in the free energy of binding between the wild type and mutant enzymes for aspartic acid lies between 3 and 4 kcal/mol, indicating the loss of such an interaction (Cronin & Kirsch, 1988).

Hydrogen Bonds. Traditionally, hydrogen bonds in proteins are treated as a separate noncovalent interaction. However, the hydrogen bond is actually an electrostatic interaction of the form charge-dipole (salt bridge) or the form dipole-dipole, and can be modelled as such by applying Coulomb's law to the Lennard-Jones potential. The electron shell surrounding the H-atom is shifted toward the atom to which it is covalently linked, thus giving the two atoms unlike partial charges. The hydrogen bond is formed by the favourable orientation of such partial electronic charges. The stabilization energy of such an interaction depends on several factors, including whether the atom to which the H-bond is directed is charged (3-5 kcal/mol) or uncharged (0.5-1.5 kcal/mol) (Fersht et al., 1985). In addition, the energetically optimum arrangement occurs when the two dipole moments are collinear, although deviations of 20° reduce the enthalpy of the interaction by only ~10%. The length of an H-bond is observed to be less than 3.5 Å and the energy of the interaction falls off rapidly with distance ($1/r^2$ or $1/r^3$).

Hydrogen bonds in biological systems are usually thought to involve only oxygen and nitrogen as one or more of the two electronegative participants in the bond. However, hydrogen bonds involving carbon as the hydrogen donor atom have been observed by neutron and X-ray crystallography, and C-F...H-N hydrogen bonding has been characterized in small-molecule systems and enzyme-inhibitor complexes (Murray-Rust et al., 1983; Ringe et al., 1985).

The hydrogen bond is thought of most often as providing the stabilization of secondary structural elements such as pleated sheets and α-helices. However, H-bonds are observed in many other types of stabilization, such as the 13-residue loop from ala 21 to asn 34 in aspartate amino transferase, which is held in place by a single hydrogen bond from the side-chain NH_2 of asn 34 to the CO of ala 21. In the seven sequences of amino transferases which are known, asn 34 is always conserved. An example of the stabilization of a protein toward thermal inactivation by hydrogen bonds has been studied by mutagenesis in triose phosphate isomerase (Casal et al., 1987). Asparagine 78 forms two hydrogen bonds to neighbouring residues. If the residue is changed to threonine, which can only form one H-bond, the thermal stability of the protein drops by 3°, which is equivalent to about 1 kcal/mol. When the asparagine residue is changed to isoleucine, which does not form any H-bonds, the thermal stability of the protein drops by 6° (2 kcal/mol).

Polar Interactions: Charge-Dipole. The distribution of partial electronic charges within an amino acid gives rise to various permanent electronic dipole and higher order multiple moments in both the backbone and the side chains. An electronic dipole is formed when two (partial) charges of opposite sign are linked. The dipole moment is a function of the charge magnitude and the separation distance. In a well-characterized example, the α-helix, the permanent electronic dipoles of the amino acid residues become aligned and their contributions sum to give a large dipole moment, which can span a substantial portion of the protein. The effect of this dipole moment is equivalent to a 1/2 positive charge of the N-terminus of the helix. Analyses of helix dipoles in proteins have demonstrated that they play a role in stabilizing charge. The energy of interaction of a charge with such a dipole has been estimated to be ~4 kcal/mol and depends critically on the precise orientation of the electronic dipole moment with respect to the charge.

Stabilization of phosphate groups in proteins provides a good example of this type of interaction. If the phosphate group is not involved in the chemistry of an enzyme, most often a helix dipole is used to help stabilize its position. The N-terminus of a helix (residues 109-124) in aspartate amino transferase is used to help stabilize the phosphate of the pyridoxal phosphate cofactor. Triose phosphate isomerase contains a helix (residues 95-103) whose termini each have a histidine residue. The histidine at the C-terminus (His 103) is used to stabilize the helix, since its +ve charge interacts favourably with the -ve end of the dipole. The histidine at the N-terminus (His 95) is part of the active site and has very unusual properties due to the fact that it is located at the positive terminus of the helix dipole. This histidine does not titrate over the pH range 4-9, determined by NMR. In addition, it does not react with any of the known reagents which alkylate the N^E of histidines. This residue appears to play a vital role in the catalytic mechanism (Nickbarg et al., 1988).

Polar Interactions: Dipole-Dipole Interactions. Dipole-dipole interactions are weaker than charge-dipole interactions and of shorter range. The interaction depends on the inverse cube of the distance between centres of the two dipoles and on the relative orientation of the two dipole moments. The optimum arrangement occurs when the two dipole moments are collinear and the positive end of one dipole is facing the negative end of the other. However, even an angle of 25° between dipoles only reduces the energy of the interaction by ~10%. The best characterized dipole-dipole interaction is the hydrogen bond, which has already been described, and which also involves other electronic interactions. Another interaction of this type is the common antiparallel arrangement of adjacent α-helices in proteins. The estimated enthalpy of this type of stabilization lies between -5 and -7 kcal/mol.

A good example of this arrangement is the four helix bundle in hemerythrin (Stenkamp & Jensen, 1978). A similar arrangement occurs in aspartate amino transferase between helix (109-124) and helix (278-294).

London (Dispersion) Interactions. The term London or van der Waal's interaction traditionally refers to attractive interactions involving induced electronic dipoles, and short range repulsive interactions due to spatial overlap of electron orbitals. The energy of interaction between such induced dipoles depends on the polarizabilities of the interacting species and is proportional to $1/r^6$ or higher terms. The short range repulsion is usually modelled as a $1/r^{12}$ distance dependence. The combination of these models, the Lennard-Jones 6-12 potential function, is a computationally easy way of estimating the effect of the combination of these two interactions. The "van der Waal's potential" combines Coulomb's law and the 6-12 potential function, thereby lumping electron shell repulsion, London forces and electrostatic interactions under one term. The only meaningful parameter which comes from the above analysis is the van der Waal's distance which occurs at the minimum of the Lennard-Jones potential function.

An example of the London interaction is the incorporation of xenon in the hydrophobic cavities found in myoglobin (Tilton et al., 1985). These cavities are formed by packing restraints and are generally empty. Xenon, which is a hydrophobic atom, can be intercolated into some of these spaces, forming a complex that is stabilized by induced-dipoles.

Weakly Polar Interaction. An electronic quadrupole moment is formed when two dipole moments are aligned in a collinear fashion with the like-charged ends of the dipole moments closest to one another. This charge arrangement is neutral and has no net electronic dipole. Electronic quadrupole moments of the aromatic amino acid side chains are substantial, and quadrupole-quadrupole interactions and charge-quadrupole and dipole-quadrupole interactions occur frequently in proteins and involve one or more aromatic moieties. The distance dependence of these three types of interactions with electronic quadrupole moments ranges from the inverse cube to the inverse fifth power of the separation.

The magnitudes and signs of these interactions also depend on the spatial arrangement of the charges, with enthalpically optimal arrangements bringing unlike charges close to one another and separating like charges.

These quadrupolar interactions define a group of weakly polar, enthalpically favourable interactions which exploit the characteristic segregation of positive and negative electronic charges found in aromatic moieties. The four types of interactions which are discussed here are aromatic-aromatic, oxygen-aromatic, sulfur-aromatic, and nitrogen-aromatic interactions. These four are both ubiquitous and numerous and the distribution of observed geometries for each type of weakly polar interaction differs from that expected from random close packing. Results of ab initio quantum mechanical calculations suggest that such an interaction contributes between -1 and -2 kcal/mol to the three-dimensional structural stability of a protein.

Aromatic-aromatic interactions occur when the hydrogen atom of one aromatic ring with its partial positive charge comes close to the π electron cloud of another aromatic ring (Burley & Petsko, 1984). The geometric arrangements of the two aromatic rings do not reflect random close packing of planar molecules in the interior of a protein. Of 33 protein structures examined, such interactions account for 60% of all aromatic rings. The typical separations between centroids fall between 3.4 and 6.5 Å.

An extreme example of such an interaction network occurs in the iron binding site of iron superoxide dismutase. The iron atom is held in place by ligands from three histidine and one aspartic acid residue. The histidine ligands, in turn, are held in place by a set of aromatic interactions, one for each histidine, and this second level of aromatic residues makes interactions with a third level of aromatic residues. The whole system forms a stable network of aromatic-aromatic interactions which does not collapse when the iron is removed from its centre.

Oxygen-aromatic interactions have been described by Thomas et al. (1982). There is a statistical preference for an oxygen atom to be found in the plane of the neighbouring aromatic ring near the hydrogen atoms. A typical distance is ~ 2.5 Å. An example of this type of interaction occurs between tyr 171 and the carbonyl oxygen of ser 223 in γ-chymotrypsin.

A similar type of interaction has been reported by Reid et al. (1985) between sulfur atoms and aromatic side chains. A typical distance is ~3 Å from the nearest H-atom. An example of such an interaction occurs between met 192 and the phenyl group of the phenylalanine side chain in the complex between γ-chymotrypsin and acetyl-leucyl-phenylalanyl trifluoromethyl ketone.

Finally, amino-aromatic interactions occur in proteins, involving the side chains of lysine, arginine, histidine, glutamine, and asparagine (Burley & Petsko, 1986). The side chain amino group is preferentially found near the π electron cloud of an aromatic ring, avoiding the partially positive hydrogen atoms, and with a preferred separation of 3 to 6 Å between the nitrogen atom and the centroid of the aromatic ring. In the 33 data sets which have been analysed, 50% of the aromatic residues were involved in such an interaction with 25% of the lysines, 50% of the arginines, 40% of the histidines, and 30-40% of the asparagines and glutamines.

The data presented above suggest that the three-dimensional structure of a protein and the interaction of a protein with a ligand is stabilized, in part, by a wide variety of electrostatic interactions that occur between amino acid (ligand) constituents.

Although the precise contribution of any given electrostatic interaction in a protein to its free energy of stabilization is uncertain because it is a function of charge distribution, separation and interaction geometry, and of the local solvent environment, enthalpy estimates for the various electrostatic interactions can be made. Attractive charge-charge interactions with optimal separation are said to contribute an enthalpy of about -5 kcal/mol, charge-dipole interactions are estimated to contribute an enthalpy of about -4 kcal/mol, the hydrogen bond, a dipole-dipole interaction, contributes an enthalpy of about -3 kcal/mol, and the weakly-polar interactions involving quadrupole moments typical contribute enthalpies of between -1 and -2.5 kcal/mol. Each of these individual enthalpy terms is small when compared to the sum of all enthalpic contributions to the stabilizations of a protein's tertiary structure, but they are substantial when compared to the free energy of protein tertiary structure stabilization, which is between -10 and -20 kcal/mol (Privalov, 1979). For example, analyses of thermostable proteins suggest that the additional free energy of stabilization required to convert a mesophilic protein into a thermophilic protein is on the order of -1 to -2 kcal/mol, which would be readily facilitated by the insertion of a single charge-charge interaction into the protein's interior. These same types of electrostatic interactions are involved in stabilizing protein-ligand complexes, and in determining the precise geometry of ligand binding. Finally, these interactions are certainly involved in the process of protein folding. Since it is now generally agreed that intermediates occur in the folding process, and that such intermediates may differ in structure from the final folded form of the protein (Baldwin, 1975; Creighton, 1978), interactions that are essential for folding (because they stabilize a necessary intermediate) can occur between groups in the protein that do not interact in the final native structure. Thus, even a residue that seems unimportant when the native structure is examined may be vital for the kinetics of the folding process that forms the native state (Yu & King, 1984).

REFERENCES

1. Baldwin, R.L. (1975) **Ann. Rev. Biochem.**, 44, 453-475.
2. Barlow, D.J., Thornton, J.M. (1983) **J. Mol. Biol.**, 168, 867.
3. Burley, S.K., Petsko, G.A., (1985) **Science**, 229, 23-28.
4. Burley, S.K., Petsko, G.A. (1986) **FEBS Letters**, 203, 139-143.
5. Casal, J.I., Ahern, T.J., Davenport, R.C., Petsko, G.A., Klibanov, A.M., (1987) **Biochemistry**, 26, 1258-1263.
6. Creighton, T.E. (1978) **Prog. Biophys. Mol. Biol.**, 33, 231-297.
7. Cronin, C.N., Kirsch, J.F., **Biochemistry**, in press.
8. Fersht, A.R., Shi, J.-P., Knill-Jones, J., Lowe, D.M., Wilkinson, A.J., Blow, D.M., Brick, P., Carter, P., Waye, M.M.Y., Winter, G., (1985) **Nature**, 314, 235.
9. Gould, R.O., Gray, A.M., Taylor, P., Walkinshaw, M.D. (1985) **J. Am. Chem. Soc.**, 107, 5921.
10. Murray-Rust, P., Stallings, W.G., Monti, C.T., Preston, R.K., Glusker, J.P. (1983) **J. Am. Chem. Soc.**, 105, 3206.
11. Nickbarg, E.B., Davenport, R.C., Petsko, G.A., Knowles, J.R. (1988) **Biochemistry**, in press.

12. Privalov, P.L., (1979) **Adv. Protein Chem.**, 33, 167.
13. Rashin, A.A., Honig, B.H. (1984) **J. Mol. Biol.**, 173, 515.
14. Reid, K.S.C., Lindley, P.F., Thornton, J.M. (1985) **FEBS Letters**, 190, 209.
15. Ringe, D., Seaton, D.B., Gelb, M., Abeles, R.H. (1985) **Biochemistry**, 24, 64.
16. Stenkamp, R.E., Jensen, L.H. (1978) **Biochemistry**, 17, 2499.
17. Singh, J., Thornton, J..M. (1985) **FEBS Letters**, 190, 1.
18. Thomas, K.A., Smith, G.M., Thomas, T.B., Feldman, R.J. (1982) **Proc. Natl. Acad. SCi. USA**, 79, 4843.
19. Tilton, Jr., R.F., Kuntz, Jr., I.D., Petsko, G.A., (1985) **Biochemistry**, 12, 2849-2857.
20. Yu, M.-H., King, J. (1984) **Proc. Natl. Acad. Sci. USA**, 81, 6584-6588.

TRANSFER OF INFORMATION THROUGH PROTEIN/LIPID INTERACTIONS

Charles M. Deber[1], G. Andrew Woolley[1], Raisa B. Deber[2], and Christopher J. Brandl[3][†]

[1]Research Institute, Hospital for Sick Children, Toronto M5G 1X8; and Departments of [1]Biochemistry and [2]Community Health, and [3]Best Department of Medical Research, University of Toronto, Toronto M5S 1A8, Ontario, Canada.

ABSTRACT Transmembrane (TM) regions in receptor proteins which span the bilayer once are proposed to function only to anchor receptors to their cellular membranes rather than be directly responsible for signal transduction. This suggestion results from a survey in which the amino acid compositions of the TM regions of receptor proteins (n = 11) could not be distinguished statistically from corresponding regions of integral membrane proteins with a functional external domain attached to a hydrophobic membrane-spanning anchor segment (n = 16), whereas both differed from the compositions of TM regions in transport proteins (n = 10). Since penetration of membranes by peptides is well-established experimentally - several aqueous-soluble peptide hormones and neuropeptides having segments with hydrophobic character are shown by spectroscopic methods to bind and penetrate phospholipid membranes, often with accompanying induction of secondary structure - the transfer of molecular information across membranes by some receptors could conceivably involve, in part, direct intramembranous contact between receptor external and cytoplasmic aqueous domains.

INTRODUCTION

Transfer of information through the conformational changes associated with movement of a protein (or segment thereof) from

---

[†] C.J.B.'s present address: Department of Biological Chemistry, Harvard Medical School, Boston, MA. 02115.

aqueous to membrane domains is widely observed in biochemistry. For example, the lethal effects on cells of diptheria toxin, a 58.3 kDa protein, depend, in part, upon the movement of a charged hydrophilic soluble protein across a lipid bilayer (1); regions proposed to be involved in conformational change and membrane interaction have been identified (2). As well, soluble proteins become membrane-associated in the biogenesis of biological membranes; in this complex process, which may well involve multiple mechanisms (3), the role of membrane interactions in the "membrane-trigger hypothesis" is stressed (4). Other examples include the membrane-attack complex of complement (5), the biosynthesis of M13 coat proteins (6), and the bee venom peptide mellitin which can insert into membranes and lyse cells (7).

In this report, we consider receptors as a class of membrane proteins, and specifically, the molecular mechanism by which they might transfer a signal across a membrane. Since receptors are membrane-embedded proteins, and the signal must cross a bilayer, receptor function must at some stage involve interaction of amino acids with lipids. We present here evidence that the transmembrane (TM) domains of many receptors may not be involved in signalling per se, and suggest instead membrane interaction of external (and/or internal) receptor aqueous domains as a mechanistic feature of transduction by some receptors. In experimental support of this hypothesis, we describe the specific conformational changes which membrane environments can induce in some biologically-active peptides, including the pain transmission neuropeptide substance P (8).

Compositions of Transmembrane Regions of Receptors

The amino acid sequences of a number of receptor proteins have now been determined, from which many receptors have been predicted to possess a single transmembrane (TM) chain of hydrophobic amino acids which separates their independent cytoplasmic and external domains (9-21). In the absence of any other transducing protein, it has been presumed that this TM passage must be responsible in molecular terms for signal transduction across the membrane bilayer, particularly where function depends on rapid activation of an intracellular domain (e.g., as in receptor kinases).

We have earlier hypothesized that if certain residues in a TM region play a specific role in protein structure or function, their distribution should differ systematically between protein categories (22). This assumption held in comparisons between a group of transport (T) proteins and a corresponding group of membrane-anchored proteins (MAP's) [i.e., integral membrane proteins which possess an external functioning domain attached to a single hydrophobic TM segment that acts primarily to "anchor" the protein within the membrane (23)]: statistically-different occurrences were found for several potentially functional residues in TM regions, notably proline (24). Thus, TM regions of receptor proteins should similarly have unique characteristics if these regions contain residues functional in signal transduction.

Data bases were constructed for 37 integral membrane proteins (11 receptors (R), 16 MAP's, 10 transporters (T)) whose function and primary sequence were known, and for which the distribution of their residues between membrane and aqueous domains has been deduced by the various cited authors using a combination of experimental methods and predictive criteria (22,24); the reader is referred to these articles for details of proteins chosen for inclusion and identification of TM residues. Mean % residue occurrences in each domain were then calculated separately by protein category. Statistical tests were performed using the MINITAB program (25). Unpaired t-tests were used to compare differences in % residue occurrences between protein categories. The t-test computes the number of standard errors between the observed value and the value "expected" if the null hypothesis of no difference between the two distributions being compared were true, and computes the associated p-value (i.e., the probability that the observed difference could have occurred by chance). The t-test is thus sensitive to both absolute difference, and to variations within each distribution; to be statistically significant, variation between distributions has to be large relative to such internal variance. A p-value ≤ 0.05 is deemed the conventional standard of statistical significance, while p-values ≤ 0.01 are deemed highly significant.

Receptor (R) proteins which function autonomously and for which a TM segment has been identified displayed the sequences for their TM segment(s) and adjacent residues shown in Table 1. MAP protein TM sequences are also given ((22) and references therein). Because this group of receptors is heterogeneous with respect to functions and mechanism of actions [six function primarily to internalize bound substrate (26,27), three are kinases involved in the rapid relay of transmembrane signals (28-30), while two are bacterial chemoreceptors that undergo methylation in response to ligand binding (31)], the residue composition of each receptor TM segment was also compared individually with the average receptor TM composition (data not shown). We found that no amino acid displayed an aberrantly high or low percentage (i.e., differing by greater than two standard deviations) in more than one receptor. Thus, the amino acids in receptor TM regions do not cluster on the basis of specific receptor function.

Using the data bases and the methodology described above, TM segments of these receptors were then found to have an average composition which could not be distinguished statistically from the corresponding group of MAP proteins [i.e., in the comparison R vs

TABLE 1
RESIDUES IN TRANSMEMBRANE AND ADJACENT REGIONS OF MEMBRANE PROTEINS

RECEPTOR (R) PROTEINS	EXTERNAL	TRANSMEMBRANE	CYTOPLASMIC
Asialoglycoprotein receptor, H1	LLQRLCSGPR	LLLSLGLSLLLLVVVCVI	GSQNSQLQEE
Asialoglycoprotein receptor, H2	LAQRLCSMVC	FSILALSFNILLLVVICVTG	SQSEGHRGAQ
Aspartate receptor	TFDQSAHDYR	FAQWQLGVLAVVLVLILMVVWFGI	RHALLNPLAR
	MFNRIR	VVTMLMMVLGVFALLQLVSGGLLF	SSLQHNQQGF
Epidermal growth factor (EGF) receptor	CPTNGPKIPS	IATGMVGALLLLVVALGIGLFM	RRRHIVRKRT
Insulin receptor, β subunit	YLDVPSNIAK	IIGPLIFVFLFSVVIGSIYLFL	RKRQPDGPLG
Interleukin-2 receptor	ETSIFTETEYQ	VAVAGCVFLLISVILLSGL	TWQRRQRKSR
LDL receptor, human	GNEKKPSSVR	ALSIVLPIVLLVFLCLGVFLLW	KNWRLKNINS
Serine chemoreceptor	DRLHDIAVSD	NNASYSQAMWILVGVMIVVLAVIFAVWFGI	KASLVAPMNR
T-cell receptor	GVLSATLYE	ILLGKATLYAVLVSTLVVMAMV	KRKNS
Transferrin receptor	TKANVTKPKR	CSGSICYGTIAVIVFFLIGFMIGYLGYC	KGVEPKTECE
Tyrosine kinase receptor	AEQRASPLTS	IVSAVVGILLVVVLGVVRGILI	KRRQQKIRKY

MEMBRANE-ANCHORED (MAP) PROTEINS	EXTERNAL	TRANSMEMBRANE	CYTOPLASMIC
Avian sarcoma virus glycoprotein	IGVDSDLIGS	WLRGLPGGIGEWAVHLLKGLLLGLVVILL—LVVCLPCCLQMLCG	NRRKMINNSI
Filamentous bacterial virus coat protein	FDSLTAQATE	MSGYAWALVVLVVGATVGI	KLFKFVSRA
Glycophorin A^M	QLAHHFSEIE	ITLIIFGVMAGVIGTILLISYGI	RRLIKKSPSD
Isomaltase subunit of small intestinal sucrase-isomaltase	VNAFSGLEIT	LIVLFVTVFIIAIALIAVLA	...PAV
Influenza virus agglutinin	PVKLSSGYKD	VILWFSFGASCFLLLAIAVGLVFICV	KNGNMRCTIC
Membrane bound immunoglobulin μ-chain	EVNAEEEGFE	NLWTTASTFIVLFLLSLFYSTTVTLF	KVK
Mouse H-2^d (partial) histocompatibility Ag	EEPPSSTKTN	TVIIAVPVVLGAVVILGAVMAFVM	KRRRNTGGKG
Murine transplantation antigen	DEPPSTVSNM	ATVAVLVVLGAAIVTGAVVAFVM	KMRRRATGGK
Neuraminidase	HNPNQK	IITIGSICLVVGLISILIQIGNISIWIS	HSIQTGSQNH
Rabies virus glycoprotein	VDLGLPNWGK	YVILSAGALTALMLIIFLMTCC	RRVNRSEPTQ
Rat leukocyte-common antigen (T200)	KPQSTSYNSK	ALIFLVFLIIVTSIALLVVLT	KIYDLRKKRS
Semliki forest virus membrane glycoprotein, E1	SGTALSWVQK	ISGGLGAFAIGAILVLVVVTCIGL	RR
Semliki forest virus membrane glycoprotein, E2	IVQYYYGLYP	AATSAVVGMSLLALISIFASCYMLVAA	RSKCLTPYAL
Sindbis virus structural protein, E1	DQERQAAISK	TSWSWLFALFGGASSLLIIGLMIFACSMMLTST	RR
Sindbis virus structural protein, E2	PHEIVQHYSH	HPVYTILAVASATVAMMGVTVAVLCAC	KARRECKTPY
Vesicular stomatitis virus glycoprotein	LVEGWFSSWK	SSIASFFFIGLIGLFLVL	RVGIHLCICKL

MAP, p > 0.05 for all 20 residues (32)]. In contrast, several residues displayed statistically-significant differences in the corresponding comparison of R- vs T-proteins (6/20 in each instance), a result which paralleled differences already found between MAP- and T- proteins (22). This analysis therefore implies that <u>the linear sequences of the TM segments in receptor proteins contain only the structural information necessary for anchoring the receptor to the membrane bilayer</u>. These results substantiate statistically a similar suggestion by Marchesi (33). High occurrences of positively-charged (Lys, Arg) residues observed in regions adjacent to the cytoplasmic side of the TM segments of both R- and MAP-proteins (Table 1), which mitigate against the vertical displacement of these segments, also support the notion of TM regions as membrane anchors (34).

Although the structural representations of most receptors depict the membrane as a barrier to direct interaction between cytoplasmic and external domains, this does not necessarily have to be the case: regions in one (or both) aqueous domains could penetrate partially into the membrane, and thereby form a second transmembrane passage via direct intraprotein contact between the two domains. In such a mechanism, ligand binding to residues in the exterior domain of an R-protein could induce a positional and/or conformational change in that domain which would then contact and influence activity in the protein's cytoplasmic domain. This situation is a logical extension of mechanism operative in receptors that act through G-proteins, where signal transduction similarly requires that protein-protein interactions occur within the membrane (35). The applicability of these proposals to receptors which function through oligomerization is considered explicitly elsewhere (32).

Transmembrane contact between receptor protein aqueous domains should, in principle, be facilitated by their inherent capacity to insert segments into the bilayer other than the established TM passage - a function which could be served by contiguous or non-contiguous conformation-dependent regions of high local hydrophobicity. For example, linear segments of 7-10 consecutive amino acids having high hydrophobicity (36) can be identified in both aqueous domains of several receptors. Domain-domain contact could ensue via membrane penetration of such hydrophobic "epitopes", and/or via perturbation of intervening lipids.

The feasibility of these postulated events cannot at present be tested directly in the receptors themselves, but is readily demonstrated in the case of the transfer to membranes of aqueous-soluble peptide hormones whose conformational behaviour can be examined in a variety of membrane environments.

Membrane Binding and Penetration of Peptides

Substance P (Arg-Pro-Lys-Pro-Gln-Gln-Phe-Phe-Gly-Leu-Met-NH_2), best known as a neuromodulator in the transmission of pain information, is also found in the gut and is involved in a number of central and peripheral processes (8). We have been studying the behaviour of substance P and related peptides of the tachykinin family in a variety of model membrane systems (37,38). Micelle-forming lipids lysophosphatidylcholine (LPC), lysophosphatidylglycerol (LPG), and sodium dodecyl sulphate (SDS) were chosen initially for these studies since the relatively small size of the micellar particles makes spectroscopic studies simpler: solutions for circular dichroism (CD) are clear, and NMR lines are not highly broadened by short relaxation times. Substance P was found to bind to these lipids, with accompanying major conformational changes (vs. aqueous phase) as judged by CD spectra (Fig. 1). Combined data from 360 MHz ^{1}H NMR, UV spectroscopy, and potentiometric titration (37), showed that the Phe residues interact with the hydrophobic region of the micelle, with the N-terminal amino group located the micelle head-group region. CD spectra suggested that substance P in micellar membranes is largely helical, due probably to intramolecular H-bond formation upon penetration of the C-terminal portion of the molecule into the membrane.

Bilayer-forming lipids (e.g., phosphatidylcholine (PC), phosphatidylglycerol (PG)), the natural components of biological membranes, clearly represent more realistic models of the *in vivo* situation than are micelles. Although light scattering by vesicles prevented measurement below 210 nm, CD spectra of substance P could be obtained upon the addition of PG, PG/PC or PC, as shown in Fig.2. A qualitative comparison with results in micellar lipids *(vide infra)* indicates similar CD spectral changes in the two systems, except that the bilayer-forming lipids showed smaller changes on a w/w basis than the corresponding micellar lipids. Weaker interactions with bilayer-forming lipids are likely attributable to a tighter-packed structure for the vesicle which more strongly resists deformation in its surface; micelles are comparatively looser aggregates which are in equilibrium with relatively high concentrations of monomer (8 mM for SDS) (39).

While secondary structure is induced in several tachykinins, including substance P, when they are complexed to model membrane systems, CD spectra of other tachykinins appear insensitive to addition of lipids (38). This diversity of conformational behaviour did not seem to be simply correlated with any particular feature such as net charge or helix-forming potential (40), and must somehow be specified by the amino acid sequence. A similar property might be expected for selected sequences (or "epitopes") of receptor soluble domains.

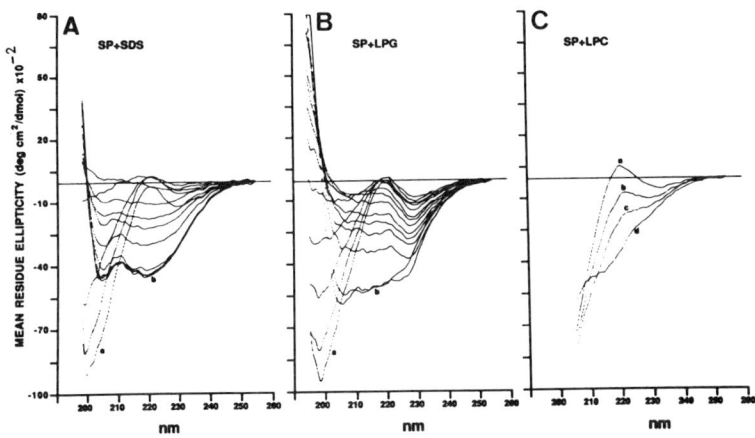

FIGURE 1. Circular dichroism spectra of the titration of substance P and (A) sodium dodecyl sulfate (SDS), (B) lysophosphatidylglycerol (LPG), and (C) lysophosphatidylcholine (LPC). Spectra of the lipid have been subtracted by computer in all cases. (A) SP = [0.38 mM] in 14.3 mM phosphate buffer, pH 7.3. Curve a, SP alone. Small volumes of a concentrated SDS solution were added to give final SDS concentrations (mM) of 0.19, 0.47, 0.94, 1.9, 2.8. 3.6. 4.5, 6.2, 8.6, 12.3, 15.6 and 53 (curve b). No further change was noted above 12.3 mM SDS. (B) SP = [0.22 mM] in 14.3 mM phosphate buffer, pH 7.3. Curve a, SP alone. Small volumes of a concentrated LPG solution were added to give final LPG concentrations (mM) of 0.1, 0.2, 0.3, 0.38, 0.48, 0.57, 0.65, 0.74, 0.83, 0.91, 1.1, 1.5, and 1.8 (curve b). No further change was noted above 1.5 mM LPG. (C) SP = [2.3 mM] in distilled water, pH 6.7 throughout. (a) SP alone. LPC concentrations (mM) = 19 (b), 34 (c), and 140 (d). Diagram from Woolley and Deber (37). Used by permission.

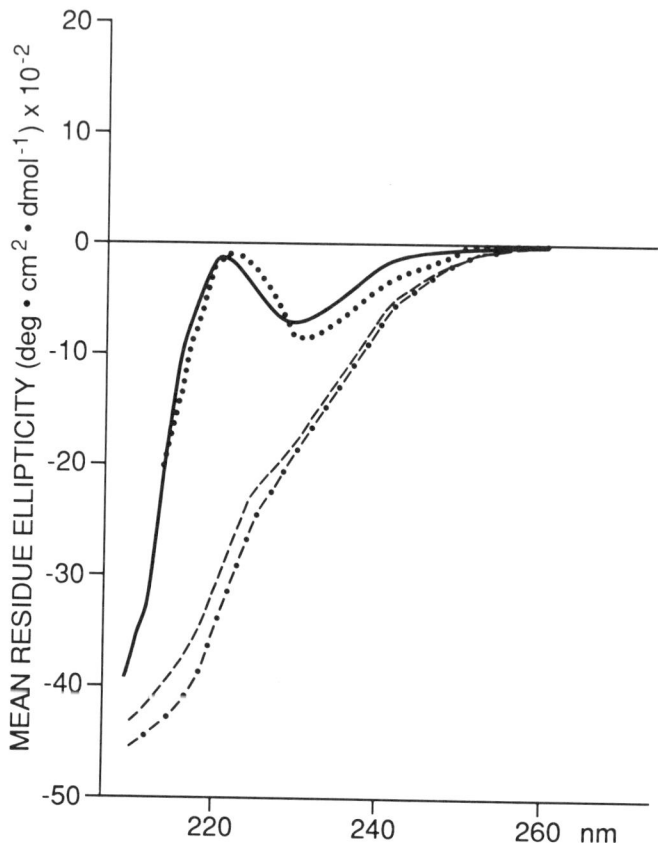

FIGURE 2. Circular dichroism spectra of Substance P (SP) in unilamellar phospholipid vesicles: The spectra are: SP (——); SP + phosphatidylcholine (PC) (·····); SP + phosphatidylglycerol (PG) (—·—·); and SP + PC/PG (50/50 w/w) (-----). Conc. of SP in each spectrum, 0.58 mM; conc. of total lipids in each of the latter three spectra, 12 mM. See text for a further discussion.

Induced Conformation Is Characteristic of Pepide Rather Than Lipid Type

Experiments suggest that peptides tend to adopt specific conformations when transferred into lipid environments but that the exact nature of the lipid component is not critical. While there are clear effects of membrane surface charge on the strength of the interaction (*viz.* SDS, LPG and PG experiments), this seems not to have a major effect on the resulting conformational ensemble adopted by membrane-bound peptide. Similar results have been obtained (41,42) in comparing interactions of substance P and other peptides (ACTH, dynorphin) with neutral PC versus negative phosphatidylserine (PS) lipids. In an extensive characterization of mellitin-lipid interactions, Lauterwein *et al* (43) similarly found the membrane-bound conformation of this small protein to be relatively insensitive to lipid head-group type. Studies on enkephalins in membranes reached similar conclusions (44). Yet, the variation in lipid type is in fact quite large: in the present work, for example, we are comparing conditions as diverse as SDS micelles (aggregates of 12-carbon aliphatic tails and sulphate head-groups) to bilayers of PC with two hydrocarbon tails of varying length and double bond character and a zwitterionic phosphocholine head-group (see also (45)). These findings suggest that the peptides are not participating in a lipid-recognition event *per se* but instead are responding to the more general properties of an interface between hydrophobic and aqueous domains. Thus, it may not be necessary to duplicate the membrane environment of receptor proteins exactly to observe relevant interactions.

CONCLUSION

Membranes can influence topology by restricting the peptide conformational ensemble. Since receptor transmembrane regions are deduced to be essentially membrane anchors, ligand-activated membrane penetration of segments located in receptor aqueous domains may be a feature of the signalling process. The present observations add to our understanding at the molecular level of peptide conformational behaviour in membranes, and suggest that lipid-mediated conformational changes may be applicable to the dynamics of the more complex interactions involved in signal transduction by receptor proteins.

ACKNOWLEDGEMENT

This work was supported, in part, by grants to C.M.D. from the Medical Research Council (MRC) of Canada, and the Natural Sciences and Engineering Research Council (NSERC) of Canada.

REFERENCES

1. Neville, DM, Jr, Hudson, TH (1986) Ann Rev Biochem 55:195.
2. Brasseur, R, Cabiaux, V, Falmagne, P, Ruysschaert, J (1986) Biochem Biophys Res Commun 136:160.
3. Wickner, WT, Lodish, HF (1985) Science, 230:400.
4. Wickner, W (1979) Ann Rev Biochem 48:23.
5. Muller-Eberhard, HJ (1975) Ann Rev Biochem 44:697.
6. Wickner, W (1977) Biochemistry 16:254.
7. Eisenberg, D (1984) Ann Rev Biochem 53:595.
8. Pernow, B (1982) Pharmacol Rev 35:85
9. Spiess, M, Schwartz, AL, Lodish, HF (1985) J Biol Chem 260:1979.
10. Spiess, M, Lodish, HF (1985) Proc Natl Acad Sci USA 82:6465.
11. Russo, AF, Koshland, Jr, DE (1983) Science 220:1016.
12. Ullrich, A, Coussens, L, Hayflick, JS, Dull, TJ, Gray, A, Tam, AW, Lee, J, Yarden, Y, Libermann, TA, Schlessinger, J, Downward, J, Mayes, ELV, Whittle, N, Waterfield, MD, Seeburg, PH (1984) Nature 309:418.
13. Ullrich, A, Bell, JR, Chen, EY, Herrera, R, Petruzzelli, LM, Dull, TJ, Gray, A, Coussens, L, Liao, Y-C, Tzubokawa, M, Mason, A, Seeburg, PH, Grunfeld, C, Rosen, OM, Ramachandran, J (1985) Nature 313:756.
14. Ebina, Y, Ellis, L, Jarnagin, K, Edery, M, Laszio, G, Clauser, E, Ou, J-H, Masiarz, F, Kan, YW, Goldfine, ID, Roth, RA, Rutter, WJ (1985) Cell 40:747.
15. Leonard, WJ, Depper, JM, Kanetisa, M, Kronke, M, Peffer, NJ, Svetlik, PB, Sullivan, M, Greene, WC (1985) Science 230:633.
16. Yamamoto, T, Davis, CG, Brown, MS, Schneider, WJ, Casey, ML, Goldstein, J, Russell, DW (1984) Cell 39:28.
17. Boyd, A, Kendall, K, Simon, MI (1983) Nature 301:623.
18. Hedrick, SM, Nielsen, EA, Kavaler, J, Cohen, DI, Davis, MM (1984) Nature 308:153.
19. Yanagi, Y, Yoshikai, Y, Leggett, K, Clark, SP, Aleksander, I, Mark, TW (1984) Nature 308:145.
20. McClelland, A, Kuhn, LC, Ruddle, FH (1984) Cell 39:267.

21. Coussens, L, Yang-Feng, TL, Liao, YC, Chen, E, Gray, A, McGrath, J, Seeburg, PH, Libermann, TA, Schlessinger, J, Francke, V, Levinson, A, Ullrich, A (1985) Science 230:1132.
22. Deber, CM, Brandl, CJ, Deber, RB, Hsu, LC, Young, XK (1986) Arch Biochem Biophys 251:68.
23. Guidotti, G (1977) J Supramolecular Struct 7:489.
24. Brandl, CJ, Deber, CM (1986) Proc Natl Acad Sci USA 83:917.
25. Ryan, TA, Jr, Joiner, BL, Ryan, BA (1982) In "Minitab References Manual" Boston: Duxbury Press.
26. Goldstein, JL, Brown, MS, Anderson, RGW, Russell, DW, Schneider, WJ (1985) Ann Rev Cell Biol. 1:1.
27. Brown, MS, Anderson, RGW, Goldstein, JL (1983) Cell 32:663.
28. Cobb, MH, Rosen, OM (1984) Biochim Biophys Acta 738:1.
29. Downward, J, Parker, P, Waterfield, MD (1984) Nature 311:483.
30. Carpenter, G, King, L, Jr, Cohen, S (1979) J Biol Chem 254:4884.
31. Koshland, Jr, DE (1981) Ann Rev Biochem 50:765.
32. Brandl, CJ, Deber, RB, Hsu, LC, Woolley, GA, Young, XK, Deber, CM (1988) Biopolymers 27:000 (in press).
33. Marchesi, VT (1986) Adv Exp Med and Biol 205:107.
34. Deber, CM, Brandl, CJ, Deber, RB, Young, XK (1988) In Marshall, GR, (ed) "Proc 10th Amer Pept Symp", Leiden, Netherlands: ESCOM Science Pub., p. 330.
35. Gilman, AG (1984) Cell 36:577.
36. Kyte, J, Doolittle, RF (1982) J Mol Biol 157:105.
37. Woolley, GA, Deber, CM (1987) Biopolymers 26:S109.
38. Woolley, GA, Deber, CM (1987) in Theodoropoulos, D (ed), "Peptides 1986", Berlin: W De Gruyter, p. 439.
39. Tanford, C. (1980) "The Hydrophobic Effect," 2nd ed, New York: Wiley & Sons.
40. Chou, PY, Fasman, GD (1978) Adv Enzymol 47:45.
41. Rolka, K, Erne, D, Schwyzer, R, (1986) Helv Chim Acta, 69:1798.
42. Schwyzer, R, Erne, D, Rolka, K (1986) Helv Chim Acta 69:1789.
43. Lauterwein, J, Bosch, C, Brown, L R, Wuthrich, K (1979) Biochim Biophys Acta 556:244.
44. Deber, CM, Behnam, BA (1984) Proc Natl Acad Sci USA 81:61.
45. Erne, D, Rolka, K, Schwyzer, R (1986) Helv Chim Acta 69:1807.

Synthetic Peptides: Approaches to Biological Problems, pages 109-123
© 1989 Alan R. Liss, Inc.

CONFORMATIONS OF THREE PEPTIDES DEDUCED FROM EXPERIMENTS AND MOLECULAR ENERGETICS

Vincent Madison, Ziva Berkovitch-Yellin[1],
David Fry, David Greeley, and Voldemar Toome

Roche Research Center, Hoffmann-La Roche Inc.,
Nutley, New Jersey 07110

ABSTRACT We have utilized molecular graphics in conjunction with the energy minimization and molecular dynamics modules of the CHARMM program package to optimize conformations of peptides subject to constraints from experimental results. This methodology will be illustrated by means of structural results obtained for synthetic peptide analogs in three systems. Based on functional and energetic considerations, we have developed a model for the ion-pore formed by a 22-residue peptide *in vitro*. For two polypeptide hormones, conformational families have been derived based on molecular dynamics and energy minimization constrained by interproton distances determined from nuclear magnetic resonance spectra. A vasoactive intestinal peptide (VIP) analog and analogs of growth hormone releasing factor (GRF) have been investigated. For the VIP analog, there are a variety of backbone conformations which are all consistent with the experimental interproton distances. In the absence of solvent and counter ions, calculations on the peptide required screening of formal charges on acidic and basic side chains to prevent collapse of secondary structure due to formation of intramolecular ion pairs. GRF is nearly completely helical in 75% methanol:water, but has much less helix in aqueous solution. Structural effects of amino acid substitutions will be discussed.

INTRODUCTION

The sodium channel protein mediates voltage-dependent modulation of sodium ion permeability in electrically excitable membranes. The protein has a molecular weight of 260,000 Daltons and contains four homologous internal repeats. A cluster of four homologous, acid-rich, helical segments has been postulated to be the ion pore (1). Based on the sequence of this segment in the rat brain protein, a 22-residue peptide, P14 (Table 1), has been shown to form sodium channels *in vitro* (2). We have used constrained energy optimization to construct models of the channel formed

[1]Present address: Department of Structural Chemistry, Weizmann Institute of Science, Rehovot, 76100 Israel.

by P14.

We are investigating the structures for analogs of two polypeptide hormones which have therapeutic potential. Vasoactive intestinal peptide (VIP) is a 28-residue peptide which acts as a bronchodilator and may be useful in treating asthma. Growth hormone releasing factor (GRF) has a 29-residue fragment which causes release of growth hormone and may be effective in animal health areas such as increasing milk production and in human health areas such as promoting wound healing (3).

TABLE 1
PEPTIDE SEQUENCES

Name	Sequence
P14	Asp-Pro-Trp-Asn-Trp-Leu-Asp-Phe-Thr-Val-Ile-Thr-Phe-Ala-Tyr-Val-Thr-Glu-Phe-Val-Asp-Leu
VIP	His-Ser-Asp-Ala-Val-Phe-Thr-Asp-Asn-Tyr-Thr-Arg-Leu-Arg-Lys-Gln-Met-Ala-Val-Lys-Lys-Tyr-Leu-Asn-Ser-Ile-Leu-Asn-NH_2
GRF(1-29)NH_2	Tyr-Ala-Asp-Ala-Ile-Phe-Thr-Asn-Ser-Tyr-Arg-Lys-Val-Leu-Gly-Gln-Leu-Ser-Ala-Arg-Lys-Leu-Leu-Gln-Asp-Ile-Met-Ser-Arg-NH_2
Analogs:	
VIP'	[N^1-Ac,Lys^{12},Lys^{14},Nle^{17},Val^{26},Thr^{28}]VIP
[Ala^{15}]GRF(1-29)NH_2	
[Sar^{15}]GRF(1-29)NH_2	
cyclo(Asp^8-Lys^{12})[Asp^8,Ala^{15}]GRF(1-29)NH_2	

VIP and GRF are similar in size and amino acid sequence. For GRF(1-29)NH_2 in 30% trifluoroethanol:water, a structure based on measured nuclear Overhauser effects (NOE's) and constrained molecular dynamics has been published (4,5). To date, there have been no detailed conformational studies reported for GRF analogs, GRF in aqueous solution, nor VIP and its analogs. Herein we report conformational studies in aqueous alcohol solutions for the VIP and GRF analogs listed in Table 1. Circular dichroism (CD) spectra were used to evaluate the secondary structure (especially α-helix) content of the peptides. After making sequence-specific assignments of the nuclear magnetic resonance (NMR) spectra, NOE's were the basis for further defination of the conformation for the peptides. The pattern of NOE's along the primary sequence yielded sequence-specific assignments of secondary structure in an initial structural model. Pairwise interproton distances derived from the NOE's were utilized for constrained optimization of the peptide structures via energy minimization and molecular dynamics simulation to provide a detailed conformational model as well as a means for assessing the uniqueness of the derived structural features. The optimization protocol and computational parameters were investigated. A procedure was found which facilitates the derivation and evaluation of peptide conformers.

METHODS

Peptide structures were built and adjusted using the Roche Interactive Molecular Graphics (RIMG) package (6). The CHARMM program system (version 19, 7) was used for constrained energy minimization and molecular dynamics optimization of the structures. The CHARMM version 18 parameters and energy functions were used except that the van der Waals radii of O and N atoms in neutral groups were increased by 0.1 Ångstrom (Å) and those in charged groups by 0.2 Å, and the explicit 10-12 hydrogen-bonding term was neglected. With these modifications satisifactory hydrogen-bonded geometries and energies were obtained for both neutral and charged groups. As appropriate for the various calculations, constraints were used for symmetry, dihedral angles or interproton distances. Generally, for pairwise electrostatic energies, the dielectric constant was r, the distance between the charges. Throughout, minimization was conducted using 200 steepest descent steps followed by assumed-basis Newton-Raphson steps until the energy converged to 0.01 kcal/mol (up to 5000 steps).

For the sodium channel model of peptide P14, an array of three or four parallel helices was symmetrically placed about an axis through the center of the array parallel to the helices. The array was constrained to C_3 or C_4 symmetry for the trimer and tetramer respectively. The backbone dihedral angles were also constrained to maintain regular α-helices. The energy of the peptide array was optimized in four steps: minimization, 15 ps dynamics at 300 K, 9 ps dynamics while gradually cooling to near 0 K, and reminimization.

For the VIP and GRF analogs, the experimentally determined NOE's were converted to pairwise distance constraints utilizing covalently fixed protons for calibration and assuming a uniform rotational correlation time for the entire peptide. Distances between 2.0 and 3.6 Å were observed with an estimated error of 0.3 Å. For the longest distance (weakest NOE), the upper error limit was increased to 0.6 Å (i.e., for a nominal interproton distance of 3.6 Å, the range of distances within the estimated error was 3.3 to 4.2 Å).

The experimentally determined distances were converted to constraint energies: $E = \text{WNOE} (kT/2) [(R-R_0)/\delta R]^2$, with WNOE = a weighting factor, k = the Boltzmann constant, T = the absolute temperature, R the distance between two protons in the structure or the $< r^{-6} >^{1/6}$ average distance between groups of protons in topologically equivalent groups, R_0 the distance derived from experimental NOE's, and δR the experimental uncertainty in the distance (different values of δR can be used for $R < R_0$ and $R > R_0$). The values of WNOE ranged between 0.01 and 25.0 as specified below.

For most of the calculations on the VIP and GRF analogs, a dielectric constant of r and reduced formal charges of ± 0.25 for the non-neutral Asp, Glu, Arg and Lys sidechains were used. For test cases, formal charges of ± 0.0, 0.5, or 1.0 were also employed. In one instance a constant dielectric constant of one was utilized.

The optimization protocol for the VIP and GRF analogs using the NOE-derived distance constraints consisted of the following steps: 1) minimization, 2) 10 ps dynamics with gradual heating from 0 K to 1000 K, 3) 300-600 ps dynamics at 1000 K, 4) at 20 ps intervals along the dynamics trajectory, 15 ps dynamics with gradual cooling from 1000 K to near 0 K followed by minimization, 5) at 20 ps intervals along the dynamics trajectory direct minimization. The energy was usually lower for the structures from quenched dynamics (step 4) than for those from direct minimization (step 5). The lower energy structure from a particular point on the trajectory was used for subsequent analysis. Peptide groups were constrained to be planar-trans and guanidino groups to be planar throughout the optimization, except that after the penultimate minimization the structures were reminimized without torsional constraints.

Optimizations were performed utilizing either an initial model with secondary structure assigned from the pattern of NOE's and tertiary structure adjusted graphically or with a fully-extended peptide chain. In the first case, WNOE was 25.0 throughout the optimization, except that for a final reminimization when WNOE was 1.0. For initial minimization utilizing the fully-extended peptide chain, the NOE-constraints were phased in gradually using WNOE of 0.01, 0.1, and 1.0 for successive minimizations and WNOE of 1.0 for the subsequent dynamics and minimizations. Alternatively, WNOE values of 0.01 and 0.1 were used in successive initial minimizations and WNOE of 0.1 for subsequent optimization.

For each peptide, the 15-30 conformations optimized using distance constraints were analyzed for global and local similarities based on RMS fits of α-carbons in the peptide backbone. Local structures (turns, helices, *etc.*) were identified by 1) aligning each segment of four consecutive α-carbons with the corresponding segment in the lowest energy conformer, 2) computing the average coordinates for each segment, 3) realigning each segment with the average segment and computing the RMS fit. A low RMS (<0.5 Å) indicated a defined conformation for that segment while a high RMS (>0.5 Å) indicated local disorder, that is, the NOE constraints were not sufficient to define a unique structure for this segment. Global comparisons were made using all α-carbons in the peptide backbone. First the conformers were sorted in order of increasing energy. The lowest energy conformer was the first member of the first family. The next conformer was added to the first if it fitted within the specified RMS cutoff (generally 1-5 Å) otherwise it became the first member of the second family. Each subsequent conformer was compared to the defined families in order and became a member of the first family which it fitted within the cutoff, otherwise it was placed in a new family. The resultant family classification has ordered the lowest energy representatives of each structural type. These representatives were compared graphically.

RESULTS AND DISCUSSION

Model for the Sodium Channel

A bundle of four α-helices was built from four peptide P14 monomers using as a template the bundle of helices observed crystallographically in several globular proteins (8). The polar residues (especially the acidic sidechains) were placed at the interior of the assembly and the apolar residues (especially the aromatic side chains) at the exterior surface of each helix. Optimization (see Methods) of the helical bundle with no symmetry nor dihedral angle constraints, resulted in close-packed occluded structures for either parallel or antiparallel strands. However, optimization with the parallel array constrained to C_4 symmetry, the helix axis aligned with the symmetry axis, and the helices constrained to idealized dihedral angles, produced a structure with a pore of the size inferred for the sodium channel protein (Figure 1). The optimized structure maintains the segregation of polar and non-polar residues of the original model and has two clusters of acidic groups (Asp^{7}s and Glu^{18}s) lining the pore. In contrast, no pore is obtained for a parallel trimeric array optimized with C_3 symmetry, aligned helix axes, and idealized dihedral angles.

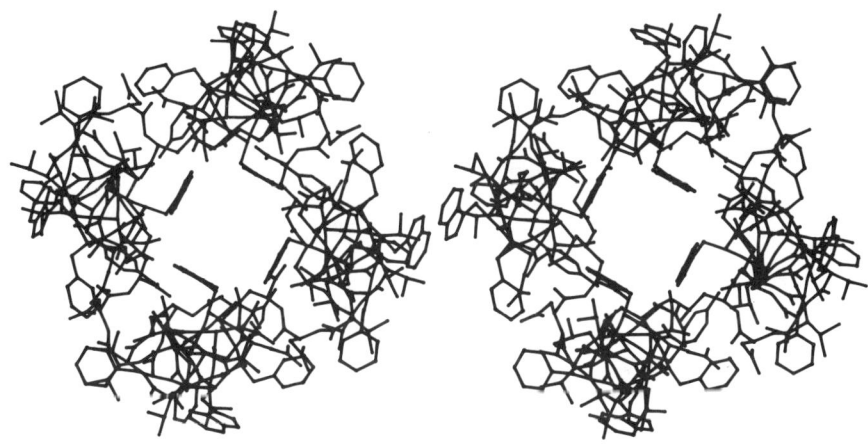

FIGURE 1. Stereo, end-view of model pore formed by a C_4-symmetric tetramer of peptide 14.

The bundle of four helices formed by the tetramer of the synthetic peptide adopts a structure that appears to be suited to form the lining of the sodium channel. The model differs from the four-helix bundle of globular proteins by containing an interior pore. For the model, the strands are parallel, the polar residues are on the inside, and the helix axes are aligned. For globular proteins, the strands are antiparallel, the polar residues are on the outside, and the helix axes cross at an angle. The last factor appears to be the most important in giving efficient packing within the globular protein, while alignment of the helix axes can give a stable bundle with an internal pore when there are at least four strands in the complex.

VIP Conformers in Aqueous Methanol; Testing of Optimization Protocols

CD spectra of a VIP analog (VIP', Table 1) in various methanol:water solutions show no pH dependence within the range 4-7. In aqueous solution, little or no regular secondary structure is indicated for VIP'. In methanol:water solutions, CD spectra indicate that the percentage α-helix (9,10) increases roughly linearly with percentage methanol.

Numerous two-dimensional nuclear magnetic resonance (NMR) spectra were obtained for VIP' in 25% methanol:water, pH 4; and in 50% methanol:water pH 6. Assignments of resonances and NOE's to specific protons were made using established methods (11). Our NMR data and its interpretation will be reported in detail elsewhere. An initial assessment of secondary structure was made from the pattern of NOE's: a type I β-turn at residues 5-8 and α-helical segments at residues 10-14 and 22-26 in 25% methanol:water, and an α-helical segment for residues 9-24 and possible β-turns at residues 3-6 and 24-27 in 50% methanol:water.

The effects of various computational parameters in the constrained dynamics and minimization protocol (Methods) were explored. Starting from the initial conformational model for VIP' in 50% methanol which has type I β-turns at residues 3-6 and 24-27, α-helix from residues 9-24, and the remainder fully extended, the protocol was applied. Due to the dominance of interactions with the charged sidechains, the α-helical structure indicated by the CD and NMR experiments (50% helix content) collapsed when the full formal charges (± 1) were used for the two Asp and five Lys sidechains. To approximate the attenuation of interactions of the charged groups due to solvent, calculations were performed with reduced formal charges. Visual comparison and RMS fitting of the α-carbons, revealed similar VIP' structures for formal charges (FC) of ± 0.0, 0.25, or 0.50 used with a dielectric constant of r. Subsequent calculations were performed with formal charges of ± 0.25 and these structures will be used for reference.

The reference optimization (FC = 0.25, optimization A) gave 27 VIP' structures which were analyzed for global similarities to each other and to the structures obtained using other formal charges. Using a RMS cutoff of 3 Å, 12 of the 27 structures were contained in the four lowest energy families with an energy range for the first family members of -69.4 to -58.3 kcal/mol. Similarly four families each were obtained for the simulations with formal charges of 0.0 and 0.5. Four of these families fall within one of the reference families (FC = 0.25) while the other four are close with RMS deviations of 3.3 to 4.6 Å. In contrast the structures obtained with full unit formal charges differ from the closest reference family by RMS deviations of 5.1 to 10.6 Å.

The 27 optimized VIP' structures (FC = 0.25) were further analyzed for constant structural elements involving at least four residues. The average RMS along the sequence is shown in Figure 2. Regions with average RMS < 0.5 Å correspond to constant structural segments, while the larger values indicate multiple structures within the ensemble. The common structural features are 1) an α-helix for residues 10-20, 2) a type III β-turn for residues 3-6, and 3) a type III β-turn for residues 24-27. With some irregularities near residue 20, residues 10-27 are essentially helical. Consistent with the data, there are a variety of conformations for residues 7-9 so that the N-terminal segment projects in various directions from the helical segment as apparent for the four lowest energy families (Figure 3A).

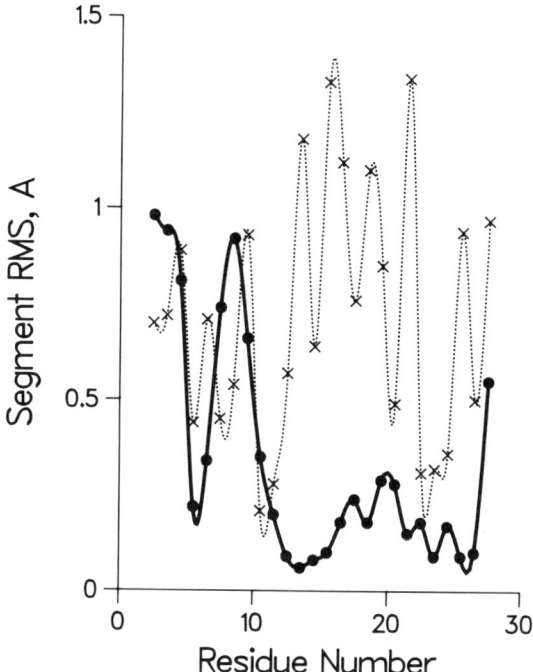

FIGURE 2. Average RMS for fit of consecutive segments of four C^α to average segment for VIP'. The optimizations started from helical models and used WNOE = 25. Optimization A, 27 conformers, NOE's for VIP' in 50% methanol:water - **filled circles**; Optimization E, 22 conformers, NOE's for VIP' in 25% methanol:water - **X's**.

Starting from a fully-extended conformation, the optimization protocol (Methods) was applied to VIP' utilizing NMR data from solutions in 50% methanol: water with WNOE=1.0 (optimization B), WNOE=0.1 (optimization C), and WNOE=0.1 using additional NOE's from reprocessed NMR data (optimization D). For optimization B, the unique structural feature was the shortening of the helical segment to residues 14-21. Even though the segment 10-14 was ordered it contained a non-helical kink. For optimization C, the unique structural feature was the absence of the turn at residues 3-6. There is a continuous helical segment at residues 10-27, but no turn at residues 3-6 in optimization D. See Table 2 for a comparison of the conserved structural features for each optimization.

The helical segment from residues 10-20 and the type III turn at residues 24-27 (one turn of helix) are present in the VIP' structures from each of the optimizations. The kink at residues 13-14 for optimization B appears to be an artifact arising from too large a value of WNOE preventing flexibility of the peptide. The absence of the turn at residues 3-6 in optimizations C and D shows that this structural feature is not unique even though it was formed in optimization B when starting from a fully-extended conformation. Optimization D uses the most complete NMR data (Table 3) and should be the most reliable. Peptide groups in the helical segment (10-27) comprise 60% of the total in the backbone which is comparable to the percentage helix calculated from CD spectra.

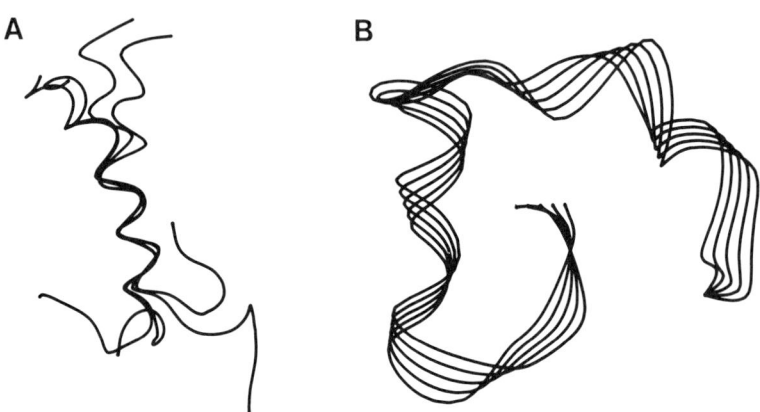

FIGURE 3. Backbone tracings for low-energy, optimized conformations for VIP'. **A.** A single line for each of the four lowest-energy families from optimization A for VIP' in 50% methanol:water. The N-termini are at the bottom of the figure, the C-termini at the top. **B.** Five-stranded ribbon for family two from optimization E for VIP' in 25% methanol:water. The C-terminus ends in an arrowhead.

TABLE 2
ORDERED STRUCTURAL SEGMENTS FOR VIP'

Opt.	% MeOH	Starting Conform.	WNOE	Segments		
A	50	Helical Model	25.0	TIII 3-6	Helix 10-20	TIII 24-27
B	50	Fully-Extended	1.0	TIII 3-6	Helix 14-21	TIII 24-27
C	50	Fully-Extended	0.1	----	Helix 9-19	TIII 24-27
D	50	Fully-Extended[a]	0.1	----	Helix 10-27	----
E	25	Helical Model	25.0	Turns 3-6&5-8	Helix 8-12&20-25	
F	25	Fully-Extended	0.1	----	Helix 8-12	
G	25	Fully-Extended[a]	0.1	----	Helix 8-13&15-19	

[a]Optimization used additional NOE's from recomputation of NMR data.

TABLE 3
NUMBER OF OBSERVED NOE'S FOR GRF AND VIP ANALOGS

Peptide	Solvent	Number of NOE's		
		Total	Long-Range[a]	i-(i+3)
GRF(1-29)NH$_2$	75% MeOH	165	24	18
[Ala^{15}]GRF(1-29)NH$_2$	75% MeOH	182	31	21
c(8-12)[Asp^8,Ala^{15}]GRF(1-29)NH$_2$	75% MeOH	202	37	25
[Sar^{15}]GRF(1-29)NH$_2$	75% MeOH	107	2	0
GRF(1-29)NH$_2$	Water	139	9	6
[Ala^{15}]GRF(1-29)NH$_2$	Water	131	14	7
c(8-12)[Asp^8,Ala^{15}]GRF(1-29)NH$_2$	Water	140	12	7
VIP'	50% MeOH	119	9	8
VIP'[b]	50% MeOH	189	35	19
VIP'	25% MeOH	103	1	0
VIP'[b]	25% MeOH	139	10	8

[a]NOE's which are not intraresidue nor sequential, but including i-(i+3).
[b]Additional NOE's from recomputation of NMR data.

In 25% methanol:water, CD measurements indicated about 25% helix. Optimization (E) was started using a conformational model in which residues 1-5, 9, 19, 22, and 26-28 were fully extended; residues 5-8 were in a type I β-turn; and residues 10-18, 20-21, and 23-25 were α-helical. The results of optimization indicated that the VIP' structure was considerably less constrained by the NOE distances in 25% methanol than in 50% methanol. Each of the 22 optimized structures differed in its global RMS fit of α-carbons by more than 4.0 Å from each of the other structures. With a 5.0 Å cutoff, 12 families were obtained with 9 conformations in the 5 lowest-energy families. The lowest energy members of these latter 5 families had energies from -96.1 to -61.4 kcal/mol. The high local RMS's for segments of four residues (Figure 2) also reflected the large number of conformations in the ensemble. Constant structural elements were discernable for the four segments with an average RMS less than 0.5 Å, but none were apparent for the regions with a RMS of 0.5 or greater. The constant elements were 1) consecutive γ-turns at residues 4 and 5 (ordering residues 3-6), 2) a non-classical β-turn at residues 5-8, 3) about 1 1/2 turns of irregular α-helix for residues 8-12, and 4) about 1 1/2 turns of irregular α-helix for residues 20-25. A backbone structure for one of the low-energy families is shown in Figure 3B.

For VIP' in 25% methanol:water, only short segments of secondary structure were predicted from the optimization even though the high value of WNOE (25.0) and starting from an ordered structure both favor retaining the original structural elements. Only the helical segment from residues 8-12 was conserved when a fully-extended conformer was the starting point (optimization F). When additional NOE's were obtained from reprocessing of the NMR data (Table 3, optimization G), two helical segments separated by a kink at residue 14 were predicted (Table 2). This latter optimization should be the most reliable.

GRF Conformers in Aqueous Methanol

The CD spectra of the GRF analogs in methanol:water are independent of pH within the 3 to 7 range. The percentage α-helix estimated (9,10) for these peptides in aqueous solution and in 75% methanol:water is given in Table 4. While only cyclo(Asp^8-Lys^{12})[Asp^8,Ala^{15}]GRF(1-29)NH$_2$ had substantial helix in aqueous solution, three of the analogs are essentially fully-helical in 75% methanol:water.

TABLE 4
HELIX CONTENT DEDUCED FROM CD SPECTRA FOR GRF ANALOGS

Peptide	% Helix	
	Water pH 3	75% MeOH pH 6
GRF(1-29)NH$_2$	20	90
[Ala^{15}]GRF(1-29)NH$_2$	25	90
c(8-12)[Asp^8,Ala^{15}]GRF(1-29)NH$_2$	45	90
[Sar^{15}]GRF(1-29)NH$_2$	15	50

NMR spectra for the GRF analogs were assigned and interpreted using established methods (11); the results will be presented in detail elsewhere. For GRF(1-29)NH$_2$ and its lactam analog, the NOE's observed in 75% methanol:water do not change significantly between pH 3 and pH 6. For the three highly-helical GRF analogs in 75% methanol:water, pH 6, the pattern of NOE's suggests an initial model with residues 1-5 extended and the remainder helical. Starting from this initial model, structural optimizations (Methods, WNOE=25.0) confirmed that these three peptides are highly helical (Figures 4A and 5) and that the structures are well determined by the NMR data (note the small RMS values in Figure 6). Cyclo(Asp^8-

Lys^{12})[Asp^8,Ala^{15}]GRF(1-29)NH$_2$ readily folds to the helical structure starting from a fully- extended conformer (WNOE=1.0). Judging from the number of long-range NOE's (Table 3), native GRF(1-29)NH$_2$ and its [Ala^{15}] analog also should readily fold from the fully-extended conformer, but these optimizations were not done. The residues in helical segments for each of the peptides are given in Table 5. The helix for the lactam analog is the most regular, that for the [Ala^{15}] analog bends near residue 18, and the native peptide has a major kink at residues 21-22 between the two helical segments. The results obtained for GRF(1-29)NH$_2$ are similar to those which have been reported by Clore and co-workers for [Nle^{27}]GRF(1-29)NH$_2$ in 30% trifluoroethanol:water, pH 4. For the latter peptide, CD spectra had indicated about 80% helix and optimizations constrained by NOE data had shown that residues 6-13 and 16-29 were helical (4,5).

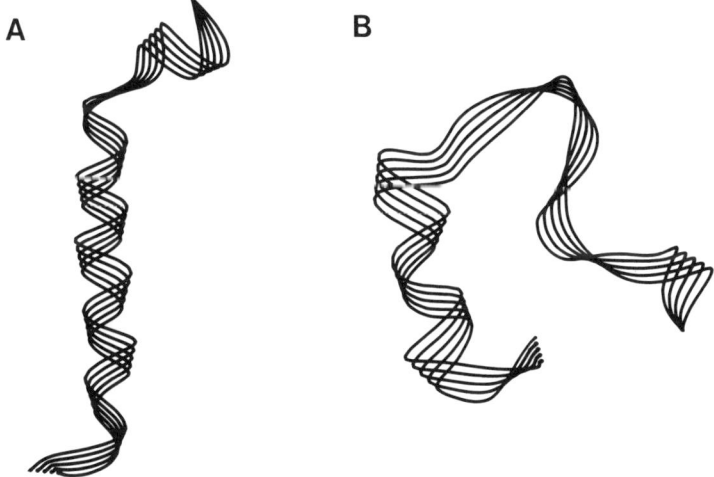

FIGURE 4. Five-stranded ribbons for backbone conformations of a low-energy family from optimizations for GRF(1-29)NH$_2$. The C-termini end in arrowheads. **A.** Family 1 for 75% methanol:water. **B.** Family 2 for water.

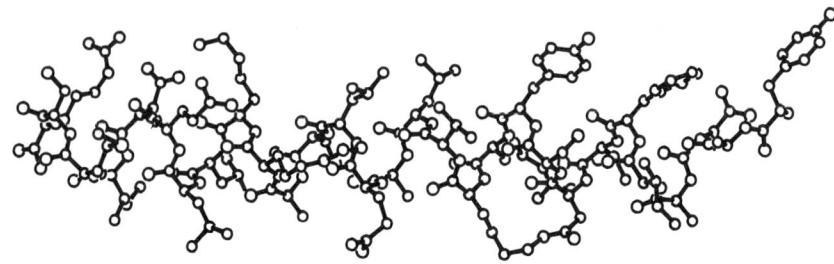
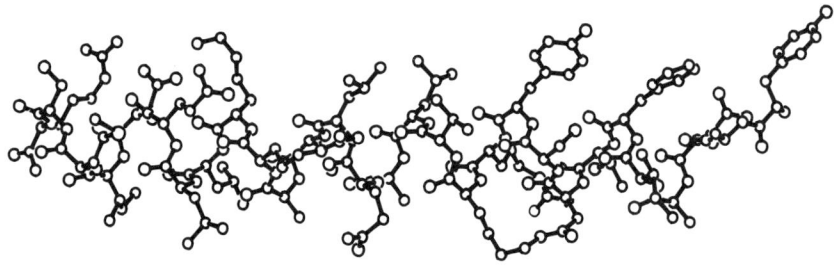

FIGURE 5. Stereo, ORTEP plot for family 1 from optimization for cyclo(Asp^8-Lys^{12})[Asp^8,Ala^{15}]GRF(1-29)NH$_2$ in 75% methanol:water. The N-terminus is on the right.

TABLE 5
ORDERED STRUCTURAL SEGMENTS FOR GRF ANALOGS

Peptide	Solvent	Segments	
GRF(1-29)NH$_2$	75% MeOH	Helix 4-21&22-28	
[Ala^{15}]GRF(1-29)NH$_2$	75% MeOH	Helix 4-29	
c(8-12)[Asp^8,Ala^{15}]GRF(1-29)NH$_2$	75% MeOH	Helix 2-29	
[Sar^{15}]GRF(1-29)NH$_2$	75% MeOH	Turn 3-5	TIII 18-21&25-28
GRF(1-29)NH$_2$	Water	Helix 10-14	----
[Ala^{15}]GRF(1-29)NH$_2$	Water	Helix 10-15	TIII 22-25
c(8-12)[Asp^8,Ala^{15}]GRF(1-29)NH$_2$	Water	Helix 7-18	TIII 22-25

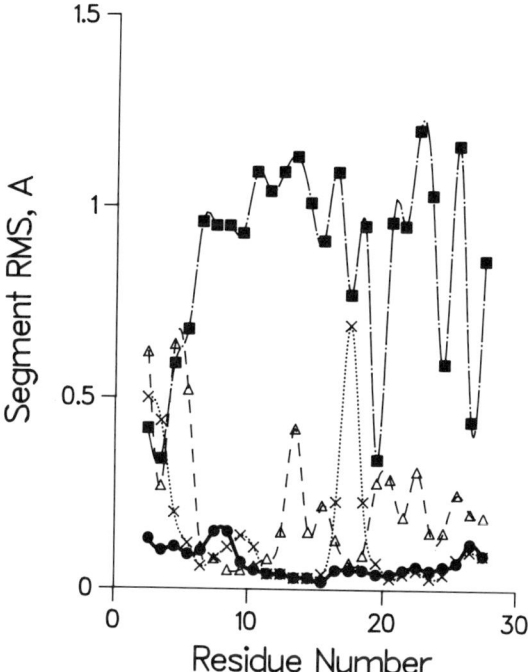

FIGURE 6. Average RMS for fit of consecutive segments of four C^α to average segment for GRF analogs in 75% methanol:water. For three analogs, the optimizations started from helical models and had WNOE = 25 : GRF(1-29)NH$_2$, 19 conformers - **triangles;** [Ala^{15}]GRF(1-29)NH$_2$, 14 conformers - **X's;** cyclo(Asp^8-Lys^{12})[Asp^8,Ala^{15}]GRF(1-29)NH$_2$, 13 conformers - **filled circles.** For the fourth analog, the starting conformer was fully-extended and WNOE = 1 : [Sar^{15}]GRF(1-29)NH$_2$, 22 conformers - **filled squares.**

In contrast to the above peptides, NMR data for [Sar^{15}]GRF(1-29)NH$_2$ in 75% methanol:water, pH 3 showed little evidence for ordered secondary structure. Optimizations (WNOE = 1.0) starting either from the above helical model or a fully-extended conformation gave similar results. Analysis for local structure gave only three points, centered at residues 3, 19 and 26, with a RMS < 0.5 Å. Graphical examination of the structures revealed reverse turns at these points, but the turn type differed in the two optimizations. Two optimized conformers (out of 34) had an α-helical segment at residues 18-27. Six additional conformers fit the α-carbons of this segment with a RMS < 2.5 Å. The eight conformers had an average of nine helical residues. For this segment, α-helix is only one of the many conformers which are consistent with NMR data since there are only two long-range NOE's (Table 3). Nevertheless, CD spectra indicated a high helical content (50% or 15 residues) for [Sar^{15}]GRF(1-29)NH$_2$.

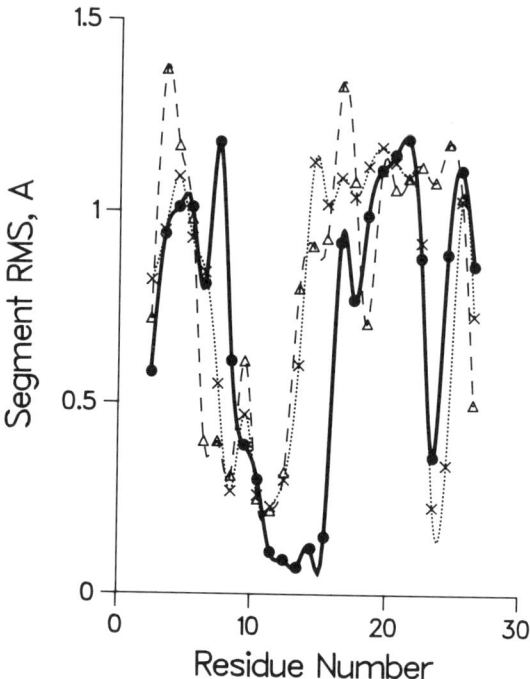

FIGURE 7. Average RMS for fit of consecutive segments of four C^α to average segment for GRF analogs in water. The optimizations started from fully-extended conformers, had WNOE = 0.1 and produced 20 optimized conformers. GRF(1-29)NH$_2$ - **triangles**; [Ala^{15}]GRF(1-29)NH$_2$ - **X's**; cyclo (Asp^8-Lys^{12}) [Asp^8,Ala^{15}] GRF (1-29) NH$_2$ - **filled circles.**

In aqueous solution at pH 3, the GRF analogs have an ordered segment beginning near residue 10 and two of the three peptides have a short ordered segment centered at residue 24 (Figure 7). Classification and examination of the optimized conformers reveals the secondary structure outlined in Table 5. GRF(1-29)NH$_2$ and its Ala^{15} analog also have a disordered (high RMS) type III turn at residues 6-9, but there is a kink between residues 9 and 10 so that this turn does not merge with the helical segment. The disorder for these two peptides compared to the ordered helical segment for the lactam analog is apparent in Figure 7 and in RMS plots for seven and eleven consecutive α-carbons (not shown). The number of helical residues corresponds well with the estimates from CD spectra in aqueous solution.

In conclusion, the optimization protocol readily folds the peptides to give conformers in accord with the CD and NMR data. Long-range NOE's are essential to uniquely define the conformation; a variety of conformers are consistent with intraresidue and sequential NOE's. A small weighting factor (WNOE) permits sampling numerous conformational states and optimization of the resulting structures.

Slowly phasing-in the NOE-based constraints prevents structural distortions.

ACKNOWLEDGEMENTS

We are grateful to M. Montal and S. Oiki for helping formulate and construct the model for the sodium channel protein; to D. Bolin, A. Felix and E. Heimer for synthetic peptides; to B. Wegrzynski for CD spectra; and to J. Blount, A. Krohn, A. Meade, K. Mueller, H. Ammann, D. Doran, P. Gerber, G. Schrepfer, T. J. O'Donnell, and A. Olson for software.

REFERENCES

1. Greenblatt RE, Blatt Y, Montal M (1985). The structure of the voltage-sensitive sodium channel. Inferences derived from computer-aided analysis of the *Electrophorus electricus* channel primary structure. FEBS Letters 193:125.
2. Oiki S, Danho W, Montal M (1988). Channel protein engineering: A synthetic 22-mer peptide from the primary structure of the voltage-sensitive sodium channel forms ionic channels in lipid bilayers. Proc Natl Acad USA (in press).
3. Heimer EP, Felix A, Ahmad M, Lambros T, McGarty T, Wang CT, Mowles T, Davidovich D (1988). Synthesis and biological evaluation of growth hormone releasing factor, structural linear and cyclic analogs (this volume).
4. Clore GM, Martin SR, Gronenborn AM (1986). Solution structure of human growth hormone releasing factor. Combined use of circular dichroism and nuclear magnetic resonance spectroscopy. J Mol Biol 191:553.
5. Brünger AT, Clore GM, Gronenborn AM, Karplus M (1987). Solution conformations of human growth hormone releasing factor: comparison of the restrained molecular dynamics and distance geometry methods for a system without long-range distance data. Protein Engineering 1:399.
6. Mueller K, Ammann HJ, Doran DM, Gerber P, Schrepfer G (1986). Aspects and applications of computer-assisted molecular modeling to drug design (CAMMDD). In Harms AF (ed) "Innovative Approaches in Drug Research," Amsterdam: Elsevier, p 125.
7. Brooks BR, Bruccoleri RE, Olafson BD, States DJ, Swaminathan S, Karplus M (1983). CHARMM: A program for macromolecular energy, minimization, and dynamics calculations. J Comp Chem 4:187.
8. Weber PC, Salemme FR (1980). Structural and functional diversity in 4-α-helical proteins. Nature 287:82.
9. Bewley TA, Brovetto-Cruz J, Li CH (1969). Human pituitary growth hormone. Physicochemical investigations of the native and reduced-alkylated protein. Biochemistry 8:4701.
10. Jibson MD, Li CH (1981). β-Endorphin. Circular dichroism of synthetic human analogs with various chain lengths in methanol solutions. Int J Peptide Protein Res 18:297.
11. Wüthrich K (1986) "NMR of Proteins and Nucleic Acids," New York: John Wiley & Sons, Inc. (and references therein).

III. SYNTHETIC PEPTIDE-BASED VACCINES

THE NEXT GENERATION OF FOOT-AND-MOUTH DISEASE VACCINES

F. Brown

Department of Virology, Wellcome Biotech Ltd
Langley Court, Beckenham, Kent, BR3 3BS

ABSTRACT The vaccines to control foot-and-mouth disease are prepared by chemically inactivating virus grown in a variety of tissue culture cells. Although these are effective if properly applied, there are still problems associated with their use, chief of which are the need to maintain a cold chain to ensure their stability and the considerable antigenic variation which the virus displays. Studies of the structure of the virus have shown that a short peptide will elicit levels of neutralizing antibody which protect experimental animals against challenge infection. The response to the peptide can be increased several fold if it is presented as part of a fusion protein with β - galactosidase or the hepatitis B core protein. Moreover, the peptide elicits antibody which is more cross-reactive than the corresponding anti-virion antibody. Preliminary observations indicate that appropriate amino acid substitutions can increase the level of cross reactivity and could lead to a vaccine which protects against viruses of all serotypes.

INTRODUCTION

Foot-and-mouth disease is the most economically important virus disease of farm animals, with productivity losses usually estimated to be of the order of 25%. Together with the indirect losses due to embargoes on trading, an outbreak of the disease can be devastating, particularly to those countries which depend heavily on the export of farm animals and their products.

The disease occurs in many countries, particularly in South America, Africa and Asia. Control is by vaccination, almost exclusively using inactivated vaccines prepared from virus grown in a variety of tissue culture cells including tongue epithelium fragments, primary pig and calf kidney cells and the baby hamster kidney cell line BHK21. Comprehensive programmes using vaccines prepared in this way have to all intents and purposes eradicated the disease from Western Europe over the last 30 years. However the situation in the rest of the world is not so encouraging. This relative lack of success probably stems partly from the lower level of efficiency of the veterinary services in those countries where the disease still occurs but there are other reasons such as the more difficult terrain which make administration of the vaccine more difficult. Moreover, the higher ambient temperatures make it important to maintain a cold chain which will ensure that the vaccine retains its potency until it is injected into the animals.

In addition to these difficulties, it is necessary to provide multivalent vaccines because of the antigenic diversity of the virus. There are seven serotypes of the virus, O, A, C, Southern African Territories SAT1, SAT2 and SAT3 and Asia1 and an animal recovered from infection with virus of one serotype is still susceptible to infection with viruses of the remaining serotypes. Moreover there is considerable antigenic variation within each serotype so that immunisation with a vaccine prepared from one isolate need not necessarily protect against infection with viruses belonging to the same serotype. For this reason it is necessary to monitor outbreak strains for their antigenic relatedness to the available vaccines so as to ensure that the vaccines can provide the protection. This relatedness is usually determined by comparing the neutralizing activity of sera from animals which have received the available vaccine against the outbreak strain and the virus used for preparing the vaccine. The ratio between the two titres (the 'r' value) gives a measure of the relatedness between the two strains. A high ratio (approaching unity) indicates that the strains are closely related antigenically and that the available vaccine would be suitable for use against the new outbreak. A low value (r approaching 0.1) indicates that the available vaccine may not be suitable for use in the outbreak.

There is thus a clear need to provide a stable vaccine with a wide antigenic spectrum. Recent work has provided information on the detailed structure of the virus and RNA sequencing studies have allowed comparison to be made between viruses belonging to different serotypes and to different sub-types within a serotype. The chemical basis for antigenic variation which has emerged from these studies should allow us to develop vaccines which can overcome the problem.

The results presented in this paper show that peptides corresponding to a linear sequence of the protein VP1 of the virus can be presented in such a way that they elicit levels of neutralizing antibody approaching those obtained with conventional inactivated vaccines. Moreover, preliminary results indicate that antigenic variation can be overcome by manipulating the sequence of amino acids in the antigen site (1). Consequently the objective of a stable, chemically synthesised peptide vaccine seems attainable, particularly as we learn more about the immune response to antigens and peptides in particular. This article summarises the current state of our knowledge.

THE STRUCTURE OF THE VIRUS IN RELATION TO ITS IMMUNOGENIC ACTIVITY

Foot-and-mouth disease virus belongs to the genus aphthovirus of the family Picornaviridae. The RNA genome of the virus has a mol.wt. of 2.6×10^6 and there are 60 copies of each of four proteins VP1-VP4. The three surface proteins VP1-VP3 have mol.wt. of c 24×10^3 and the internal protein VP4, which is myristylated at its N-terminus, has a mol.wt. of c 10×10^3. The crucial property of the virus in the context of vaccination is that when VP1 is cleaved in situ by trypsin, the immunogenicity of some strains is reduced considerably (2). This indicated that VP1 was critical in determining the immunogenic activity of the virus particle and this view was confirmed by the demonstration by Laporte and his colleagues (3) that the isolated VP1 would elicit neutralizing antibody when injected into pigs. Consequently, studies on the possibility of a sub-unit vaccine have been focussed on this protein.

One of the inherent problems, however, was the very low intrinsic potency of the isolated protein, compared with that of the virus particle. This has been presumed to be due to the altered configuration of the protein once it had been released from the virus particle and there are no convincing reports to suggest that this problem has been overcome.

IMMUNOGENIC SITES ON VP1

Pioneering studies by Strohmaier and his colleagues in Germany (4) had shown that fragments of VP1 would elicit the formation of neutralizing antibody in mice. Extension of these studies by Strohmaier's group (5) and by Bittle et al. (6) and Pfaff et al. (7) in 1982 led to the finding that a sequence of amino acids within the region 141-160 would elicit levels of neutralizing antibody which were high enough to protect guinea pigs against challenge infection. This protection was achieved with a single inoculation. Moreover, the level of response could be boosted considerably with a second injection of the peptide, thus providing the basis for a system of multiple injections using a triggered release system, in which the vaccine is a molecule which is stable at body temperature.

These results have focussed attention on the 141-160 sequence, with particular emphasis on methods for enhancing the neutralizing antibody response to it. The initial experiments had shown that about 50μg of peptide coupled to a protein such as keyhole limpet haemocyanin, were required to afford protection in guinea pigs. This amount compares with 0.01μg of this sequence which is present on the 60 copies of VP1 present in the 1μg of virus which is sufficient to protect guinea pigs. Several approaches have been made to enhance the response, based on the reasoning that the configuration of the peptide when it forms part of the virus particle is likely to be important in determining the specific response.

CONFORMATION OF THE PEPTIDE SEQUENCE ON THE VIRUS PARTICLE

The only method for solving this problem is by X-ray diffraction studies of virus crystals. The virus has been crystallised (8) and some preliminary data are now available (9). The crystals of virus of serotype O-1, strain BFS belong to the space group I23 with cell dimensions a=b=c = 345Å, $\alpha = \beta = \gamma = 90°$. There are two virus particles per unit cell, which means that the asymmetric unit is one-twelfth of a complete particle. These diffract to high resolution and an 83% complete data set is now available at a resolution of 4.5Å. Given the space group there are only two possible orientations of the particle in the unit cell. A rotation function has been calculated which shows the correct orientation; this has been confirmed by placing models of rhino virus (10) and Mengo virus (11) in both orientations, calculating the diffraction that would be observed from these viruses and comparing this with the observed scattering obtained with the FMDV crystals.

Phases obtained by Fourier inversion of a composite virus 'rhengo' provided a starting point for phase refinement and extension, using the methods and programmes developed initially by Bricogne (12) and further developed by Rayment (13) and Rossmann (11) and their colleagues. Such a strategy of phase determination has already been used in the case of Mengo virus (11). The outcome of the studies with FMDV, which is awaited with considerable interest, should provide valuable information on the configuration of the peptide sequence in situ and perhaps indicate better methods of presenting it to the host's immune system.

PRESENTATION OF THE PEPTIDE

In the absence of information on the configuration of the immunogenic peptide on the virus particle, we have studied this problem in a rather more empirical way. Most of the initial experiments used conjugates of the peptides and carrier proteins. This conjugation was achieved with a variety of reagents, none of which could be regarded as providing a product of known structure. In an attempt to achieve an ordered coupling of the peptide to a carrier protein we have expressed it as part of a fusion protein in E.coli cells. In our first experiments the peptide sequence was fused to β-galactosidase at its N-terminus. Such a construction has the potentially additional advantage that the antigenic sites on β-galactosidase which are recognised by helper T cells have been identified. This fusion protein was no more immunogenic for mice or guinea pigs than the chemically linked peptide (14,15). Surprisingly, fusion proteins consisting of two or four copies of the peptide linked to β-galactosidase have a much greater immunogenic activity than the single copy construct and as little as 40μg of peptide in such a construct, given as a single inoculation, will protect pigs against challenge inoculation with 60000 ID50 of virus (16). We are unable to account for this greatly enhanced activity but it seems possible that the repetitive sequence is influencing the response.

An even greater response is obtained when the peptide is expressed as part of the hepatitis B core particle (17). A construct with the DNA sequence coding for the peptide attached to the 5' end of the gene coding for the core protein poisoned the E.coli expression system. However, it could be expressed in vaccinia virus and the expressed protein formed core particles which could be purified by conventional centrifugation methods. The hybrid cores reacted with core protein antibody and with antibody against both FMDV particles and the 141-160 peptide. Crucially, the core particles elicited high levels of neutralizing antibody and protection of guinea pigs could be obtained with a single inoculation of as little as 2µg of protein (equivalent to 0.2µg of peptide). The level of neutralizing antibody obtained with 2µg of the hybrid core protein was similar that obtained with 1µg of virus particles.

Clearly presentation of the peptide in a defined configuration leads to enhanced responses and it will be interesting to determine whether the low responses to potentially immunogenic peptides from other viruses can be increased in the same way.

THE IMPORTANCE OF T CELLS IN THE IMMUNE RESPONSE TO THE PEPTIDE

In most experiments to test its immunogenicity the peptide has been coupled to a carrier protein either by chemical linking or by expression as a fusion protein. However, we have found that the uncoupled peptide will evoke protective neutralizing antibody in guinea pigs, provided it is injected with incomplete Freund's adjuvant or in a liposome (18). However, the response in cattle and pigs to the uncoupled peptide was much lower.

This limitation of the host immune response is clearly a problem for the development of peptides as vaccines but the growing understanding of the importance of the T cell response and how it functions at the molecular level now allows a rational approach to its study. It has become clear that a peptide will only be immunogenic in those recipients whose histocompatibility proteins recognise it. The requirements for a peptide vaccine, therefore, consist of at least two factors, one recognised by B cells and the second by T cells. Indeed, the helper T cell is the focal point of the immune response because it is necessary for the antibody response and in particular for an anamnestic response. These facts have led us to consider whether the factors which determine the response to the FMDV peptide are related to MHC restriction.

In preliminary experiments, we have studied the response to the uncoupled peptide in a number of well defined mouse haplotypes (19). We have found that mice belonging to the $H-2^k$ haplotype respond well to the peptide whereas $H-2^d$ haplotype mice do not respond (Fig. 1). However, by using a peptide which provides T cell help in the $H-2^d$ mice in combination with the FMDV peptide a response was obtained similar to that obtained in the $H-2^k$ mice (Fig. 2). The peptides used were those defined by Berzofsky and Grey and their colleagues (20,21,22) namely two sequences from sperm whale myoglobin (SWMI 132-148 and SWMII 105-121) and one sequence from ovalbumin (OVA 323-339). However, the response to the three hybrid peptides differed in one significant respect in that neutralizing antibodies were evoked only in those $H-2^d$ mice receiving the OVA and SWMI sequences. This result suggests that T-helper cell epitopes may control antibody production of specific B cell clones.

To characterise the nature of the virus neutralizing and non-neutralizing antibody populations in those mice which had received the hybrid peptides with the OVA and SWMII T cell sites, the 28 days post-inoculation sera were tested with a range of peptides located within the 141-160 sequence of the FMDV peptide. The antibodies to the peptide with the added OVA T cell site (i.e. those which neutralize the virus) recognised peptides from the 147-156 region, which is known from our previous work to be critical for the induction of virus neutralizing antibody. In contrast, the antibodies to the peptide with the added SWMII T cell site reacted only with those peptides from the N-terminus i.e. they did not react with those from the region known to be critical for the induction of neutralizing antibody.

The hybrid peptides which we used were synthesised so that the B cell epitope was N terminal to the T cell epitope. However, at this stage of the work we have no knowledge that this is the best arrangement of the individual epitopes or even whether it is necessary for them to be covalently linked.

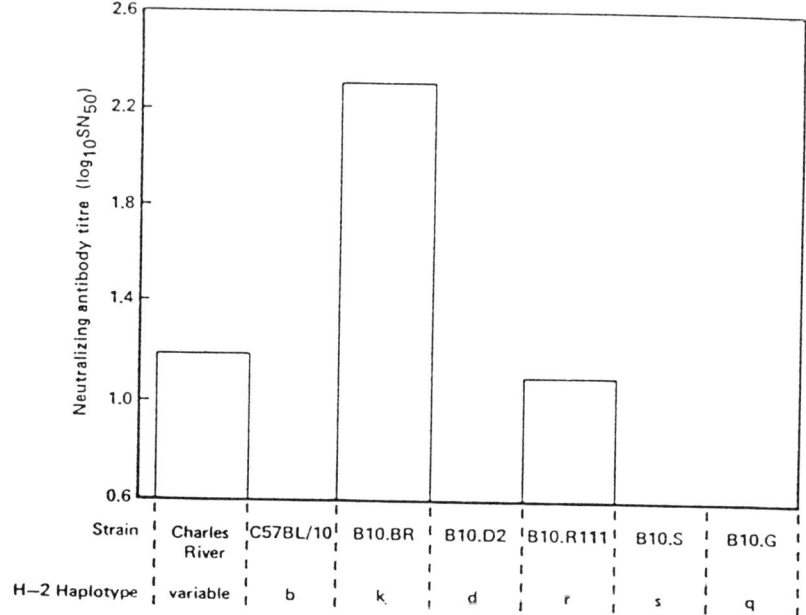

Fig 1 Neutralizing antibody response of mice of different haplotypes to the uncoupled 141-160 peptide of FMDV. The sera were collected 56 days after inoculation of the uncoupled peptide with Freund's incomplete adjuvant and the neutralizing activity measured against 100 tissue culture ID_{50} in a micro-neutralization test.

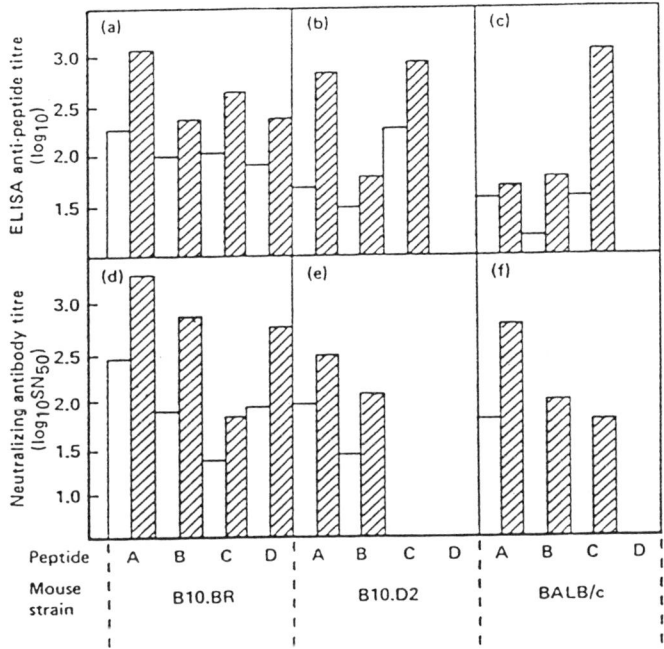

Fig 2 Anti-peptide and neutralizing antibody response of mice belonging to the $H-2^k$ (B10.BR) and $H-2^d$ (B10.D2 and BALB/C) haplotypes to the 141-160 peptide of FMDV linked at its C terminus to (A) OVA; (B) SWMI; (C) SWMII or (D) 161-177 from FMDV-VP1 and inoculated with Freund's incomplete adjuvant. The open columns give the values for sera collected 28 days after a single inoculation and the hatched columns give the values 28 days after a second inoculation 63 days after the first.

OVERCOMING ANTIGENIC VARIATION

As referred to above, the occurrence of FMDV as seven serotypes poses severe problems in the control of the disease. Consequently it is necessary to have available vaccines against all the serotypes or at least the capacity to produce them. It was expected that peptides would be even more strain specific than conventional vaccines. Unexpectedly, we found that antisera prepared against the 141-160 peptide were more cross-reactive than the sera prepared against conventional vaccines (23). The reason for this cross-reactivity may be related to the conserved region of the peptide sequence between amino acids 145 and 151 (Fig. 3). The viruses which were not neutralized by the anti-peptide antiserum had an amino acid change at position 148. Some of our early work with closely related viruses of serotype A also showed the importance of the residue at this position in determining antigenic specificity (24).

In preliminary experiments we have further demonstrated that the cross-neutralizing activity of the anti-peptide sera extends to viruses of other serotypes. It seems that if the structural basis for this cross-reactivity can be determined precisely, peptides which will elicit neutralizing antibody against all serotypes could be designed. Of particular interest in this connection is the presence of the sequence Arg Gly Asp at positions 145-147. This sequence, which was first pointed out by Geysen et al. (25) occurs, with one exception, in all the FMDV isolates studied so far. Moreover, a peptide of sequence Asn Leu Arg Gly Asp Leu Glu, linked to keyhole limpet haemocyanin, elicited antibody which neutralized not only viruses which had the same sequence at those positions but also viruses belonging to other serotypes. These observations indicate that by presenting the Arg Gly Asp sequence to the host in different configurations it may be possible to enhance the cross-reactivity of the response.

ARE PEPTIDES A PRACTICAL OPTION FOR FOOT AND MOUTH VACCINATION?

The concept of a totally synthetic vaccine against FMD, largely unacceptable a few years ago, is now beginning to attract more support as the results of a variety of experimental approaches begin to accumulate. The major

Virus	141									150
Kaufbeuren B64	Val	Pro	Asn	Leu	Arg	Gly	Asp	Leu	Gln	Val
Kaufbeuren B7	.	.	.	.	.	.	.	.	.	.
BFS 1848	.	.	.	.	.	.	.	.	.	.
BFS 1860	.	.	.	.	.	.	.	.	.	.
O-VI (Subtype 6)	.	.	.	Val	.	.	.	Thr	.	.
Hong Kong	Met	Ser	.	Val	.	.	.	.	.	.
Indonesia 7/83	Thr	Thr	.	Val	.	.	.	.	.	.
Thailand 1/80	Leu	Thr	.	Val	.	.	.	Arg	.	.

	151									160
Kaufbeuren B64	Leu	Ala	Gln	Lys	Val	Ala	Arg	Thr	Leu	Pro
Kaufbeuren B7	.	.	.	.	.	.	.	.	.	.
BFS 1848	.	.	.	.	.	.	.	.	.	.
BFS 1860	.	.	.	.	.	.	.	.	.	.
O-VI (Subtype 6)	.	Asp	.	.	.	Ser	.	Ala	.	.
Hong Kong	.	Thr	.	.	Ala	Ser	.	Ala	.	.
Indonesia 7/83	.	.	.	.	Ala	Ala	.	.	.	.
Thailand 1/80	.	.	.	.	Ala	.	.	Pro	.	.

Point signifies no change from Kaufbeuren B64 sequence.

Fig 3   Amino acid sequences of the 141-160 region of VP1 of several isolates of FMDV, serotype O

argument against a peptide vaccine initially was the conceived amount of peptide which would be necessary to evoke a protective immune response. Moreover, the excellent responses to the peptide in guinea pigs appeared not to be reflected in cattle. Two groups of experiments have now shown that these problems can be overcome. Firstly, two or more copies of the peptide linked to the N terminus of β-galactosidase elicit much higher neutralizing antibody responses than the fusion protein containing only one copy (16). The reason for this increased activity is not understood and warrants close study. The improved immune response with the hepatitis B core construct is even more dramatic. Milich and his colleagues (26) have provided considerable information on the immunogenic properties of the hepatitis B core particle and this method for presenting peptides may hold the key to their delivery. More detailed investigations of the T cell sites of the core particle are proceeding and the results of these studies are eagerly awaited. It would clearly be of importance to measure the activity of the FMDV peptide core construct in cattle and pigs.

The second approach follows from the recognition that to realise their full potential as vaccines, uncoupled peptides must contain domains which react with helper T cell receptors and Ia antigens in addition to binding sites for anti-protein antibodies. The contrast between the rather disappointing responses in cattle, and the encouraging observations in guinea pigs clearly required investigation of the genetic restriction involved. The results of the preliminary experiments in mice, in which it was shown that the genetic restriction in mice of the $H-2^d$ haplotype can be overcome by linking the FMDV peptide to a defined helper T cell determinant from ovalbumin or sperm whale myoglobin (19), provide grounds for optimism that the restriction in cattle can be overcome similarly. This rational approach to peptide vaccines based on a detailed knowledge of the immune response should provide the information necessary to design synthetic immunogens with B and T cell epitopes that will overcome the "within species" and "between species" variations.

In addition to these two approaches it is becoming apparent that correct targetting of the peptide could lead to enhanced activity. Although many suggestions have been made, this aspect of antigen presentation has only been investigated empirically. As more details of antigen processing become available this extra dimension in antigen presentation should become fully understood at the molecular level.

Finally, the vulnerability of the peptide bond clearly needs to be considered when peptide vaccines are under discussion. If we accept that the shape of the epitope is of major importance, there seems to be no reason why the shape of a peptide, which can be established by direct analysis, should not be mimicked either by peptides consisting of D-amino acids or even by other invulnerable molecules. Irrespective of the approach, however, the urgent need is to understand antigen processing at the molecular level so that some general rules can be drawn up.

Even with our largely empirical approach, we have shown that the activity of the FMDV peptide can be enhanced by several orders of magnitude. With the refinements outlined in the preceding paragraphs it seems reasonable to expect that potent immunogens based on amino acid sequences will emerge as the vaccines of the future. If, in addition the structural basis for the cross-reactivity of the anti-peptide antisera can be elucidated, there would be the added bonus of a vaccine which could be used against viruses of all serotypes.

REFERENCES

1. Brown F (1985). Antigenic structure of foot-and-mouth disease virus. In: Immunochemistry of viruses. Eds. M.H.V. van Regenmortel and A.R. Neurath. Elisvier Science Publishers B.V.

2. Wild TF, Burroughs JN Brown F (1969). Surface structure of foot-and-mouth disease virus, J gen Virol 4: 313.

3. Laporte J, Grosclaude J, Wantyghem J, Bernard S Rouze P (1973). Neutralisation en culture cellulaire du pouvoir infectieux du virus de la fievre aphteuse par des serums provenant de porcs immunises a l'aide d'une proteine virale purifiee C, r hebd Seanc Acad Sci Paris 276: 3399.

4. Kaaden OR, Adam KH, Strohmaier K (977). Induction of neutralizing antibodies and immunity in vaccinated guinea pigs by cyanogen bromide peptides of VP3 in foot-and-mouth disease virus. J gen Virol 197: 397.

5. Strohmaier K, Franze R, Adam KH (1982). Localisation and characterisation of the antigenic portion of the foot-and-mouth disease virus protein. J gen Virol 59: 295.
6. Bittle JL, Houghten RA, Alexander H, Shinnick TM, Sutcliffe JG, Lerner RA, Rowlands DJ, Brown F. (1982). Protection against foot-and-mouth disease by immunization with a chemically synthesised peptide predicted from the viral nucleotide sequence, Nature, 298: 30.
7. Pfaff E, Mussgay M, Bohm HO, Schulze GE, Schaller, H (1982). Antibodies against a pre-selected peptide recognise and neutralize foot-and-mouth disease virus. EMBO J 1: 869.
8. Fox G, Stuart D, Acharya R, Fry E, Rowlands D, Brown F (1987). Crystallization and preliminary X-ray diffraction analysis of foot-and-mouth disease virus. J Mol Biol 196: 591.
9. Stuart G, Fry E, Acharya R, Fox G, Rowlands DJ, Brown F (unpublished data).
10. Rossman MG, Arnold E, Erickson JW, Johnson JE, Kamer G, Luo M, Mosser AG, Rueckert RR, Sherry B, Vriend G (1985). Structure of a human common cold virus and functional relationship to other picornaviruses. Nature 317: 145.
11. Luo M, Vriend G, Kamer G, Minor I, Arnold E, Rossmann MG, Boege U, Scraba DG, Duke GM, Palmenberg AG (1987). The atomic structure of Mengo virus at 3.0A resolution. Science 235: 182.
12. Bricogne G (1974). Geometric sources in redundancy in intensity data and their use for phase determination. Acta Crytallogr Sect. A30: 395.
13. Rayment I, Baker TS, Caspar DLD (1983). A description of techniques and application of molecular replacement used to determine the structure of polyoma virus capsid at 22.5A resolution. Acta Crystallogr Sect. B39: 505.

14. Winther MD, Allen G, Bomford RH, Brown F (1986). Bacterially expressed antigenic peptide from foot-and-mouth disease virus capsid elicits variable immunologic responses in animals Journal of Immunology 136: 1835.

15. Broekhuijsen MP, Blom T, Kottenhagen M, Pouwels PH, Meloen RH, Barteling SJ, Enger-Valk BE (1986). Synthesis of fusion proteins containing antigenic determinants of foot-and-mouth disease virus, Vaccine 4: 119.

16. Broekhuijsen MP, Van Rijn JMM, Blom AJM, Pouwels PH, Enger-Valk BE, Brown F, Francis MJ (1987). Fusion proteins with multiple copies of the major antigenic determinant of foot-and-mouth disease virus protect both the natural host and laboratory animals J gen Virol 68: 3137.

17. Clarke BE, Newton SE, Carroll AR, Francis MJ, Appleyard G, Syred AD, Highfield PE, Rowlands DJ, Brown F (1987). Improved immunogenicity of a peptide epitope after fusion to hepatitis B core protein, Nature 330: 381.

18. Francis MJ, Fry CM, Rowlands DJ, Bittle JL, Houghten RA, Lerner RA, Brown F (1987). Immune response to uncoupled peptides of foot-and-mouth disease virus Immunology 61: 1.

19. Francis MJ, Hastings GZ, Syred AD, McGinn B, Brown F, Rowlands DJ (1987). Non-responsiveness to a foot-and-mouth disease virus peptide overcome by addition of foreign helper T-cell determinants, Nature 330: 168.

20. Berkower I, Matis LA, Buckenmeyer GK, Gurd FRN, Longo DL Berzofsky JA (1984). Identification of distinct predominant epitopes recognised by myoglobin specific T cells under the control of different Ir genes and characterization of representative T cell clones, J Immunol 132: 1370.

21. Streicher HZ, Berkower JJ, Busch M, Gurd FRN, Berzofsky JA (1984). Antigen conformation determines processing requirements for T-cell activation Proc natn Acad Sci USA 81: 6831.

22. Shimonkevitz R, Colon S, Kappler JW, Marrack P, Grey H (1984). Antigen recognition by H-2 restricted T cells. II. A tryptic ovalbumin peptide that substitutes for processed antigen, J Immunol 133: 2067.
23. Ouldridge EJ, Parry NR, Barnett PV, Bolwell C, Rowlands DJ, Brown F (1986). Comparison of the structures of the major antigenic sites of foot-and-mouth disease viruses of two different serotypes Ed F Brown, RM Channock, RA Lerner, Vaccines 86: 45.

24. Rowlands DJ, Clarke BE, Carroll AR, Brown F, Nicholson BH, Bittle JL, Houghten RA, Lerner RA (1983). Chemical basis of antigenic variation in foot-and-mouth disease virus Nature 306: 694.

25. Geysen HM, Meloen RH, Barteling SJ (1984). Use of peptide synthesis to probe viral antigens for epitopes to a resolution of a single amino acid Proc Natl Acad. Sci USA 81: 3998.

26. Milich DR, McLachlan A (1986). The nucleocapsid of hepatitis B virus is both a T-cell-independent and a T-cell-dependent antigen Science 234: 1398.

BIOLOGICAL ROLE OF PRE-S SEQUENCES OF THE HEPATITIS B VIRUS (HBV) ENVELOPE PROTEIN

A. Robert Neurath[1], Nathan Strick[1], Stephen B.H. Kent[2], Karen Parker[2], Chin Sook Kim[2], Marc Girard[3], Harold E. Ralph[1] and Jay Valinsky[1]

[1]The Lindsley F. Kimball Research Institute of The New York Blood Center, New York, New York 10021, [2]Div. of Biology, California Institute of Technology, Pasadena, California 91125 and [3]Pasteur Vaccins, B.P. no. 10, 92430 Marnes-la-Coquette, France

ABSTRACT Studies with synthetic peptides permitted the mapping of sites within the HBV envelope (env) proteins involved in virus attachment to cell receptors and in eliciting virus-neutralizing (VN) antibodies (Ab). Such Ab are elicited by each of the three regions of the HBV env protein: S-protein, preS2 and preS1 sequences. We describe here: 1) The localization of epitopes within the preS sequence involved in antiviral Ab binding (B-cell epitopes) and in immunogenicity (T-cell epitopes); 2) Immunological properties of hybrid synthetic peptides consisting of B- and T-cell epitopes derived from distinct portions of the HBV env protein; 3) The role of carriers and adjuvants in eliciting immune responses to synthetic peptides; 4) The effect of glycosylation in recognition of native HBV env proteins by antipeptide antisera; 5) The role of interaction between distinct domains of HBV env proteins in their immunologic recognition; 6) Immunogens prepared by combining recombinant DNA products with synthetic peptides; and 7) The safety of vaccines containing preS sequences.

INTRODUCTION

The HBV env consists of three distinct but related proteins. Due to their differential glycosylation, six components can be discerned by polyacrylamide gel electrophoresis (PAGE): P25, GP29 (S-Protein); GP33, GP36 (M-protein); and P39, GP41 (L-protein) (1). Work with synthetic peptides having sequences predicted from the nucleotide sequence of the HBV env gene confirmed that M- and L-protein correspond to S-protein + the preS2 sequence, and S-protein + the preS2 and preS1 sequences, respectively (1). Selected peptides from the preS region meet the following criteria: 1) They are recognized by Ab to the native env protein; 2) Antipeptide antisera recognize the native env protein well; 3) The peptides without a carrier elicit high levels of Ab reacting with the native protein. This indicates that they contain both B- and T-cell epitopes; 4) Some T-cell epitopes on the peptides mimic the corresponding T-cell epitopes on the native protein; 5) Selected peptides prime animals for a response to the native env protein; 6) Peptides derived from the preS1 sequence mimic the HBV receptor binding site; 7) The peptides elicit VN Ab (1,4).

Unlike the preS region, S-protein contains cysteine residues which are involved in the formation of disulfide bonds within the native env protein. Reduction and alkylation of S-protein leads to the formation of an antigen partially cross-reacting with the native protein (2). Ab elicited by native S-protein are VN and protective (3), while the biological properties of Ab to reduced and alkylated S-protein remain undefined. For this reason, peptides from the sequence of S-protein, to be effective as components of synthetic hepatitis B vaccines, should mimic conformational epitopes. Peptides from the S(110-150) region seem the most promising in this respect (2,3).

RESULTS

Delineation of Determinants Essential for Antigenicity and Immunogenicity of the HBV env protein.

Peptides derived from the N-terminal 1/2 and 1/3 of the preS2 and preS1 sequences, respectively, meet the criteria listed in the Introduction. The length of these peptides corresponds to 21-36 amino acid residues [preS(120-145), preS(120-153), preS(12-32), preS(12-47), preS(21-47)] (1,5). In order to describe the design of synthetic peptides containing B- and T-cell epitopes from nonadjacent regions of the HBV env protein, and of complex immunogens containing several synthetic peptides, it is necessary to summarize results (reviewed in refs. 1-5) of experiments designed to localize contiguous B- and T-cell epitopes on the preS sequence. A dominant preS2-specific epitope is localized within the sequence preS(132-139) (Fig. 1). The immunodominant B-cell epitope within the N-terminal half of the preS1 sequence has not been sufficiently defined and is likely to be localized between residues 21-47. A short peptide, preS(26-32) is recognized by anti-HBV at a relatively high dilution (1/1600).

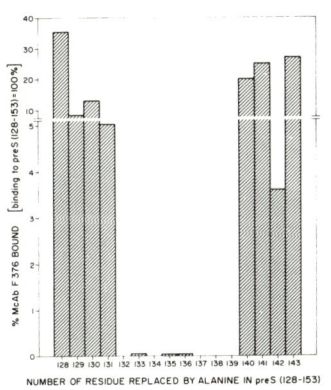
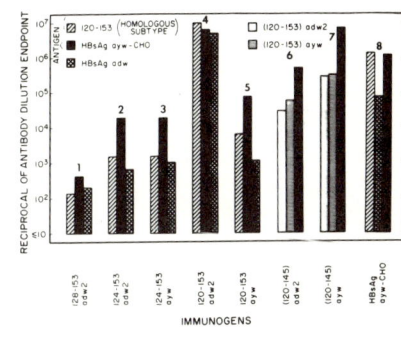

Fig. 1 (left). Binding of monoclonal Ab (McAb) F376 to peptides preS(128-153) having amino acid residues at indicated positions replaced by alanine.

Fig. 2 (right). Anti-preS2-specific Ab responses elicited by HBsAg and by peptides from the preS2 region. The peptides were used in unconjugated form with complete and incomplete Freund's adjuvant. Antigens described in the insert were used for coating of polystyrene plates. Two rabbits were used per immunogen in this and in all subsequent experiments. Geometric mean Ab titers are given.

It has been claimed that peptides are much less immunogenic than the native protein from which they are derived and that their immunogenicity can be enhanced by linking to protein carriers. These assertions are not applicable to selected peptides from the preS region of the HBV env protein. Thus, the levels of Ab elicited by the peptides preS(120-153)adw2 and preS(120-145)ayw were significantly higher than that elicited by intact HBsAgayw (7) (compare columns 4, 7 and 8 in Fig. 2). Relatively high levels of antibodies were also elicited by peptides preS(120-153)ayw and preS(120-145)adw2 (columns 5 and 6). The dominant B-cell epitope in the preS2 sequence is located C-terminally from residue ~128, and the sequence preS(120-123) is not essential for recognition of the preS2 sequence by anti-HBV. Therefore, shorter peptides, truncated at the N-terminus [preS(128-153) and preS(124-153)] were tested for immunogenicity in free form using complete Freund's adjuvant. These peptides (columns 1-3, Fig. 2) were much less immunogenic than the full length peptide preS(120-153). Thus, residues preS(120-127) are essential for immunogenicity of preS2 peptides, in agreement with results obtained in mice (6).

Peptides from the preS1 sequence, preS(1-21) and preS(12-32) elicited in free form high levels of Ab against the homologous peptide and the preS1 sequence of native HBV (columns 1 and 2, Fig. 3). The peptide preS(21-47)adw2 was poorly immunogenic (column 3). The same peptide derived from the ayw sequence elicited antipeptide Ab at reasonably high levels, but only low levels of Ab recognizing the native preS1 sequence. Therefore, longer peptides preS(12-47), preS(13-47), preS(15-47), preS(17-47) and preS(19-47) (subtype adw2) were tested for immunogenicity. They were about equally immunogenic. Results obtained with the peptide preS(15-47) are shown in Fig. 3 (column 5). The peptide preS(15-47)ayw also elicited reasonable levels of anti-preS1 specific Ab (column 6). In conclusion, N-terminal residues preS(12-20) are important for immunogenicity of preS1 specific peptides, indicating the localization of a dominant T-cell epitope within this sequence.

Antigenicity and Immunogenicity of Hybrid Peptides Derived from Nonadjacent Regions of the HBV env Protein.

To assess the usefulness for vaccination of hybrid peptides containing B- and T-cell epitopes from nonadjacent regions, peptides preS(12-25)-preS(132-139) and preS(120-131)-preS(26-32) were synthesized. The first peptide was preferentially recognized by anti-preS(120-153) as compared with anti-preS(15-47) (Fig. 4). The second peptide was preferentially recognized by anti-preS(15-47) as compared to anti-preS(120-153). This confirms that the dominant B-cell epitopes recognized by the anti-peptide sera were localized within the preS(132-139) and preS(26-32) segments of the two respective hybrid peptides. Anti-HBV recognized the two

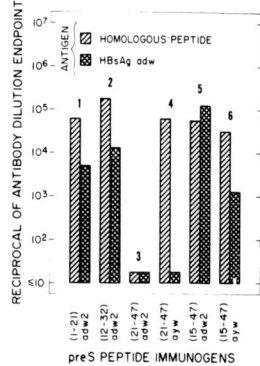 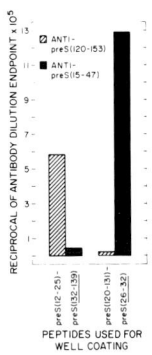

Fig. 3 (left). Immunogenicity of selected synthetic peptides from the preS1 region of the HBV env protein.

Fig. 4 (right). Recognition by anti-preS(120-153) and anti-preS(15-47), respectively, of hybrid peptides having B- and T-cell epitopes derived from nonadjacent portions of the preS sequence. Segments corresponding to B-cell epitopes are underlined. Unlike in the following figures, the scale on the ordinate is linear.

hybrid peptides by about one order of magnitude less efficiently than the respective full length peptides preS(15-47) and prcS(120-153) (data not shown). This indicates the importance of sequences flanking the preS(132-139) and preS(26-32) segments for recognition by anti-HBV. The hybrid peptides elicited much lower levels of Ab recognizing the native HBV env in comparision with the respective full length preS1 and preS2 peptides (Fig. 5). The level of antipeptide Ab elicited by preS(12-25)-preS(132-139) was also much lower in comparison with Ab elicited by the peptide preS(120-153). On the other hand, the level of Ab recognizing the preS(12-47) sequence was similar in sera of rabbits immunized with this peptide and with the hybrid peptides preS(120-131)-preS(26-32) and preS(120-131)-preS(21-47), respectively (Fig. 5). However, the latter peptide failed to elicit Ab recognizing the native preS1 sequence. Peptides preS(120-131)-preS(26-32) and preS(120-131)-preS(21-47) elicited Ab recognizing the preS(120-153) sequence, indicating that the N-terminal half of the latter sequence also contains B-cell epitopes. In conclusion, only relatively long peptides can optimally mimic B-cell epitopes on the native preS1 and preS2 sequences.

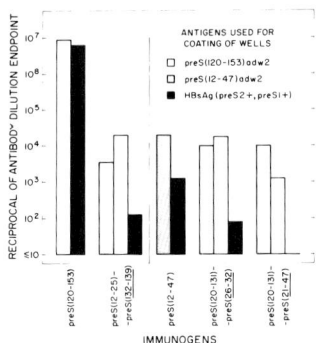 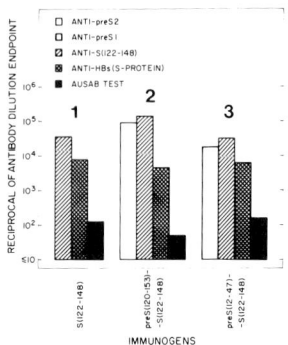

Fig. 5 (left). Comparative immunogenicities of peptides from the preS2 and preS1 sequences, respectively, and of hybrid peptides described in Fig. 4.

Fig. 6 (right). Comparative immunogenicities of the peptide S(122-148) used for immunization either in free form or covalently linked to preS(12-153) and preS(12-47), respectively. The AUSAB test measures predominantly Ab to S-protein, but preS2 sequences are also detectable with a sensitivity varying from lot to lot of the test kits.

A peptide derived from the S-protein sequence, S(122-148), having 2 out of the 5 cysteine residues in positions 137 and 138 replaced by aminobutyric acid was shown to be highly immunogenic when linked to a protein carrier (8). Since this peptide appears to contain T-cell epitopes (2,6,9), a similar peptide having Gly-Gly-Tyr at the C-terminus, was oxidized by ferricyanide and tested for immunogenicity in free form. The peptide elicited Ab recognizing not only the peptide itself but also native S-protein (Fig. 6, column 1).

In order to: 1) develop immunogens containing epitopes from distinct regions of the HBV env protein; and 2) assess the influence of preS2 and preS1 specific peptides on the immunogenicity of the S-peptide, peptides preS(120-153) and preS(12-47), respectively, were covalently linked to the peptide S(122-148) by disulfide bonds. This was accomplished by derivatizing the respective preS peptides having Gly-Gly-Cys residues at the C-terminus with 2,2'-dithiodipyridine. The derivatized peptides were reacted with the peptide S(122-

148). The resulting hybrid peptides elicited Ab against the appropriate preS region of the HBV env protein and against native S-protein (Fig. 6, columns 2,3). However, coupling of the preS peptides to the S-peptide did not enhance the immune response to S-protein.

These results suggest that cross-linking of the described peptides derived from nonadjacent regions of the HBV env protein may not be the method of choice to prepare complex immunogens eliciting Ab against multiple epitopes.

Immunogenicity of Synthetic Peptides Linked to Nonimmunogenic Carriers Administered Without Oil-Containing Adjuvants.

To realize the potential of synthetic peptides as vaccines, it is preferable to design peptide derivatives which are immunogenic when administered with alum, the only approved adjuvant for human use, or without any adjuvant. Peptides linked to iscoms are immunogenic without any additional adjuvant (1). Some peptides which are nonimmunogenic in free form, even with complete Freund's adjuvant, become immunogenic when linked to liposomes (10). The peptide preS(120-145) was linked to liposomes and injected into rabbits without any adjuvant. The peptide-liposome complex elicited Ab against the peptide and against preS2 sequences of the native HBV env protein (Fig. 7, column 1). The liposomes also elicited anti-preS2 specific Ab (dilution endpoint $\sim$ 1/500) in chimpanzees (unpublished results of collaborative studies with Dr. A.M. Prince). New methods for the preparation of synthetic peptide derivatives which adsorb to alum were also explored. Dextran 2000 (mol. weight 2×10^6) was converted into an amino derivative. Sulfhydryl-reactive groups were introduced into the dextran derivative by activation with m-maleimidobenzoylsulfosuccinimide-ester (MBS). Alternatively, the amino-dextran was cross-linked with glutaraldehyde (= GA-dextran) and the aldehyde group was utilized for linking peptides to the carrier. Peptides were linked to the GA-dextran either directly or through diamino bridges using polylysine. Sulfhydryl groups were introduced into the polylysine-GA-dextran by treatment with N-succinimidyl-3-(2-pyridyldithio) propionate (SPDP). The peptides preS(120-153) and preS(15-47), each with a Gly-Gly-Cys linker at the C-terminus, were attached to the distinct dextran derivatives either through $-NH_2$ or -SH groups. The resulting complexes were adsorbed quantitatively to alum and used to immunize rabbits. The derivatives based on GA-dextran were more immunogenic than those consisting of MBS-activated dextran (Fig. 7, columns 2-7). The dextran derivatives are suitable for the simultaneous attachment of multiple peptides for the production of polyvalent immunogens. Additional improvements are needed to increase the immunogenicity of dextran conjugates.

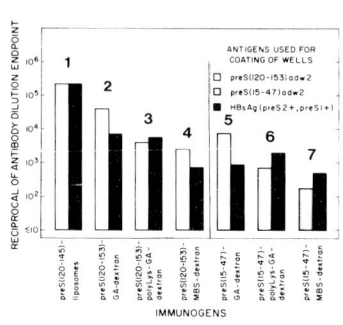 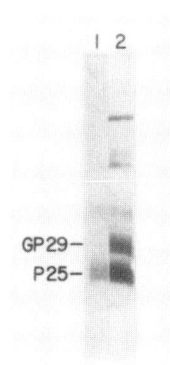

Fig. 7 (left). Effect of different carriers on the immunogenicity of preS2 and preS1 peptides. In the case of GA-dextran, rabbits immunized with five 100-200 µg doses of immunogens with alum adjuvants were further hyperimmunized with the same immunogens using two additional doses with an adjuvant from Syntex. The results on the graph correspond to the antisera obtained after the final bleeding of rabbits. Hyperimmunization with immunogens in combination with the Syntex adjuvant resulted in a further 2- to 5-fold increase of Ab levels as compared with the last bleedings obtained after immunization with the respective immunogens in combination with alum.

Fig. 8 (right). Detection of HBsAg polypeptides after PAGE under reducing conditions by: 1) immunoblotting with anti-S(135-155); 2) staining with silver. (Results obtained in collaboration with A. Mohamad, P. Price and G. Acs.)

The Effect of Glycosylation of HBV Env Proteins on Their Recognition by Antipeptide Antisera.

Carbohydrate moieties of glycoproteins can strongly influence the interaction of Ab to protein antigenic determinants of a glycosylated protein (11). S-protein and preS2 sequences are present in HBV in both glycosylated and non-glycosylated forms. The N-glycosylation sites are at positions S(146) and preS(123) (1,2). These sites are within regions corresponding to important B- and/or T-cell epitopes. It is important to know whether or not antisera against peptides derived from these regions recognize both the glycosylated and non-glycosylated forms of HBV env proteins. PreS2 sequences within the HBV env M-protein (GP33 and GP36) are recognized by Ab to peptides preS(120-145) or preS(120-153) (1,3). However, T-cell epitopes on the preS2 sequence of M-protein and on the peptide preS(120-145) appear to be unrelated (6). This observation may be explained by the glycosylation of residue preS(123). The recognition of S-protein by Ab to peptides containing the Asn residue at position S(146) is inhibited by glycosylation (Fig. 8). Since non-glycosylated HBsAg is used for vaccination against hepatitis B, the failure of synthetic peptides to elicit Ab recognizing the glycosylated form of HBsAg S-protein might not represent an impediment to the potential application of these peptides for vaccination.

Synthetic Peptides as Probes for Studying Interdomain Interactions within HBV Env Proteins.

Although very little is known about the tertiary structure of native HBV env proteins, the following results suggest that the preS1 and preS2 sequences interact with each other: 1) The preS2 sequence is always glycosylated in M-protein. However, the presence of preS1 sequences within L-protein prevents the glycosylation of Asn 123 in the preS2 sequence (1); 2) The preS2 sequence becomes more resistant to proteolytic cleavage if present in L-protein as compared with M-protein (12,13); 3) Tyrosines of the preS2 sequence in L-protein cannot be labeled with ^{125}I (13); 4) Recombinant L-protein failed to elicit anti-preS2 Ab, although anti-preS1 Ab was elicited (14).

In order to directly asses the interaction between preS1 and preS2 sequences, the peptide preS(12-47) was attached to recombinant HBsAg particles containing S-protein and the preS2 sequence (see below). The immunologic properties of the resulting hybrid particles were compared with the original recombinant particles (CHO-HBsAg). Attachment of myristilated preS(12-47) strongly inhibited the accessibility of preS2-specific epitopes to anti-preS(120-153) (Fig. 9).

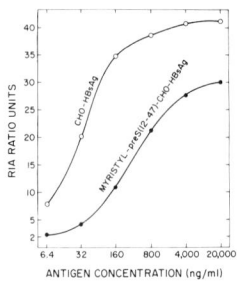 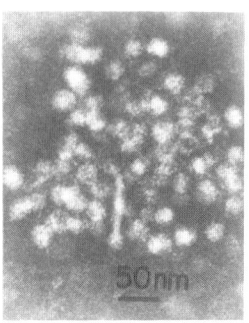

Fig. 9 (left). Detection of preS2 sequences in recombinant CHO-HBsAg particles containing the S and preS2 sequence (7) and in hybrid particles to which the myristylated peptide preS(12-47) was attached. Serial dilutions of the respective antigen preparations were added to wells of polystyrene plates coated with anti-preS(120-153). The attachment of HBsAg was detected with ^{125}I-labeled Ab to S-protein.

Fig. 10 (right). Electron micrograph of an immune precipitate generated by reacting HBsAg particles complexed with myristylated preS(12-47) with anti-preS(12-47).

The interaction between distinct domains of HBV env proteins has to be considered in any strategy for developing vaccines consisting of peptides derived from distinct regions of the HBV env protein.

Hybrid Immunogens Consisting of Recombinant S-protein with Attached Synthetic Peptides from the PreS Sequence.

The preS1 and preS2 regions of HBV env proteins are usually underrepresented in preparations of HBsAg obtained from plasma of HBV carriers. PreS1 sequences inhibit the secretion of HBsAg from eukaryotic cells. Therefore, the content of preS1 sequences in recombinant HBsAg is relatively low (1). A combination of peptides derived from the preS1 and possibly from the preS2 sequence with HBsAg particles consisting either of S-protein or of S-protein + preS2 sequences is expected to result in improved immunogens eliciting protective VN Ab of the broadest possible specificities.

Epitopes on the HBV Envelope Protein

The addition of petides to HBsAg can theoretically be accomplished by chemical linkage. This requires active groups on both the peptide and on the surface of HBsAg, for example, SH-groups and NH_2-groups. HBsAg S-protein does not have available free SH-groups, since all cysteine residues are involved in the formation of disulfide bonds. The $\leq$-NH_2 groups on the surface of HBsAg also seem to play an important role in immunogenicity and antigenicity (2). For this reason, methods have to be designed which would allow the attachment of peptides to HBsAg not involving a chemical linkage but strong enough to allow the HBsAg protein to function as a carrier. S-protein has on its surface exposed hydrophobic regions which interact strongly with different hydrophobic adsorbents (15). It should be possible to adsorb synthetic peptides having a covalently attached hydrophobic tail to HBsAg S-protein. To accomplish this, the peptide preS(12-47) was myristylated and the resulting myristyl-preS(12-47) was mixed with an equivalent weight amount of recombinant CHO-HBsAg containing preS2 sequences. Anti-preS(12-47) precipitated the modified HBsAg, demonstrating that the peptide has become attached to the particles (Fig. 10). HBsAg premixed with the myristylated peptide was quantitatively recovered in the immune complex (Fig. 11). To determine whether or not

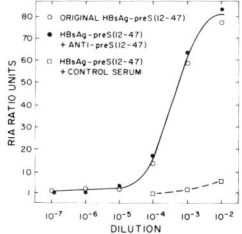
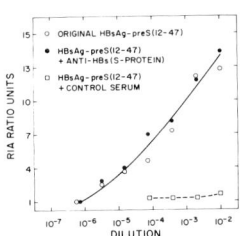

Fig. 11 (left). Evidence that anti-preS(12-47) quantitatively precipitates HBsAg S-protein present in complexes generated by adding myristylated preS(12-47) to HBsAg-CHO particles. Serial dilutions in 3M NaSCN of the immune complex or of HBsAg-CHO-preS(12-47) were added to polystyrene beads. Attached HBsAg was detected by ^{125}I-labeled Ab to S-protein.

Fig. 12 (right). Evidence that anti-HBs (S-protein) quantitatively precipitates myristylated preS(12-47) added to HBsAg-CHO particles.

all myristyl-preS(12-47) became attached to HBsAg, the derivatized HBsAg was mixed with anti-HBs (S-protein), the immune complex was separated by centrifugation, dissolved in 3M NaSCN and tested for the presence of preS1 sequences by a double antibody RIA using rabbit anti-pres(12-47). All myristyl-preS(12-47) was recovered in the precipitate (Fig. 12).

Myristyl-preS(12-47)-CHO-HBsAg elicited high levels of Ab to preS1, preS2 and S sequences and was more immunogenic than CHO-HBsAg (compare columns 4 and 5, Fig. 13). The augmentation of the antibody response to S-protein was remarkable. The latter finding was further confirmed in experiments with mice using alum as adjuvant. On the other hand, covalent linking of preS(12-47) to CHO-HBsAg (50 µg of peptide per mg HBsAg) resulted in an immunogen which failed to elicit anti-preS1 specific Ab (Column 3). The preS(12-47) peptide, either before or after myristylation, without attachment to any carrier, elicited anti-preS1 specific Ab recognizing the native HBV env protein (columns 1 and 2, Fig. 13). The addition of preS1-specific peptides to HBsAg containing S-protein ± preS2 sequences is anticipated to result in immunogens which will: 1) elicit a broader spectrum of protective Ab; and, 2) augment significantly the immune response to S-protein. Thus, a combination of synthetic peptides with recombinant HBsAg presents a feasible approach for the development of hepatitis B vaccines before all problems related to the design of fully synthetic vaccines can be solved.

Safety of Hepatitis B Vaccines Containing PreS Sequences.

Based on observations that: 1) The preS2 sequence reacts with glutaraldehyde-polymerized human serum albumin (GA-HSA); and, 2) Both anti-HSA and anti-preS2-specific Ab were allegedly detected in sera of HBV-infected humans, it was concluded that the preS sequence carrying the HBV receptor for hepatocytes may elicit an autoimmune response not only to HSA, but also to the liver cell membrane, resulting in hepatocellular damage (20). Although the native preS2 sequence as well as the peptide preS(120-153), but not the peptide preS(120-145), specifically recognize HSA (Fig. 14), the arguments concerning an autoimmune response to the liver, all of which are amenable to experimental verification, are unproven: 1) The specificity of assays for anti-HSA in human

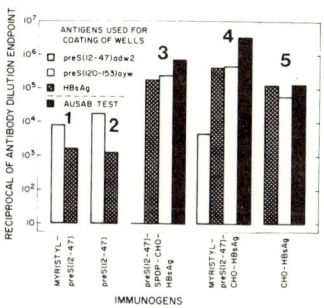
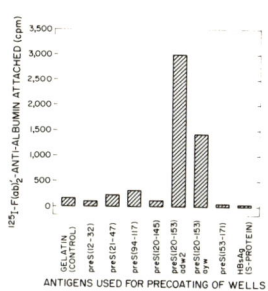

Fig. 13 (left). Comparative immunogenicity of preS(12-47) administered with complete Freund's adjuvant; of myristylated preS(12-47) administered with incomplete Freund's adjuvant; of HBsAg-CHO particles with covalently linked preS(12-47) and of complexes formed by addition of myristylated preS(12-47) to HBsAg-CHO (the latter two preparations were administered with complete Freund's adjuvant). Five doses of antigen, 200 µg each, were administered in bi-weekly doses to rabbits.

Fig. 14 (right). Reaction of peptides from the preS sequence and of native HBsAg S-protein with human serum albumin (HSA). Wells of polystyrene plates were coated with the distinct antigens and were subsequently reacted with HSA (1 mg/ml) followed by ^{125}I-labeled F(ab)$'_2$ anti-HSA. Counts were corrected for nonspecific attachment of ^{125}I-labeled anti-HSA attached to HBsAg-(synthetic peptide)-coated wells to which HSA had not been added.

sera has not been convincingly demonstrated; 2) The claimed simultaneous appearance of anti-HSA and anti-preS-specific Ab in HBV-infected humans does not prove a causal relationship between immunization with preS-specific immunogens and the appearance of anti-HSA; 3) No evidence exists that a postulated autoimmune response to HSA is in any way related to liver damage. In contrast, direct experimental data do not support any of these arguments: 1) Immunization with native preS2 sequences or with the corresponding synthetic peptides (in free form not linked to a carrier) does not lead to an anti-HSA response (17); 2) Neither Ab to preS1 or preS2 sequences, nor anti-GA-HSA recognize proteins exposed on the continuous human liver cell line HepG2 (Fig. 15). Thus, it seems very unlikely that any of these antibodies would recognize proteins exposed on the human liver.

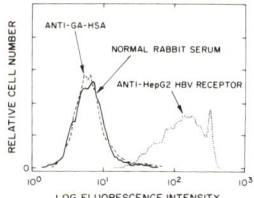

Fig. 15. Lack of reactivity of anti-GA-HSA with human hepatoma HepG2 cells. Similar negative results were obtained with anti-HSA, anti-preS(120-153) and anti-preS(12-47). A polyclonal Ab to the receptor for HBV (1) was used as a positive control antiserum. Cells were reacted with 1:50 diluted sera followed by 400-fold diluted fluorescein isothiocyanate-labeled goat anti-rabbit IgG and analyzed by cytofluorography.

DISCUSSION

Research with synthetic peptides derived from the HBV env protein has represented a profitable approach to the solution of several biological problems related to HBV infection and to the immune response to this virus, listed in the summary. This research has also led to approaches with more general significance: 1) application of peptides for elucidation of interdomain interactions within proteins; and 2) preparation of synthetic peptide-HBsAg complexes based on hydrophobic interactions offering a general opportunity to prepare multi-specific vaccines against several infectious agents by combining HBsAg with peptides derived from the infectious agents of choice.

Concerns regarding the possibility that immunogens containing virus receptor binding sites will elicit anti-idiotypic Ab deleterious to host cells (16), although not substantiated by any experimental data, or by any experience with vaccines containing whole virus, can be theoretically avoided by utilizing as immunogens peptides not containing the respective virus receptor binding sites. However, in the case of HBV, there are no compelling reasons to limit the arsenal of protective immunogens to products lacking the hepatocyte receptor binding site.

ACKNOWLEDGEMENTS

This research was supported, in part, by grants #CA43315 from the National Cancer Institute and #311 from the American Cancer Society. We thank P. Sproul and E. Seda for technical assistance and T. Huima and D. Varlese for manuscript preparation.

REFERENCES

1. Neurath AR, Kent SBH (1987). The preS region of hepadnavirus envelope proteins. In Maramorosh K, Murphy FA, Shatkin AJ (eds): " Advances in Virus Research." Orlando, Florida: Academic Press, **32** (in press).
2. Neurath AR, Kent SBH, Strick N (1988). Synthetic hepatitis B virus (HBV) peptides. In Kurstak E, Marusyk RG, Murphy FA, Van Regenmortel MHV (eds): "Applied Virology Research." New York: Plenum Pub Corp (in press).
3. Neurath AR, Jameson BA, Huima T (1987). Hepatitis B virus proteins eliciting protective immunity. Microbiol Sci **4**:45.
4. Neurath AR, Kent SBH, Seto B, Strick N, Parker K, Girard M (1988). Design of synthetic peptides mimicking the immunologic and biologic functions of the preS1 sequence of the HBV envelope protein. In Chanock RM, Lerner RA, Brown F, Ginsberg H (eds): "Vaccines 88," Cold Spring Harbor Laboratory, New York (in press).
5. Neurath AR, Kent SBH, Strick N, Parker K (1988). Delineation of contiguous determinants essential for biological functions of the preS sequence of the hepatitis B virus envelope protein: its antigenicity, immunogenicity and cell receptor recognition. Ann Inst Pasteur Virol (in press).
6. Milich DR (1987). Genetic and molecular basis for T- and B- cell recognition of hepatitis B viral antigens. Immunol Rev **99**:71.
7. Michel M-L, Pontisso P, Sobczak E, Malpiece Y, Streek R.E. and Tiollais P (1984). Synthesis in animal cells of hepatitis B surface antigen particles carrying a re-

ceptor for polymerized human serum albumin. Proc Natl Acad Sci USA **81**:7708.
8. Zheng J, Hu PS, Xu LG, Liu ZP, Huang WT (1986). Serotypic antigenic structure and immunological activity of the synthetic peptide vaccine for hepatitis B. Biopolymers **25**:S201.
9. Steward MW, Sisley BM, Stanley C, Brown E and Howard CR (1986). Humoral and cellular responses in man following immunisation with hepatitis B vaccine: analysis with synthetic peptides. In Peeters H (ed): "Protides of the Biological Fluids." Oxford, Pergamon Press, p. 137.
10. Neurath AR, Kent SBH, Strick N (1984). Antibodies to hepatitis B surface antigen (HBsAg) elicited by immunization with a synthetic peptide covalently linked to liposomes. J Gen Virol **65**:1009.
11. Alexander S, Elder JH (1984). Carbohydrate dramatically influences immune reactivity of antisera to viral glycoprotein antigens. Science **226**:1328.
12. Takahashi K, Kishimoto S, Ohnuma H, Machida A, Takai E, Tsuda F, Miyamoto H, Tanaka T, Matsushita K, Oda K, Miyakawa Y, Mayumi M (1986). Polypeptides coded for by the region preS and gene S of hepatitis B virus DNA with the receptor for polymerized human serum albumin: expression on hepatitis B particles produced in the HBeAg or anti-HBe phase of hepatitis B virus infection. J Immunol **136**:3467.
13. Heermann K-H, Kruse F, Seifer M, Gerlich WH (1987). Immunogenicity of the gene S and preS domains in hepatitis B virions and HBsAg filaments. Intervirology **28**:14.
14. Cheng K, Smith GL, Moss B (1986). Hepatitis B virus large surface protein is not secreted but is immunogenic when selectively expressed in recombinant vaccinia virus. J Virol **60**:337.
15. Neurath AR, Lerman S, Chen M and Prince AM (1975). Hydrophobic chromatography of hepatitis B surface antigen on 1,9-diaminononane or 1,10-diaminodecane linked to agarose. J Gen Virol **28**:251.
16. Hilleman MR (1988). Perspectives in the quest for a vaccine against AIDS. In: Gallo RC, Haseltine W, Klein G, zur Hausen H (eds): "Human Retroviruses, Cancer and AIDS: Approaches to Prevention and Therapy." New York: Allan R. Liss, p. 291.
17. Neurath AR, Strick N, Kent SBH, Parker K (1988). Antibodies recognizing human serum albumin are not elicited by immunization with preS2 sequences of the hepatitis B virus envelope protein. J Med Virol (in press).

STUDIES ON THE DEVELOPMENT OF A SYNTHETIC PEPTIDE VACCINE AGAINST FELINE LEUKEMIA VIRUS

Harish P. M. Kumar*, C. K. Grant**, and John H. Elder*

Department of Molecular Biology,
Scripps Clinic and Research Foundation,
10666 North Torrey Pines Road,
La Jolla, California 92037*
and Pacific Northwest Research Foundation,
Seattle, Washington 98104**

ABSTRACT: We have prepared synthetic peptides corresponding to sequences of the envelope genes of Gardner-Arnstein, Snyder-Theilen, Sarma, and Glasgow-1 isolates of Feline leukemia virus (FeLV). The peptides were coupled to Keyhole Limpet hemocyanin (KLH) and injected into rabbits for preparation of peptide-specific antisera. Antisera were subsequently tested for the ability to neutralize a wide range of FeLV isolates in vitro. We were able to define eight regions of the envelope gene which could serve as targets for neutralization; five of these epitopes were within the sequence coding for the major envelope protein gp70; three were in the coding region for p15E. Most of the neutralizing epitopes proved to be type-specific, justifiable given amino acid variability within certain of these epitopes. However, certain conserved epitopes in the C-terminal portion of gp70 and in p15E served as targets for neutralization in the subtype B isolates, but were less efficient in neutralizing subtypes A and C. The implication from these findings is that the availability of certain epitopes varies from one isolate to another, possibly via masking of the sites by carbohydrate sidechains on gp70. Using nested peptides and a monoclonal antibody prepared to intact virions, we were able to define a five amino acid minimal binding epitope at one broadly reactive neutralization site. Studies of two FeLv variants which were not neutralized by this monoclonal antibody revealed that changes both around and within the binding epitope can negatively influence antibody binding and thus neutralization. The overall conclusion from our studies to date is that the affinity of antibody binding is particularly critical for effective neutralization. Studies are now in progress to try to optimize antibody responses to peptide immunogens.

INTRODUCTION

The feline leukemia viruses (FeLV) are a divergent group of horizonatally transmitted retroviruses of domestic cats. FeLV is a type C retrovirus which is often found associated with malignancies of hemopoietic origin (1-5). However, the most prevalent health problem to the cat population is not leukemia, but rather an immune suppressive syndrome which often results in death from various opportunistic infections (6). This story has recently been complicated by the discovery of a lentivirus in cats, termed FTLV (7), which is highly related in structure to the human immunodeficiency virus (HIV). FeLV is only distantly related to the latter viruses and at this juncture, it is unclear whether many cases of immune suppression have in fact been caused by the lentivirus. However, studies using molecularly cloned isolates of FeLV have reported immune suppressive symptoms in experimentally treated animals (8). Thus, it is likely that both of these retrovirus groups are immune suppressive, but probably differ in the manner in which the disease is induced. In any event, both feline retrovirus groups offer excellent small animal models for the development of retrovirus vaccines. To date, our studies have centered on elucidating the immune response to FeLV, with particular emphasis on the development of an antisynthetic peptide vaccine.

Several approaches to vaccine development are currently being examined, including the use of traditional killed or attenuated whole viruses, recombinant subunit immunogens, anti-idiotypic antibodies, and synthetic peptides. Of these methods, only the whole virus vaccines have thus far proved efficacious in the field. However, the other approaches offer the possibility of dissecting the immune response so that we may fully appreciate the mechanisms involved in protection. In addition, the retroviruses offer a rather unique problem to vaccinology in that they possess the ability to produce DNA copies of their RNA genomes via reverse transcriptase. The viral DNA acts as a transposable element to integrate into the host DNA, where it may undergo potentially deleterious recombinational events. The latter characteristics have raised considerable concerns regarding any vaccine which still contains the genetic material of the virus. Thus, it is imperative to explore approaches which will lead to thoroughly characterized vaccines. The development of antisynthetic peptide vaccines is the epitomy of this approach, since it entails the use of chemically synthesized and defined peptide immunogens, dictated by the nucleotide sequence of the virus in question. The approach currently has the limitation that only "continuous" epitopes are examined, since the practical limitation of peptide synthesis is approximately 30 amino acid residues 2 (with consistent fidelity and purity). Thus, "discontinuous" epitopes, which involve residues distal in the linear sequence are excluded from analysis. However, continuous epitopes can be characterized completely using synthetic peptides and residues critical to antibody binding can be precisely defined. A secondary issue (from a scientific standpoint) is that a synthetic peptide vaccine would be infinitely more cost efficient than any of the other approaches and the product would not present risks from adverse reactions to contaminants in the immunogen.

The following is a summary of our findings to date using FeLV as a model for development of a synthetic peptide vaccine. Considerable work remains to be done before this approach results in an efficacious vaccine. However, these studies have yielded valuable information regarding the requirements for efficient virus neutralization by humoral antibody, the nature of neutralizing epitopes, and the plethora of ways the retroviruses have evolved to circumvent the immune surveillance system.

MATERIALS AND METHODS

Detailed procedures and the appropriate primary references for the methods used here have been described elsewhere (9,10). Peptides were prepared by both the solid state (11) as well as the "T-Bag" method (12). The peptides were conjugated either to keyhole limpet hemocyanin or soybean trypsin inhibitor via cysteine residues (13,14). In certain instances, peptides were also conjugated via the N-terminal primary amine (15,16). In general, we found that the anti-peptide response was biased toward the end attached to the carrier molecule. We were unable to ascribe any deleterious or advantageous effects to altering the coupling procedure, with the exception that peptides which also coupled via internal lysines in the latter method (or by glutaraldehyde coupling, 17) yielded antibody responses with high anti-peptide titers, but poor reaction with the native protein. Antisera produced upon inoculation of rabbits or cats were assayed for in vitro virus neutralizing ability using an immunoblotting procedure (18,19). Titers represent the serum dilutions at the midpoint in this assay.

RESULTS

Selection of Peptides

When we initiated these studies, there were no sequences of FeLV available. We, therefore, performed nucleotide sequence analysis of the envelope gene of Gardner-Arnstein FeLV (20). All primary peptide selections were made from this sequence. Based on the results of the primary studies, peptides were also prepared from other FeLV sequences as they became available, including isolates of subtypes A (21) and C (22), as well as other subtype B isolates (10). A composite of these sequences is shown in Figure 1. As can be seen, FeLV is structurally diverse, particularly within the envelope gene. Differences in gp70 provide the majority of specific sequences which delineate the A,B, and C subtypes. In addition, each isolate contains point mutations which distinguish it from all others. Another degree of variability is contributed by carbohydrate sidechains on gp70 (potential N-linked sites boxed in Figure 1). The carbohydrate moieties vary from one isolate to the other and can bias the host's response to the viral glycoprotein (23). On the other hand, certain regions of the envelope gene are totally conserved among all the known isolates. It is possible that such conserved regions resulted from differences in mutation frequencies. However, it seems more probable that these regions represent areas of the molecule that are particularly critical for function. Since, typically, molecularly cloned viruses are selected for competency prior to sequence analyses, variants containing changes in critical regions would not be observed. Regardless, from the standpoint of vaccine development, these constant regions can be considered as legitimate targets for broadly reactive antibody responses, since competent variants are our only concern. We thus have both constant and variable regions of the molecule to consider for peptide selection.

Neutralizing Epitopes on FeLV

In a sense, it was fortunate that the above diversity was undocumented at the beginning of these studies. Since we did not know which regions were constant and which were variable, we virtually blanketed the envelope gene, synthesizing peptides which encompassed approximately 85% of the sequences encoding gp70 and p15E (9). We prepared antisera to these peptides, conjugated to KLH, and tested for in vitro neutralization of the homologous Gardner-Arnstein FeLV. We found that eight regions of the envelope gene could serve as neutralization targets (figure 2). Five of the epitopes were within the sequence encoding gp70; three were within p15E. An immediate conclusion from these data is that neutralization can occur at sites other than the epitope involved in primary binding to a cellular receptor. Although the receptor binding epitope is still unknown for FeLV, it is highly unlikely that all these regions are involved in such an interaction. Secondly, the data indicate that antibodies to certain regions of the small envelope protein, p15E, can result in neutralization, an observation previously unreported using heteroantisera or monoclonal antibodies.

We next tested the neutralization positive anti-peptide antisera for neutralization of a variety of subtype A, B, and C isolates. We found that certain

Vaccine for FeLV 163

```
Glasgow-1         ANPSPHQIYNVTNVITNVQTNTQANATSMLGTLTDAYPTLHVDLCDLVGDTNEPIVLNPTNVKHG   65
Gardner-Arnstein           V      T    LV G K          F   MYF     II N   N SDQE ------
Rickard                           T    LV G K          F   MYF     II N   N SDQE ------
Snyder-Theilen                    T    LV G K          F   MYF     II N   N SDQE ------
λB1                               T    LV G K          F   MYF     II N   N SDQE ------
Sarma-C                    V                 SR            Y                APD ---RSW

Glasgow-1         ARYSSSKYGCKTTDRKKQQQTYPFYVCPGHAPSLGPKGTHCGGAQDGFCAAWGCETTGEAWWKPT  130
Gardner-Arnstein  ----FPG    DQPM RW   RNT      -----NRKQ   P      V         TY R
Rickard           ----FPG    DQPM RW   RNT      -----NRKQ   P      V         TY R
Snyder-Theilen    ----FPG    DHPM RW   RNT      -----NRKQ   P      V         TY R
λB1               ----FPG    DQPM RW   RNT      -----NRKQ   P      V         TY R
Sarma-C              TH                                 M   Y

Glasgow-1         SSWDYITVKRGSSQ------------------------DNSCEGKCNPLVLQFTQKGRQASWDGP  195
Gardner-Arnstein     K  VT GIYQCSGGGWCGPCYDKAVHSSTTGA EG R       I           T
Rickard              K  VT GIYQCSGGGWCGPCYDKAVHSSTTGAGEG R       I           T
Snyder-Theilen       K  VT GIYQCSGGGWCGPCYDKAVHSSTTGA EG R       I           T
λB1                  K  VT GIYQCSGGGWCGPCYDKAVHSSITGA EG R       I           T
Sarma-C                   N ----------------------         K                 R

Glasgow-1         KMWGLRLYRTGYDPIALFTVSRQVSTITPPQAMGPNLVLPDQKPPSRQSQTGSKVATQRPQTNES  260
Gardner-Arnstein       S         S        S     M                  IE R TPHHS G GG
Rickard                S         S        S     M                  IE R TPHHS G GG
Snyder-Theilen         S         S        S     M                  IC R TPHHS G GG
λB1                    S         S        S     M L                IC R TPHHS G GG
Sarma-C                          S        S     M                    K   T   ITS

Glasgow-1         ----------APRSVAPTTMGPKRIRTGDRLINLVQGTYLALNATDFPKTKDCWLCLVSRPPYY  325
Gardner-Arnstein  TPGITLVNASI   L TPV PAS    G                    R
Rickard           TPGITLVNASI   L TPV PAS    G N                  V N
Snyder-Theilen    TPGITLVNASI   L TPV PAS    G N                  V N
λB1               TPGITLVNASI   L TPV PAS    G N                  V N
Sarma-C           ----------T      SA        G

Glasgow-1         EGIAILGRYSHUJIPPPSCLSIDQHKLTISFYSGQGLCIGTVPKTHQALQNETQQGHTGAHYLAA  390
Gardner-Arnstein                                                KK  K  K T       V
Rickard              V              D                           KK  K  K T
Snyder-Theilen       V              D                           KK  K  K T
λB1                  V              D                  S        KK  K  K T
Sarma-C              V              T                           KK  K  K T

Glasgow-1         PNGTYNACNTGLTPCISMAVLNUTSDFCVLIELWPRVTYHQPEYVYTHFAKAVRFRR         447
Gardner-Arnstein                                              A
Rickard           S
Snyder-Theilen    S    I                                         D T L
λB1               S                                               D T L
Sarma-C                                                         I D
```

Figure 1. Comparison of the amino acid sequences of the envelope glycoproteins (gp70s) of six different isolates of feline leukemia virus; Glasgow-1 (28); Gardner-Arnstein (20); Ricard subtype B (30); Snyder-Theilen (25); lamda B1 (10) and Sarma-C (22). Only the sequence of the Glasgow-1 isolate is shown in its entirety. Only changes are shown for the other isolates. Dashed lines represent relative insertions/deletions. Underlined sites are probably not used due to the presence of proline (arrows) in the center or to the right of the consensus sequences.

```
                Leader                        GP70
              /                             /
              MESPTHPKPSKDKTLSWNLVFLVGILFTIDIGMANPSPHQVYNVIWTTTNLVTGTKANATSMLG
                   I10B
                   ⎯⎯⎯⎯
              TLTDAFPTMYFDLCDIIGNIWNPSDQEPFPGYGCDQPMRRWQQRNTPFYVCPGHANRKQCGGPQ

              DGFCAVWGCETTGETYWRPTSSWDYITVKKGVTQGIYQCSGGGWCGPCYDKAVHSSTTGASEGG
                                                                  I-26B
                                                                  ⎯⎯⎯⎯⎯
              RCNPLILQFTQKGRATSWDGPKSWGLRLYRSGYDPIALFSVSRQVMTTTPPQAMGPNLVLPDQK
                      C8                            C9B
                      ⎯⎯                            ⎯⎯⎯
              PPSRQSQIESRVTPHHSQGNGGTPGITTLVNASIAPLSTPVTPASPKRIGTGDRLINLVQGTYL

              ALNATDPNRTKDCWLCLVSRPPYYEGIAILGNYSNQTNPPPSCLSIPQHKLTISEVSGQGLCIG
                    C14B
                    ⎯⎯⎯⎯
              TVPKTHQALCNETQQGHTGAHYLAAPNGTYWACNTGLTPCISMAVLNWTSDFCVLIELWPRVTY
                        P15E
                       /
              HQPEYVYTHFAKAARFRREPISLTVAIMLGGLTVGGIAAGVGTGTKALIETAQFRQLQMAMHTD
                     C18B
                     ⎯⎯⎯⎯
              IQALEESISALEKSLTSLSEVVLQNRRGLDILFLQEGGLCAALKEECCFYADHTGLVRDNMAKL
                I6B          I7B
                ⎯⎯⎯          ⎯⎯⎯
              RERLKQRQQLFDSQQGWFEGWFNKSPWFTTLISSIMGPLLILLLILLFGPCILNRLVQFVKDRI

              SVVQALILTQQYQQIKQYDPDRP
```

Figure 2. amino acid sequence of Gardner-Arnstein envelope gene. Details of nucleotide sequence analyses have been described (9). Underlined regions denote peptides which elicited neutralizing antibody responses.

of the antisera were broadly neutralizing, while others only neutralized subtype B isolates. This was to be expected, since the subtype B Gardner-Arnstein FeLV was used for peptide selection. However, subsequent sequence comparisons between Gardner-Arnstein and subtype A and C isolates (24,9), which were completed during this study, led to several interesting observations (10, Figure 3). The most N-terminal neutralizing epitope, termed I-10, varied markedly among the FeLV subtypes and the neutralization response was subtype specific. Subsequent preparation of antisera to the equivalent peptides from subtypes A and C yielded antisera specific for their respective subtypes. The next region, termed I-26/C8, comprises two epitopes which are highly conserved (exceptions noted below) and neutralization occurred with all three subtypes. The C9 epitope was variable and gave subtype-specific neutralization. However, C-14, the C-terminal epitope of gp70 and the p15E epitopes (C-18, I-6, and I-7) were highly conserved in all three subtypes, yet only served as effective neutralization targets in the subtype B isolates. These results imply that the availability of certain regions of the molecule to antibody binding varies from one subtype to another, either via conformational differences or differential masking by carbohydrate sidechains. Thus, both variable and conserved epitopes may serve as targets for virus neutralization. However, sequence

PEPTIDE	VIRUS	SEQUENCE
I-10	Gardner-Arnstein	CDIIGNTWNPSDQEP----------FPGYG
	Glasgow-1	LV D E IVLN TNVKHGARYSSSK
	Sarma	LV D E IAPD ---RSWARYSSSTH
I-26/C8	Gardner-Arnstein	QVMTITPPQAMGPNLVLPDQKPPSRQSQIESRVTP
	Glasgow-1	
	Sarma	
C-9	Gardner-Arnstein	LSTPVTPASPKRIGTGDR
	Glasgow-1	R VAP TMG R
	Sarma	R VASATMG
C-14	Gardner-Arnstein	CIGTVPKTHQALCNETQQGHT
	Glasgow-1	
	Sarma	KK K K
C-18	Gardner-Arnstein	TDIQALEESISALEKSLTSLSE
	Glasgow-1	
	Sarma	
I-6	Gardner-Arnstein	AKLRERLKQRQQLF
	Glasgow-1	
	Sarma	
I-7	Gardner-Arnstein	DSQQGWFEGWFNKSPWFTTLISS
	Glasgow-1	
	Sarma	

Figure 3. Comparison of the amino acid sequences of neutralizing regions from representative subtype A, B, and C FeLV. Peptides refer to regions of Gardner-Arnstein shown in Figure 2. Sequences are shown as the single amino acid code. Gardner-Anrstein (20, subtype B) is shown in entirety; only changes are shown for Glasgow-1 (28, subtype A) and Sarma (22, subtype C). Dashed lines represent relative deletions.

conservation at a given site does not necessarily guarantee broad cross-neutralization of FeLV variants.

Characterization of a Neutralizing Epitope

We have concentrated our efforts on characterizing the I-26/C8 epitopes described above, since they represent relatively conserved epitopes which could serve as targets for a broadly protective vaccine. Further studies were aided by the availability of a monoclonal antibody which was prepared against whole virus, but reacted with the I-26 epitope. Using a different monoclonal antibody and DNAse I fragments in gt11, Nunberg and colleagues had previously identified this region as containing a neutralizing epitope (25). We prepared nested peptides around the I-26 peptide and performed immunoblot and competition binding studies to define the limits of the antibody binding epitope (Figure 4). The results of these studies indicated that the minimal binding epitope comprised the amino acid sequence Met-Gly-Pro-Asn-Leu. Removal of the methionine residue in the N-terminal reductive series or the leucine residue in the C-terminal reductive series totally abrogated antibody binding. The presence of three amino acids which are found at high frequency in reverse turns (Gly-Pro-Asn, 26) in the

PEPTIDE	SEQUENCE	Ki (nM)
I-26	QVMTTTPPQAMGPNLVLP	1
-3B	PQAMGPNLVLPDQKPPS	40
13B1	MGPNLVLPDQKPPSR	380
13B2	GPNLVLPDQKPPSRQSQ	>20,000
I-85	MTTTPPQAMGPNL	130
I-86	VMTTTPPQAMGPN	>20,000
I-87	TTTPPQAMGPNLV	30
I-26L	QVMTTTLPQAMGPNLVLP	10
I-26D	QVMTTTPPQAMGPDLVLP	>20,000

Figure 4. Inhibition of monoclonal antibody binding to intact virus by nested peptides. A neutralizing monoclonal antibody, termed C11D8 (29) was prepared against intact virus, but reacted with the I-26 peptide (9, see Figure 2). Nested peptides were synthesized in order to map the antibody binding epitope. The ability of the peptides to interfere with antibody binding to whole virus was assessed. Ki, peptide concentration which yielded 50% inhibition. Direct immunoblot assays were also performed against these peptides (9) and the same conclusions were reached. The minimal binding epitope is the sequence MGPNL (underlined in peptide I-87).

epitope suggests that such a structure may exist in this region of gp70. It has been suggested that reverse turns are likely candidates as continuous epitopes, due to spacial considerations for antibody binding in short peptide segments (31).

Neutralization-Resistant Variants

Although the I-26 region is, in general, highly conserved among FeLV isolates, we discovered and characteriazed a natural variant which was not neutralized by the monoclonal antibody reactive at this site (10). The nucleotide sequence of this variant revealed that the five amino acid binding epitope was conserved. However, a single nucleotide change resulted in sustitution of leucine for proline at a position three amino acids N-terminal to the binding epitope in the resistant variant. A peptide identical to I-26, but substituted with leucine at this position (I-26L) was synthesized and compared with I-26 for ability to compete for antibody binding to whole virus (Figure 4). The results of this study indicated that the substitution of leucine for proline caused a ten-fold reduction in the ability of the peptide to inhibit antibody binding. This result, using short synthetic peptides, was not as dramatic as the observed reduction in antibody binding to the intact glycoprotein of the resistant isolate (10). However, we assume this to be a reflection of the relative flexibility of short peptides in solution. We interpret the results to mean that the antibody binding epitope is retained in the resistant isolate, but that the conformation of the epitope is altered such that the antibody no longer binds with high affinity.

Another FeLV variant has now been reported (27) which contains an alteration in the neutralizing epitope described above. This isolate contains a single point mutation which changes the antibody binding epitope from MGPNL to MGPDL. We have not tested this isolate to determine whether the antibody will elicit neutralization. However, we have prepared a peptide identical to I-26, except that aspartic acid has been substituted for asparagine in the binding epitope. The neutralizing monoclonal antibody no longer recognizes the peptide containing this change (Figure 4). It is therefore probable that this single point mutation would facilitate escape from neutralization at this site in an animal primed to react to the predominant (MGPNL) eptiope.

DISCUSSION

The lesson we have learned from the above studies is that relatively conserved areas of the gp70 molecule can serve as targets for neutralization. However, very subtle changes around these epitopes can completely alter the immunological profile of the site, possibly facilitating survival of an FeLV variant containing such changes. Therefore, if we are to design a vaccine around a single epitope, it should be expected that variants will arise which must be dealt with as they occur. If these variations are infinite, then the task may be impossible. However, the relative conservation of the region of gp70 described above implies that competent viruses will not undergo infinite changes in such a region, leaving room for justifiable optimism.

In general, the synthetic peptide approach to vaccine development has yielded considerable information at the molecular level regarding the nature of neutralizing epitopes. This data could not have been generated by any other approach. From a vaccine standpoint, more work will be required to produce a functional protective response in cats. In vivo experiments using the peptides described above have yielded antipeptide responses, but not protection from subsequent viral challenge. Much of our current research is devoted to optimizing antipeptide responses. The antipeptide antisera we have thus far generated react with high specificity, but low affinity relative to either heteroantisera or the neutralizing monoclonal antibodies we have examined (9,10). This lower affinity correlates directly with lower relative neutralizing titers and this may explain the poor responses we have observed in vivo. We may be able to solve part of this problem by altering the presentation of the peptide immunogen so as to better mimic the epitope in the native molecule. Lower affinity may also result from poor maturation of the immune response, which may in turn be caused by the lack of appropriate T-cell help epitopes on our immunogens. We are currently investigating both of these areas in order to improve antipeptide responses.

ACKNOWLEDGEMENTS

The authors wish to thank Alexander Smart for valuable technical assistance. We also wish to thank Dr. James Bittle, Hannah Alexander, and James Hogle for valuable discussions and support. The contribution by Johnson and Johnson Biotechnology Center for many of the peptides employed in this study is gratefully acknowledged. This research was suppoted in part by grant R01-CA-43362 (J.E.) and Public Health Service Grant 5001CA43371 (C.G.) from the National Cancer Institute.

REFERENCES

1. Essex M (1975). Horizontally and vertically transmitted oncornaviruses of cats. Adv Cancer Res 21:175.
2. Essex M, Cotter SM, Sliski AH, Hardy WD, Stephenson JR, Aaronson SA, Jarrett O (1977). Horizontal transmission of feline leukemia virus under natural conditions in a feline leukemia cluster household. Int J Cancer 19:90.
3. Hardy WD, Old LJ, Hess PW, Essex M, Cotter S (1973). Horizontal transmission of feline leukemia virus. Nature (London) 144:266.
4. Jarret O, Hardy WD, Golden MC, Hay D (1978). The frequency of occurrence of feline leukemia virus subgroups in cats. Int J Cancer 21:334.
5. Rickard CG, Post JE, Noronka F, Barr IM (1969). A transmissible virus-induced lymphocytic leukemia of the cat. J Natl Cancer Inst 42:987.
6. Hardy WD (1980). Feline leukemia virus diseases. In Hardy WD, Essex M, McClelland AJ (eds): "Feline Leukemia Viruses," Amsterdam: Elsevier Biomedical Press, p 3.
7. Pederson NC, Ho EW, Brown ML, Yammamoto JK (1987). Isolation of a T-lymphotropic virus from domestic cats with an immunodeficiency-like syndrome. Science 235:790.
8. Overbaugh J, Donahue PR, Quackenbush SL, Hoover EA, Mullins JI (1988) Science 239:906.
9. Elder JH, McGee JS, Munson M, Houghten RA, Kloetzer W, Bittle JL, Grant CK (1987). Localization of neutralizing regions of the envelope gene of feline leukemia virus by using anti-synthetic peptide antibodies. J Virol 61:8.
10. Nicolaisen-Strouss K, Kumar HPM, Fitting T, Grant C, Elder JH (1987). Natural feline leukemia virus variant escapes neutralization by a monoclonal antibody via an amino acid change outside the antibody-binding epitope. J Virol 61:3410.

11. Merrifield RB (1963). Solid phase peptide synthesis. I. The synthesis of a tetrapeptide. J Am Chem Soc 85:2149.
12. Houghten RA (1985). General method for the rapid solid-phase synthesis of large numbers of peptides: specificity of antigen-antibody interaction at the level of individual amino acids. Proc Natl Acad Sci USA 82:5131.
13. Johnson DA, Elder JH (1983). Antibody directed to determinants of a Moloney virus derived MCF gp70 recognizes antigens on normal immature thymus cells, brain, testes, and murine leukemia cells. J Exp Med 159:1751.
14. Liu FT, Zinnecker M, Hamaska T, Katz DH (1979) New procedures for preparation and isolation of conjugates and a synthetic copolymer of D-amino acids and immunochemical characterization of such conjugates. Biochemistry 18:690.
15. Briand JP, Muller S, Van Regenmortel MHV (1985) Synthetic peptides as antigens: Pitfalls of conjugation methods. J Immunol Methods 78:59.
16. Staros JV, Wright RW, Swingle DM (1986). Enhancement by N-Hydroxysulfosuccinimide of water-soluble carbodiimide-mediated coupling reactions. Anal Biochem 156:220.
17. Avrameas S, Ternynck T (1969). The cross-linking of proteins with glutaraldehyde and its use for the preparation of immunoadsorbents. Immunochemistry 6:53.
18. Elder JH, Munson M (1984). Modification of Western blotting technique for detection and quantitation of infectious virus. Biotechniques 2:170.
19. Johnson DA, Gautsch JW, Sprotsman R, Elder JH (1983). Improved technique utilizing non-fat dry milk for analysis of protein and nucleic acids transferred to nitrocellulose. Gene Anal Tech 1:3.
20. Elder JH, Mullins JI (1983). Nucleotide sequence of the envelope gene of Gardner-Arnstein feline leukemia virus B reveals unique sequence homologies with a murine mink cell focus-forming virus. J Virol 46:871.
21. Neil JC, Hughes D, McFarlane R, Wilkie NM, Onions DE, Lees G, Jarrett O (1984). Transduction and rearrangement of the myce gene by feline leukaemia virus in naturally occurring T-cell leukaemias. Nature (London) 308:814.
22. Luciw P, Parkes D, Potter S, Najerian R (1985). Feline leukemia virus (FeLV), strains A/Glasgow-1 and C, env genes. In Weiss R, Teich N, Varmus H, Coffin J (eds): "RNA Tumor Viruses," Cold Spring Harbor, New York: Cold Spring Harbor Laboratory, p 1000.
23. Alexander S, Elder JH (1984). Carbohydrate dramatically influences immune reactivity of antisera to viral glycoprotein antigens. Science 226:1328.
24. Rosenberg, ZF, Pedersen FS, Haseltine WA (1980). Comparative analysis of the genomes of feline leukemia viruses. J. Virol 35:542.

25. Nunberg JH, Williams ME, Innis MA (1984). Nucleotide sequences of the envelope genes of two isolates of feline leukemia virus subgroup B. J Virol 49:629.
26. Chou PY, Fasman GD (1974). Prediction of protein conformation. Biochemistry 13:222.
27. Mullins JI, Chen CS, Hoover EA (1986). Disease-specific and tissue-specific production of unintegrated feline leukaemia virus variant DNA in feline AIDS. Nature (London) 319:333.
28. Stewart MA, Warnock M, Wheller A, Wilkie N, Mullin JI, Onions DE, Neil JC (1986). Nucleotide sequences of a feline leukemia virus Subgroup A Envelope gene and long terminal repeat and evidence for the recombinational origin of subgroup B viruses. J Virol 58:825.
29. Grant CK, Ernisse BJ, Jarrett O, Jones FR (1983). Feline leukemia virus envelope gp70 of subgroups B and C defined by monoclonal antibodies with cytotoxic and neutralizing functions. J Immunol 131:3042.
30. Elder JH, Mullins JI (1985). Feline leukemia virus (FeLV), strains B and B/Glasgow-7, env genes. In Weis R, Teich N, Varmus H, Coffin J (eds): "RNA tumor viruses," Cold Spring Harbor, New York: Cold Spring Harbor Laboratory, p 1005.
31. Barlow DJ, Edwards MS, Thornton JM (1986). Continuous and discontinuous protein antigenic determinants. Nature 322:747.

SPECIFICITY AT THE LEVEL OF SINGLE AMINO ACIDS OF ANTI-WHOLE FOOT-AND-MOUTH DISEASE VIRUS SUBPOPULATIONS PRESENT IN POLYCLONAL ANTI-PEPTIDE SERA

Rob H. Meloen, Wouter C. Puyk, Hanneke Lankhof, Jaap G. van Bekkum, Adri Thomas, and Wim M.M. Schaaper

Central Veterinary Institute, P.O. Box 65
8200 AB Lelystad, The Netherlands.

ABSTRACT Polyclonal anti-peptide sera raised against small synthetic peptides corresponding with an immunodominant site on the surface of Foot-and-Mouth-Disease Virus appear to consist of multiple antibody subpopulations. Adsorption of these sera with whole virus removes part of the anti-peptide activity but all anti-virus activity. The fine specificities of the anti-peptide/anti-virus subpopulations could be determined at the level of single amino acids. They showed to be similar to fine specificities of monoclonal and polyclonal antibodies raised against whole virus reactive with the same peptides.

INTRODUCTION

Synthetic peptides are widely being applied for the development of peptide vaccines and for the production of anti-peptide antibodies to be used as research tools (Tanaka et al., 1985; Sutcliff et al., 1983; Jemmerson and Paterson, 1986; Dardenne et al., 1985; Lerner, 1984; Houghten et al., 1984; Walter, 1986). Such anti-peptide antibodies need to crossreact with whole proteins of which the peptide is derived. Theoretically "free" peptides may have multiple conformations in contrast to the peptide "in situ" on whole protein. Because the immune system has no prior knowledge of the desired conforma-

tion, it will produce antibodies against any conformation it encounters. Therefore it is expected that antipeptide antibodies crossreactive with whole protein will belong to a subpopulation of all anti-peptide antibodies present.
The proportion of these subpopulations may vary depending on the peptide, the length of the peptide, the way the peptide is coupled to the carrier protein (Schaaper et al., in prep.) or whether the peptide conformation has been constrained (f.i. by cyclization).
In order to optimise anti-peptide sera an estimation of the proportion and fine specificity of the desired subpopulation of antibodies must be known.

In this report we describe how these aspects are determined for a number of antisera raised against peptides derived from a structural protein of Foot-and-Mouth-Disease Virus (FMDV). The peptides were predefined for their immunogenicity with the aid of monoclonal antibodies (Mab's) and polyclonal sera raised against whole FMDV.

METHODS

Overlapping nonapeptides coupled to solid supports (PEPSCAN) of VP1, 2 and 3 of FMDV type A10 were used and tested as described (Geysen et al., 1987; Meloen et al., 1987). Overlapping peptides of multiple length (3, 4, 5, 6, 7, 8 and 9 amino acids) were produced covering the 133-160 area of VP1 of FMDV type A10. (The number of each peptide corresponds with that of its N-terminal amino acid).

Anti-peptide sera were prepared (Table 2) as described (Geysen et al., 1985).

Neutralizing Mab's were raised against whole FMDV (A10) as described (Meloen et al., 1983).

Anti-viral activities were raised as described (Meloen and Briaire, 1980).

Anti-peptide sera were adsorbed with whole virus by the addition of one mg per ml of CsCl purified virus and incubation overnight at 4°C. Free virus and virus-antibody complexes were removed by centrifugation of the serum virus mixture for 1 h at 50.000 rpm on a cushion of 10% sucrose.

RESULTS

Mab's and polyclonal sera define an immunogenic peptide on VP1 of FMDV.

All (more than 600) overlapping nonapeptides of VP1, VP2 and VP3 of the outer structural protein of FMDV-A10 were tested against four Mab's and two rabbit polyclonal antibodies raised against whole virus. They reacted solely with nonapeptides between 142 to 146 of VP1. A typical example is shown in Fig. 1. The amino acid sequences containing the reactive peptides are shown in Table 1.

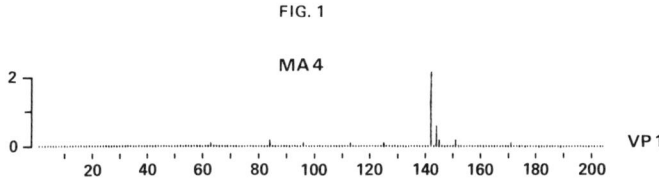

FIGURE 1. PEPSCAN result of all overlapping nonapeptides of VP1 of FMDV type A10 and monoclonal antibody MA 4.

TABLE 1.
REACTIVITIES OF MAB'S AND POLYCLONAL ANTISERA RAISED AGAINST WHOLE FMDV TYPE A10.

Antibody	Virus-ELISA	MNT	Maximal reactive sequence
			140 150
MA 4	7.0	3.0	SGDLGSIA
MA 11	7.7	2.7	RSGDLGSIA
MA 28	6.5	3.3	GDLGSIAAR
MA 30	7.5	3.6	RSGDLGSIA
R 218	5.3	2.7	DLGSIAARV
R 219	5.5	3.0	DLGSIAARV

174 Meloen et al.

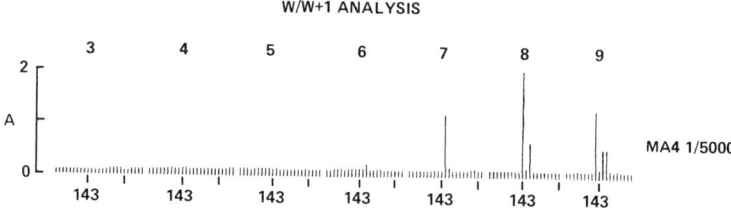

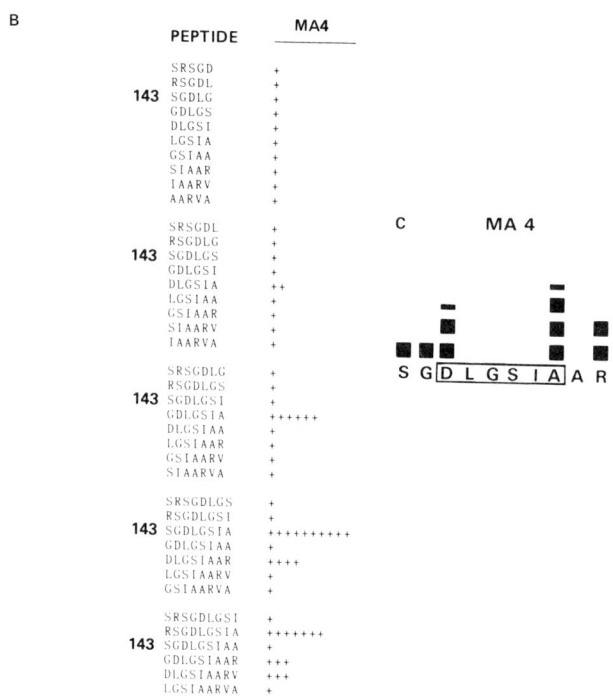

FIGURE 2. Analysis of the antibody fine specificity with the aid of multiple length peptides from the 133 to 160 area of VP1 of FMDV. In this area all overlapping 3, 4, 5, 6, 7, 8 and 9 amino acid long peptides are used. Panel A shows the results with Mab MA 4. Panel B shows the peptides written out with their reactivities schematically indicated. In panel C the schematic results of the analysis are shown when all peptides differing one amino acid are compared. The more blocks the more the amino

acid is assumed to contribute to binding. In addition the smallest reactive peptide (DLGSIA) is indicated.

Fine specificities of antibodies raised against whole virus.

The fine specificities of the antibodies was determined by their interactions with the 3, 4, 5, 6, 7, 8 and 9 long overlapping peptides in the area 133 to 160 of VP1. A typical result is shown in Fig. 2A. The reactive sequences are written out in Fig. 2B. Reactivity starts with a six amino acid long peptide, DLGSIA. This sequence is boxed in Fig. 2C. If one compares the reactivity of for instance DLGSIA and GDLSGIA, reactivity of the latter peptide is much higher than that of the former. We assume that this reflects the influence on antibody binding by the addition of the N-terminal G. This amino acid seems to contribute to antibody binding; in Fig. 2C this is indicated by a black square above the G. Comparisons are made for each pair of peptides differing by one amino acid. Any time a substantial increase in reactivity is obtained a black block is added to the appropriate amino aacid. The final result of this analysis is shown in fig. 2C for MA 4. Identical analyses were performed for the other three monoclonal antibodies (MA11, 28 and 30) and the two polyclonal sera raised in rabbits. A compilation of the results is shown in Fig. 4A.

Proportions and fine specificities of anti-viral activities present in anti-peptide sera.

We assume that anti-peptide sera raised against peptides derived from the 140-160 area of VP 1 contain anti-peptide antibodies reactive and nonreactive with whole virus. In order to study the virus specific anti-peptide sera they were adsorbed with virus. The results (Table 2) show that, although almost all anti-viral activity (as measured in an ELISA and a neutralization test) is removed, the bulk of anti-peptide activity is still present, indicating that the anti-peptide antibodies crossreactive with virus are a minor subpopulation.

TABLE 2.
REACTIVITY OF ANTI-PEPTIDE SERA RAISED IN RABBITS BEFORE
AND AFTER ADSORPTION WITH WHOLE VIRUS.

a-peptide serum		peptide-ELISA		virus-ELISA		Neutralizing activity	
		before	after	before	after	before	after
a-CSRSGDLGSAARV	(535/5)	4.6*	4.8	4.2	1.9	1.8	---
	(536/5)	4.3	4.4	4.5	2.0	1.8	---
a-CDLGSIAAC****	(418/5)	4.2	4.2	3.8	---**	1.2	---
	(441/5)	4.4	4.2	3.8	2.2	0.6	---
a-DLGSIAAC	(265/5)	2.8	2.2	3.7	---	0.6	---
	(269/5)	2.3	2.2	3.1	---	---	---
a-VIRUS	(219/5)	n.d.***	n.d.	5.5	n.d.	3.0	n.d.

* titres are -^{10}log of the final dilution
** no reactivity observed
*** not done
**** cyclic structure via disulphide bond between two C's.

The anti-peptide sera before and after adsorption were furthermore tested against the multiple length peptides. A typical example is shown in Fig. 3. Fig. 3A shows the results obtained with unadsorbed serum. The pattern is widely different from the ones obtained with anti-virus antibodies (Fig. 2A). Not only are much more peptides reactive, but shorter peptides react as well (indicative of the low specificity of some of the anti-peptide antibody subpopulations present). After adsorption with virus these latter reactivities are still present.

However, the pattern of the reactivities removed by whole virus (obtained by subtracting reactivities in panel B from those of panel A) are reminiscent of those obtained with anti-whole virus antibodies (Fig. 3C). An analysis similar to that applied to anti-whole virus antibodies is possible. The other anti-peptide sera were treated in the same way and the results are compiled in Fig. 4B.

Fine Specificity of an Immunodominant Epitope 177

FIG. 3

REACTIVITIES OF NEUTRALIZING ANTI–(C)SRSGDLGSIAARV R535/5

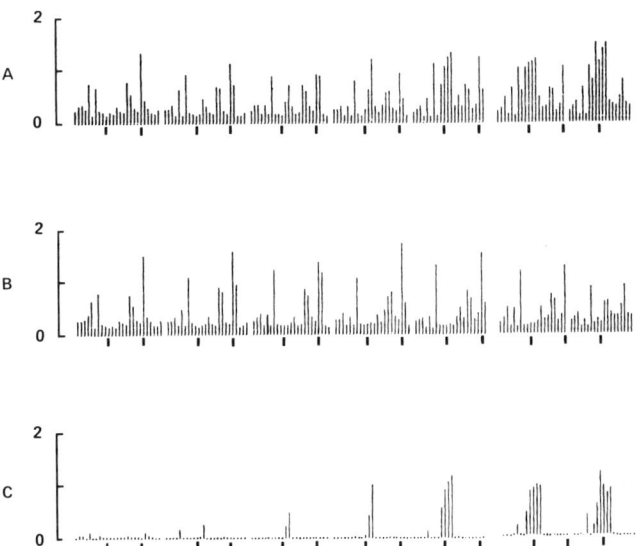

FIGURE 3. A neutralizing anti-peptide serum tested against the same multiple length peptide shown in Fig. 2.
A. Reactivities with anti-peptide serum R536/5.
B. Reactivities of the same peptides with the anti-peptide serum adsorbed with whole virus.
C. The virus specific reactivities of the anti-peptide serum. These reactivities are obtained by subtracting those of panel B from panel A.

Just as in the case of anti-whole virus antibodies the smallest reactive peptides start with the D. This amino acid always seem to make major contributions to antibody binding. In one case two subpopulations reactive with virus are observed (Fig. 4b, serum R 536/5). By comparing Fig. 3 B and C we made a rough estimation of the amount of anti-peptide antibodies cross-reactive with virus. Depending on the serum it seems to vary between 10 and 30%.

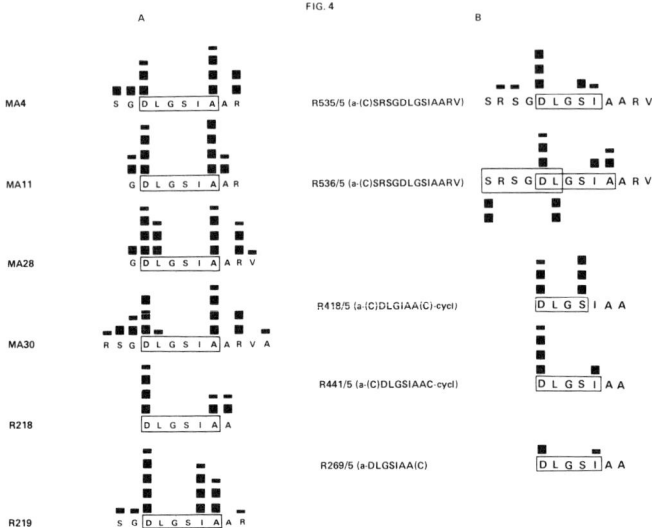

FIGURE 4. Analysis of the reactivities obtained with multiple length peptides and antibodies raised against whole virus (panel A: MA 4, 11, 28 and 30 are Mab's, R 218 and R 219 are sera raised in rabbits) and anti-peptide sera after adsorption with whole virus (panel B). R 535/5, R 536/5 and R 269/5 are antisera raised against lineair peptides as indicated. R 418/5 and R 441/5 are sera raised against cyclized peptides.
For further explanation see text; for serological data see Table 1 and 2.

DISCUSSION

As expected anti-peptide sera appeared to contain subpopulations of anti-peptide antibodies that did and did not crossreact with whole virus. The proportion of anti-peptide antibodies crossreactive with virus was estimated to be between 10 to 30%. Those that did cross react showed to have fine specificities similar to antibodies raised with whole virus with these results. We showed that the fine specificity at the level of single amino acids of the subpopulation of anti-virus antibodies appears to be independent of the species used, the length and the constraints induced by cyclization of the peptides.

ACKNOWLEDGEMENT

We like to thank Douwe Kuperus for excellent technical assistance.

REFERENCES

1. Dardenne D, Savino W, Berrih S, Bach JF (1985). A zinc-dependent epitope on the molecule of thymulin, a thymic hormone. Proc Natl Acad Sci 82:7035.
2. Geysen HM, Barteling SJ, Meloen RH (1985). Small peptides induce antibodies with a sequence and structural requirement for binding antigen comparable to antibodies raised against the native protein. Proc Natl Acad Sci 82:178.
3. Geysen HM, Rodda SJ, Mason TJ, Tribbick G, Schoofs PG (1987). Stategies for epitope analysis using peptide synthesis. J Immun Meth 102:259.
4. Houghten RA, Ostresh JM, Klipstein FA (1984). Chemical synthesis of an octadecapoptide with the biological and immunological properties of human heat stable Esterichia coli enterotoxin. Eur J Biochem 145:157.
5. Jemmerson R, Paterson Y (1986). Mapping antigenic sites on proteins: Implications for the design of synthetic vaccines. Biotech 4:18.
6. Lerner RA (1984). Antibodies of predetermined specificity in biology and medicine. Adv Immunol 36:1.
7. Meloen RH, Briaire J (1980). A study of the cross-reacting antigens on the intact Foot-and-Mouth-Disease Virus and its 12S subunit with antisera against the structural proteins. J Gen Virol 45:761.
8. Meloen RH, Briaire J, Woortmeyer RJ, Zaane D van (1983). The main antigenic determinant detected by neutralizing monoclonal antibodies on the intact Foot-and-Mouth-Disease Virus particle is absent from isolated VP1. J Gen Virol 64:1193.

9. Meloen RH, Puijk WC, Meyer DJA, Lankhof H, Posthumus WPA, Schaaper WMM (1987). Antigenicity and immunogenicity of synthetic peptides of Foot-and-Mouth-Disease Virus. J Gen Virol 68:305.
10. Sutcliffe JS, Milner RJ, Shinnick MT, Bloom FE (1983). Identifying the protein products of brain specific genes with antibodies to chemically synthesized peptides. Cell 33:671.
11. Tanaka T, Slamon DJ, Cline MJ (1985). Efficient generation of antibodies to oncoproteins by using synthetic peptide antigens. Proc Natl Acad Sci USA 82:3400.
12. Walter G (1986). Production and use of antibodies against synthetic peptides. J Immun Meth 88:149.

CONSTRUCTION OF CARRIER-FREE SYNTHETIC PEPTIDE ANTIGENS CAPABLE OF STIMULATING BOOSTABLE ANTIBODY RESPONSES TO A 75 kDa MALARIAL PARASITE PROTEIN

Sylvia J. Richman, Thomas Vedvick, Pawan Sharma, Janette Flint, Feroza Ardeshir, Mitchell Gross*, Carol Silverman*, and Robert T. Reese

The Agouron Institute, La Jolla, California 92037
*Smith, Kline & French Laboratories,
King of Prussia, Pennsylvania 19406

ABSTRACT   A 75 kDa *Plasmodium falciparum* antigen is believed to be a major component of the surface of the extracellular asexual blood stage form of this human malarial parasite. Based on the sequence of a cDNA clone which encoded ~40% of the carboxyl portion of this molecule, synthetic peptide chemistry was used to model several different parts of the molecule. A 27-residue peptide model of a highly charged helical region was found to be an excellent antigen without conjugation. It stimulated high levels of IgG which were reactive with the native parasite protein in immunoblots, ELISA, and radioimmunoprecipitation assays. It was also demonstrated that even conformationally determined epitopes could be successfully modeled with large properly constructed polypeptides. A group of carefully designed polypeptides was used to create a model of a conformationally determined region of the protein and

---

The Agouron Institute, 505 Coast Boulevard South, La Jolla, California 92037
*Smith, Kline & French Laboratories, 709 Swedeland Road, King of Prussia, Pennsylvania 19406

to dissect the epitopes against which various animals could respond. This approach has been exceedingly valuable since it demonstrated differences between the epitopes recognized by monkeys, rabbits and humans. The data question the usefulness of the owl monkey model for testing the protective capacity of this molecule but more importantly remind us that biologically reactive compounds should be engineered for man not a model.

INTRODUCTION

The surface of the extracellular, merozoite form of the human malarial parasite, *P. falciparum*, is composed of a series of glycosylated and non-glycosylated molecules (1). Based on electron microscopic and biochemical data, these proteins are believed to participate in the process in which this asexual blood form of the parasite attaches to and penetrates erythrocytes. An abundant, highly conserved 75 kDa parasite protein appears to be one of these surface molecules (1-3). The cDNA clone C7 encoding ~40% of this protein has been isolated and its cDNA sequenced (3). In order to test the potential of this molecule to induce immunologic protection against malaria, methods for producing large amounts of protein which could successfully model the native structure were devised. Data obtained on proteins produced by recombinant systems as well as synthetic chemistry will be examined. An entirely different approach was used in this work than that adopted by other members of this Symposium. Rather than stepping along the structure with small overlapping peptides, careful analysis of the predicted secondary structure was used to select specific regions for modeling. In addition, chemical studies on the native and recombinant structures were critical in directing us to the immunodominant portions of the molecule whose structures were controlled by long range interactions within the protein.

MATERIALS AND METHODS

The Honduras I/CDC isolate of *P. falciparum* was grown and harvested as previously described (4). Immune monkey sera were obtained from a series of owl monkeys which had been infected with the FVO isolate of *P. falciparum* and then drug cured (3,5). Monkey 33 had been injected with Honduras I parasitized erythrocytes after first exposure to FVO. Monkeys 37 and 39 are Bolivian karyotype VI; 33 is a Colombian type III.

The 13-, 16-, 19- and 28-residue peptides were synthesized in a Sam II peptide synthesizer (6,7); the 49-, 64- and 76-residue peptides were made in an Applied Biosystems model 430A synthesizer (Fig. 1). Before purification the large cysteine-containing peptides were refolded in a solution of 6 M guanidine·HCl containing reduced and oxidized glutathione. The 28-residue peptide, which contains an N-terminal cysteine which is not part of the original structure, was allowed to dimerize at neutral pH at a concentration of 5 mM. All peptides were purified on a semi-preparative 10 μ C18 column. Amino acid analysis was used to verify the composition of each peptide and to allow calculation of its molar absorptivity. To substantiate that the amino acid sequence of the long peptides was correct, they were analyzed on Applied Biosystems model 477A sequencer.

Reduction of 200-300 nmol of peptide was conducted in a 3 ml solution of 6 M guanidine·HCl, 0.5 mM EDTA, 0.4 M Tris HCl, pH 8.6, 20 mM dithiothreitol. The peptide was alkylated with 60 mM (25.5 mg) iodoacetamide. Sequencing of peptide 5 verified that this procedure had not detectably affected the lysine residues and that cysteine had been quantitatively converted to carboxymethyl cysteine.

The 13- and 19-residue peptides were conjugated to keyhole limpet hemocyanin with 1% glutaraldehyde and reduced with sodium borohydride, a modification of the procedure of Avrameas and Ternyck (8). Coupling ratios of 115-180 nmol of peptide per mg of protein were obtained. Ethyl (dimethylaminopropyl) carbodiimide was used to conjugate peptides to bovine serum albumin (BSA). Ratios of 9-12 peptides per BSA molecule were obtained.

Rabbits were injected with 50-100 μg of free peptide or 200 μg of peptide bound to carrier. Initial immunizations were in complete Freund's adjuvant with boosters being given in incomplete Freund's adjuvant on day 28. Immunoblots, radioimmunoprecipitation assays and enzyme linked immunoabsorbant assays were conducted as previously noted (6). High level expression of the fusion protein was achieved by subcloning into pMG27N and expressing in *E. coli* strain AR58.

RESULTS

The secondary structure based on Chou-Fasman rules of the protein produced by clone C7 is found in Figure 1. The first reason for our synthetic peptide studies was to prove that the reading frame (and thus the amino acid sequence) which was predicted from the DNA sequence was truly correct. To answer this question we initially centered our attention around two regions of the structure, the carboxyl terminal repeat region (residues 210-222), and the highly hydrophilic region extending from residues 111 to 129 which was predicted to be helical. The gly-gly-met-pro which is repeated a number of times and was predicted to induce turns was selected because repeating sequences have historically been very good antigens. In addition, we believed that if the repeats induced substantial secondary structure, then the likelihood that they would stimulate antibodies which would be reactive with the native structure would be increased. A 13-residue peptide was constructed to model the repeat region and a 19-mer to model part of the hydrophilic region. The peptides were conjugated to KLH with glutaraldehyde and both rats and rabbits immunized. We found that substantial antibody levels were produced which were reactive with the peptides conjugated to BSA. Interestingly however, only the anti-repeat antibodies reacted with the authentic parasite protein (6). This was sufficient to prove that the predicted protein sequence was correct in that region of the molecule but we were concerned that antibodies produced against the 19-residue peptide conjugated to KLH with glutaraldehyde gave no clear reaction with the native structure.

Three possibilities were obvious which could lead to this result. The protein sequence in the helical region of the molecule could be wrong. The 19-residue peptide could be too small to properly model this helical region, or alternatively, conjugation of the peptide to KLH with glutaraldehyde may have substantially altered the

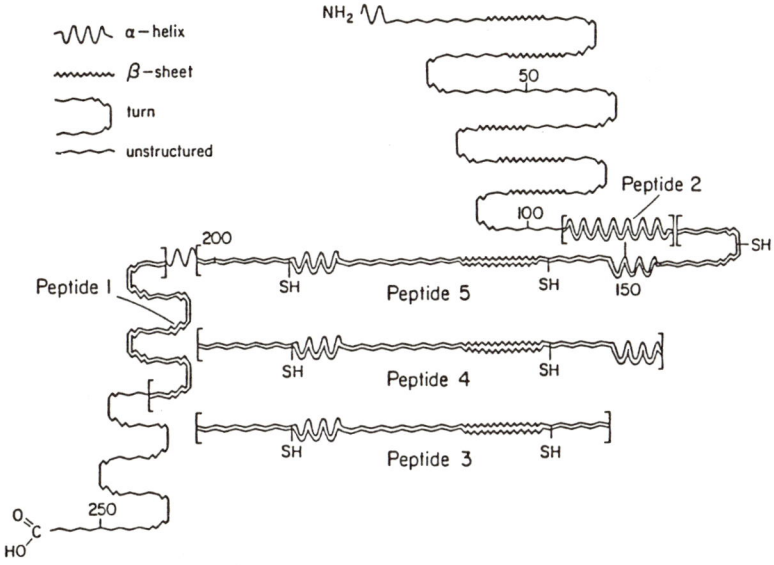

Peptide	Sequence	Residues
1	NH$_2$-MPGG-MPGG-MPGG-MPGG-CONH$_2$	212-227 (16)
2	NH$_2$-CGDEIDRMVN-DAEKYKAEDE-ENRKRIEA-CONH$_2$	104-129 (28)
3	NH$_2$-AEIETCMKTI-TTILEWLEKN-QLAGKDEYEA-KNKEAESVCA-PIMSKIYQD-CONH$_2$	156-204 (49)
4	NH$_2$-KSSLEDQKIK-FKLQPAEIET-CMKTITTILE-WLEKNQLAGK-DEYEAKNKEA-ESVCAPIMSK-IYQD-CONH$_2$	141-204 (64)
5	NH$_2$-ARNSLENYCY-GVKSSLEDQK-IKEKLQPAEI-ETCMKTITTI-LEWLEKNQLA-GKDEYEAKNK-EAESVCAPIM-SKIYQD-CONH$_2$	129-204 (76)

FIGURE 1. Predicted schematic representation of the secondary structure of the protein encoded by clone C7. The amino acid sequence was analyzed by the programs Choufasman and Plotchoufasman [Genetics Computer Group (9)] modified by Reese to reflect that β turns involve four amino acids.

peptide's structure. Examination of the DNA sequence data gave us no reason to believe there should be an error in the amino acid sequence predicted in that region of the protein. There is, however, considerable precedent for relatively small synthetic peptides not properly modeling the portion of proteins they were made to represent. In fact we will shortly provide evidence that a 49-residue peptide based on a different part of this 75 kDa parasite protein was unable to properly model the portion of the molecule upon which it was based. However, the region being modeled by this 19-mer was predicted to be helical and we had no reason to believe that it would be substantially affected by long range interactions such as disulfide bridging. The 19-residue peptide was examined by circular dichroism and found to generate a profile which was consistent with substantial helical structure even when studied in water (6). This suggested that the peptide was likely to be a good model for the part of p75 it was constructed to represent assuming that the predicted amino acid sequence was correct. If this was true then the problem probably arose from the conjugation procedure.

From the beginning we had been concerned about glutaraldehyde potentially destroying epitopes while effecting conjugation. Such a problem was likely to be magnified with this hydrophilic peptide, since 3 of the 19 residues were lysines. For this reason other conjugation methods were tried but none gave high enough coupling ratios to be useful.

To determine if we could make a complete antigen which would induce high levels of IgG without conjugation, we decided to reconstruct the antigen to include the entire portion of the molecule which was predicted to be helical (residues 104-129, Fig. 1). We hoped the additional amino acids would help stabilize the helical structures which exist in this region. Examination of this peptide by CD again substantiated that it contained considerable helical structure (6). A cysteine was then added to the amino terminus as residue-28 to allow dimer formation. This was done to increase the molecular mass up to ~6.6 kDa since molecules below 4-5 kDa have generally been poorly immunogenic. Dimer formation should also decrease the dissociation constant of this material from antibody producing cells. The idea was that if a dimer could bind to an antibody producing cell by two independent sites, then the probability that both would dissociate at the same time would be substantially reduced and the likelihood of triggering antibody production increased.

Two rabbits were immunized with this antigen and the IgM and IgG responses against the peptide measured by ELISA. Substantial IgG antibody was produced by both rabbits when immunized with the antigen in CFA. These data suggested that the molecule contained both T and B cell sites and was thus a complete antigen (6). To determine if the antibodies stimulated by the 28-

mer were capable of reacting with the authentic parasite protein, three different assays were conducted with the realization that the conditions used for specific assays heavily influence how native a protein will be and thus influence the potential for reactivity. Figure 2 demonstrates that the antibodies reacted not only in immunoblots, but in ELISA and immunoprecipiation assays as well. Thus, we have now rigorously demonstrated that it is quite realistic to synthesize entire antigens which can act as excellent models for the native protein.

Another portion of the molecule about which we were concerned was the region which extended from residue 156 through 204. This is the region between peptides 1 and 2 (Fig. 1) and is the portion of the known sequence which contains cysteine. Our particular interest in this part of the molecule came from the observation that when the recombinant gene product which was being expressed by clone C7 in *E. coli* was isolated and then reduced and alkylated it lost most of its reactivity with immune monkey serum. This suggested that much of the immunoreactivity in monkeys for this molecule was due to conformationally determined epitopes created or stabilized by disulfide bonds. As there were three cysteines in this region, how disulfide bridging occurred was not obvious. What was clear was that an approach involving small overlapping peptides would not solve the problem since the cysteines were separated in one case by over 50 residues and thus long range interactions were likely to be important.

The strategy which was adopted was to construct a 49-residue peptide (peptide 3, Fig. 1) which would include the two cysteines at 161 and 194. After removing 30% of the resin 15 more residues were added so as to include the hydrophilic region which extended from 141 to 152 and was predicted to be helical (64-mer, peptide 4, Fig. 1). Approximately 50% of the remaining resin was removed after completion of the 64-mer and the peptide extended 12 more amino acids to include the third cysteine as well as the turn which was predicted (76-mer, peptide 5, Fig. 1). By having samples of the peptide from a series of points in the synthesis, although the final structure was large, our ability to appraise the quality of the synthesis was simplified and if errors were present, we could specifically identify and perhaps correct them. Amino acid analysis was used to confirm the composition of each peptide. In addition the amino acid sequence of each was substantiated by primary structural

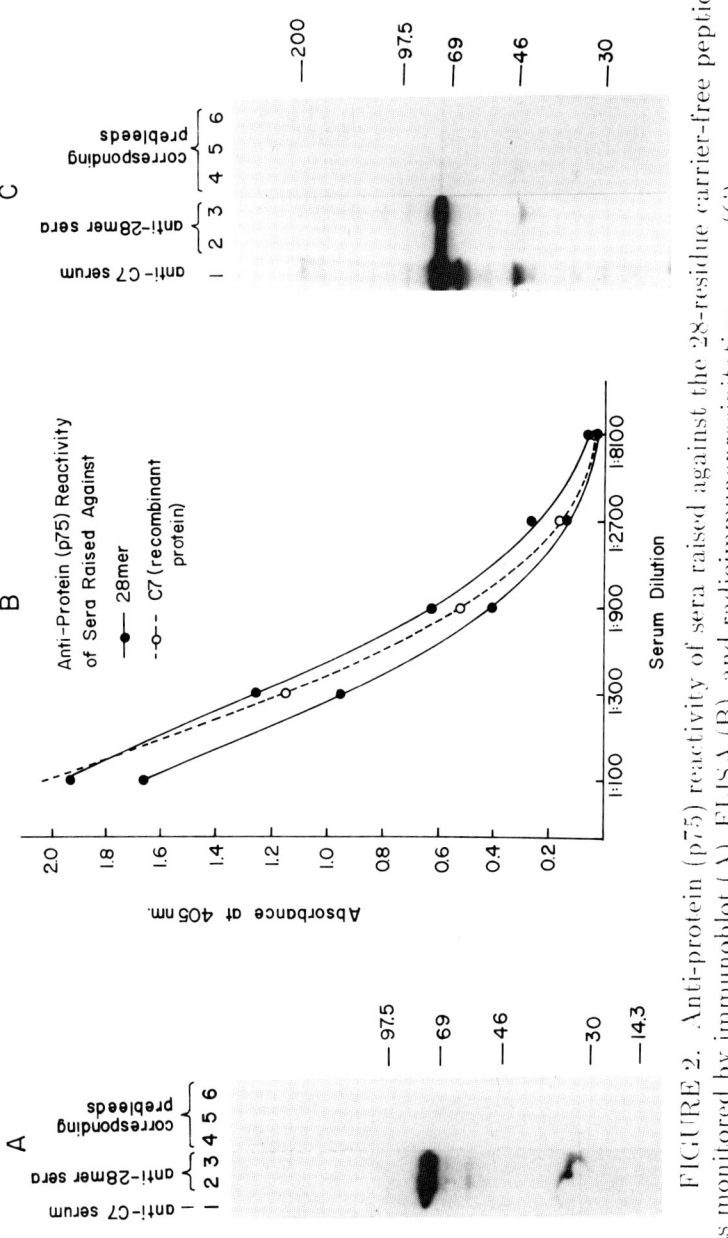

FIGURE 2. Anti-protein (p75) reactivity of sera raised against the 28-residue carrier-free peptide as monitored by immunoblot (A), ELISA (B), and radioimmunoprecipitation assay (C).

analysis.

As this was as much a test for the presence of a T and B cell site within each to test the ability of each peptide to properly model the region in question, the capacity of each peptide to directly bind antibodies from immune monkeys or to competitively inhibit its binding with the fusion protein from recombinant bacteria was measured.

In Fig. 3 we find that antibody production by owl monkeys against the carboxyl 40% of this parasite protein is largely against the region of the molecule between residues 129 and 204. Monkeys produce very little antibody against either the repeat region (16-mer) or the area modeled by the 28-mer. Fig. 3A and B both demonstrate that while the 49-mer is unable to bind efficiently with these monkey antibodies, the 64-mer and 76-mer can. However, the enhanced binding by the larger peptide cannot be attributed simply to new epitopes which are formed by the additional amino acids. If reduction and alkylation of the 64-mer is used to block dilsulfide bridging between cysteine 161 and 194, the destablized form of the peptide is only capable of binding antibody in a fashion similar to the 49-mer. Thus, these two cysteines, which are conserved phylogenetically, appear to be responsible for the critical disulfide bond which stabilizes this portion of the molecule. The reduced and alkylated form of the 76-mer binds somewhat more antibody than the same form of the 64-mer suggesting the additional 12 residues, which probably add a turn as well as the third cysteine, apparently participate in the creation of an additional epitope. Nevertheless, use of the 76-mer in its nonalkylated form again demonstrates that the major portion of this region which is capable of binding antibody is dependent on proper folding. This folding is stabilized by long range forces, two cysteines separated by 32 amino acids. Had we relied on the more typical approach of making small overlapping peptides, this conformationally determined immunodominant region would have been completely missed.

To confirm the previous inhibition data, direct studies were conducted to measure the ability of these peptides to bind antiparasite antibodies present in immune monkey serum. In these experiments, most monkey sera produced very similar patterns characterized by the data presented in Figure 4A. Here you will note that neither the 28-mer nor the 49-residue peptides were effective in binding the monkey antibodies, while the 64-mer and

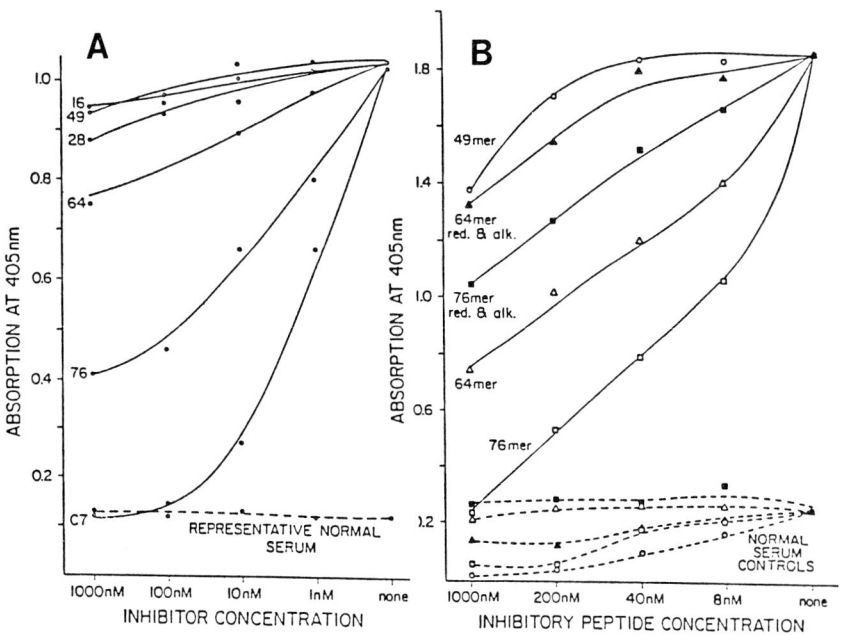

FIGURE 3. Inhibition of binding of antibodies from immune monkey 33 to the C7 fusion protein or peptide 5. A) Plates were coated with the purified C7 fusion protein and peptides 1-5 as well as the fusion protein (C7) used to inhibit antibody binding. B) Plates were coated with refolded peptide 5 and binding of antibody inhibited with peptides 3-5. Both the 64-mer and the 76-mer were tested in refolded as well as reduced and alkylated forms.

the 76-mer respectively, had increasingly high antibody binding capacities. One monkey (Aotus-M39) appeared to produce less antibody to p75 in general than did the others. Antiserum from this monkey bound to the 28-mer, < 49-mer, < 64-mer, < 76-mer. In this case, however all bound in a very similar fashion with even the greatest difference (28-mer vs. 76-mer) not being dramatic.

Immune sera from rabbits immunized with the entire fusion protein from recombinant C7 bacteria were found to inhibit the *in vitro* growth of *P. falciparum* even when used at a 1:100 dilution.

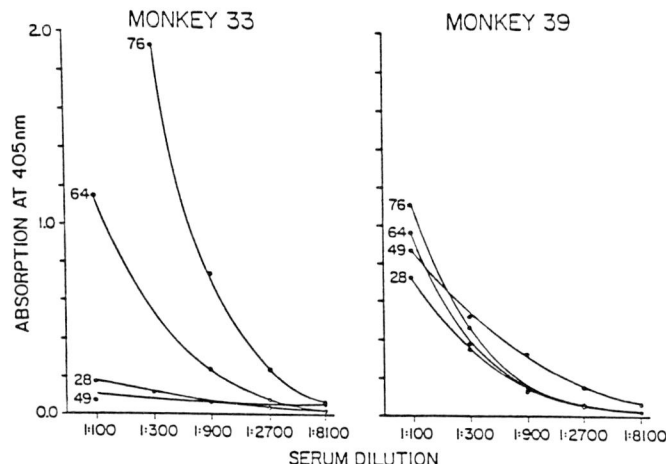

FIGURE 4. Reactivity in ELISA of sera from immune monkeys 33 and 39 with peptides 2-5. Peptides 2-5 were independently used to coat plates. The amount of antibody reactivity with each peptide was measured.

Although we have previously found that when rabbits were immunized with any of the independent peptides except for the 16-residue repeat and the 49-mer, substantial levels of antibody would be produced which could react with the native protein; nevertheless, when immunized with the entire fusion protein, the immunodominant region in rabbits based on the inhibition ELISA appears to be the helix modeled by the 28-mer. Since Aotus monkeys respond mainly to the conformationally determined epitope contained within p75, and in general poorly to the rest of the structure, it is questionable whether owl monkeys would be protected if immunized with these constructions even though they effectively model parts of p75.

To appraise the ability of humans to respond to the 28-mer and the 76-mer, sera from a series of immune humans were examined for their ability to bind directly to those peptides (Fig. 5). Individuals could clearly produce antibodies to both peptides but it appeared to occur in a reciprocal fashion. If the antibody reacted well with the 76-mer then there was substantially less reaction with the 28-mer and visa versa. These data as well as the previous evidence suggest that cysteine 161 and 194 come together forming a disulfide bridge which stacks the 2 hydrophilic helical regions found in the 76-mer with the hydrophilic helical region modeled with the 28-mer. This creates a major hydrophilic center in this portion of p75 (Fig. 6) against which animals (and man) respond according to their specific genetic potentials. Man may well be able to react with this region sufficiently to gain some degree of immunologic protection against the parasite. Owl monkeys appear to respond in a different fashion and thus may provide a poor model for testing the potential usefulness of this protein as part of a possible vaccine for man.

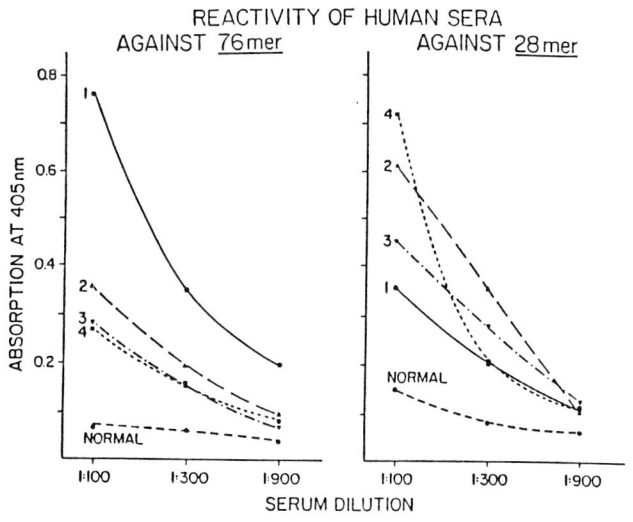

FIGURE 5. Reactivity in ELISA of sera from immune humans with peptides 2 and 5. Peptides 2 and 5 were independently used to coat plates. The amount of antibody reactivity with each peptide was measured.

DISCUSSION

In recent years peptide chemistry has played an increasingly valuable role in proving amino acid sequences predicted from

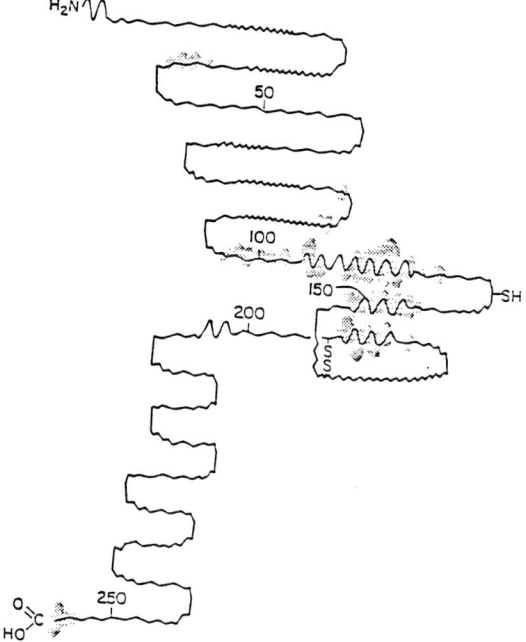

FIGURE 6. Schematic representation of a proposed folding pattern for the carboxyl terminal portion of the 75 kDa malarial parasite protein which is consistent with available data. The shaded region represents portions which are hydrophilic. The amount of shading correlates with the degree of hydrophilicity, helix, β-sheet, β-turn, unstructured.

molecular biological data. It has also been used extensively for epitope mapping. Because of its simplicity, one approach which has recently gained popularity for defining epitopes is to make a large number of small overlapping peptides which completely cover the amino acid sequence to be examined.

Epitopes on molecules are defined by two things; the structure itself and the genetics of the animal into which it is injected which allow it to be recognized as foreign. The energy of interaction between the epitope and the antibody are largely determined by the dissociation constant since the rate of association appears to be limited only by diffusion. Thus the stability of the union of antibody and epitope is wholly dependent on the number and the energy of the bonds which can be readily formed.

With this understanding, our view is that the best epitope is one which is large enough to have multiple sites of interaction with the antibody and constrained enough in its structure so that once any portion of it starts to interact with its corresponding site on the antibody, a number of other bonds will rapidly form, locking the two together. Ionic bonds provide the greatest energy and have the longest range effects. They should contribute heavily to the process of attracting the epitope into the antigen binding site in the proper orientation and holding it in place while other interactions are created. In addition, charged, highly hydrophilic, portions of molecules are generally on the surface of structures and thus accessible to interaction with antibodies. Reiteration of epitopes, which occurs when amino acid sequences are repeated, decreases the effective dissociation rate since one epitope may come loose from an antigen binding site only to be replaced by the same structure not far away on the same molecule. Epitopes which are repeated enough times and are physically arranged so that more than one antigen binding site can interact with two or more epitopes at any point in time, have an even greater impact since the probability of dissociation occurring at both sites at the same time is very small. It is the product of the energies of each of the individual binding sites.

During the last 20 years, however, it has become increasingly clear that many of the important epitopes created by proteins are conformationally determined, often by relatively long range forces. Since these conformations are sometimes stabilized by disulfide bridges, simple reduction and alkylation can provide valuable information that they exist and can suggest the region of the molecule which is likely to be involved.

In the approach used in the present work we have tried to take into account all of these factors. Regions of the molecule which were selected for modeling as potential epitopes had either helical or turn structures which the Chou-Fasman rules predict with substantial reliability. Furthermore, once synthesized, the peptide can be examined by physical methods to see if it assumes the predicted structure. Since highly hydrophilic regions may often be the most desirable parts of the molecule to model, synthesis of sufficiently large structures to be complete antigens may often provide a viable alternative to conjugation which eliminates its many negative effects.

Rabbits produce antibodies to a broad range of epitopes created by the 75 kDa malarial parasite protein. The immune response of owl monkeys appears to be limited to a much smaller portion of the molecule. Reduction and alkylation of the fusion protein made by the recombinant clone C7 demonstrated that most of the monkey antibodies reacted with conformationally determined epitopes which were stabilized by disulfide bonds. The synthetic peptides allowed us to help define where these epitopes were and, together with their reactivity with the monkey sera, to obtain an impression of how the polypeptide chain was likely to fold. A valuable result of the study was the observation that although monkeys might produce relatively large amounts of antibody to the fusion protein or some of the peptide constructions which model portions of it, these animals did *not* have the capacity to respond to the same regions of the molecule as man or rabbits do. To effectively evaluate the capacity of man to be at least partially protected against malaria by immunization with this antigen, it may be necessary to test man himself regardless of the data obtained with the monkey model. Another critical point should be remembered by all of us who are actively trying to engineer structures to induce some specific biologic effect. If a model is to be used, we must be careful in our attempts at being clever, not to optimize our structure for the model only to find that it poorly reflects the response in the host of interest. We must also be careful not to discard a product because it did not work well in the model, if there is evidence that it might still work in man.

REFERENCES

1. Reese RT, Ardeshir F, Flint JE, Howard RF, Ramasamy R, Richman SJ, Stanley HA (1985). Use of biotechnology to study *Plasmodium falciparum* asexual blood-stage antigens. In Siddiqui WA (ed): "Proceedings of the Asia and Pacific Conference on Malaria," University of Hawaii, p 309.

2. Ardeshir F, Flint JE, Reese RT (1985). Expression of *Plasmodium falciparum* surface antigens in *Escherichia coli*. Proc Natl Acad Sci USA 82:2518.
3. Ardeshir F, Flint JE, Richman SJ, Reese RT (1987). The 75 kDa merozoite surface protein of *Plasmodium falciparum*. EMBO J 6:493.
4. Trager W, Jensen JB (1976). Human malaria parasites in continuous culture. Science 193:673.
5. Reese RT, Motyl MR (1979). Immunologic inhibition of the *in vitro* growth of *Plasmodium falciparum*. J Immunol 123:1894.
6. Richman SJ, Reese RT (1988). Immunologic modeling of a 75-kDa malarial protein with carrier-free synthetic peptides. Proc Natl Acad Sci USA 85: in press.
7. Merrifield RB (1963). Solid phase peptide synthesis. I. The Synthesis of a tetrapeptide. J Am Chem Soc 85:2149.
8. Avrameas S, Ternynck T (1969). The crosslinking of proteins with glutaraldehyde and its use for the preparation of immunoadsorbents. Immunochem 6:53.
9. Devereaux J, Haeberli P, Smithies O (1984). A comprehensive set of sequence analysis programs for the VAX. Nuc Acids Res 12:387.

IV. DETERMINATIONS OF ANTIGENIC DOMAINS

USE OF MULTIPLE SYNTHETIC PEPTIDE ANALOGS TO STUDY
STRUCTURAL FEATURES OF PEPTIDES CAUSING T CELL ACTIVATION[1]

D.C. Anderson,[2] W. van Schooten, M. Barry,[2] A.A.M. Janson
and René R.P. de Vries

Department of Pathobiology, University of Washington,
Seattle WA 98119 and Department of Immunohematology and
Blood Bank, University Hospital of Leiden, Leiden
The Netherlands

ABSTRACT   Multiple analogs of a peptide stimulating helper T cell proliferation were made using solid phase peptide synthesis. These analogs have been used to find critical residues for T cell activation in this T cell epitope and to probe the importance of predicted secondary structure, amphipathicity and pattern of side chains in activation. The results suggest that neither predicted alpha helicity nor amphipathicity are necessary for activation, which appears to require specific interactions of amino acids in most positions of this peptide with cellular receptor(s). A number of analogs retain activity, suggesting the possibility of further modifications to probe molecular events in antigen presentation and T cell activation.

INTRODUCTION

The molecular basis of peptide stimulation of the cellular immune response is currently of great interest. Acti-

[1] This work was supported by the University of Washington Graduate School Research Fund, the Dutch Foundation for Medical Research, the United Nations Development Program/World Bank/World Health Organization Special Program for Research and Training in Tropical Diseases, and the Netherlands Leprosy Relief Association.
[2] Present address: Dept. of Biochemistry, NeoRx Corp., Seattle, WA 98119

vating peptides can be delineated in sequence (1) and have been shown to bind to major histocompatibility complex (MHC) coded proteins such as Ia proteins in mice or HLA-DR proteins in humans, in the process of antigen presentation (2). This is thought to usually follow proteolytic processing of the foreign protein containing the peptide inside cells such as macrophages, B cells or dendritic cells (3). Responding T helper cells are subsequently activated after formation of a presumed ternary complex of their T cell receptor and the peptide-MHC protein complex (4). Understanding the mode of action of these peptides may also be important for synthetic vaccine development for diseases such as leprosy, caused by Mycobacterium leprae (5). The severity of leprosy is at least partially a function of the type of HLA protein of the patient (6). Thus leprosy may be an important model for the study of the role in disease of human immune response genes, which code the HLA proteins. The HLA-DR2 allele is over-represented in the tuberculoid and lepromatous types of leprosy (7-8). Thus properties of HLA-DR2 restricted T cell clones may be of interest in understanding the molecular basis of this disease.

Here we are interested in studying the molecular requirements for stimulation of 2F10, an M. leprae-specific helper T cell clone which responds to the peptide LQAAPALD-KL-amide (9) from the 65 kDa protein of M. leprae. This immunodominant protein, which has a strong sequence homology to heat shock proteins from a variety of organisms (10), contains a number of B cell epitopes (11) as well as peptides activating T cells (12-14).

RESULTS AND DISCUSSION

Construction and Testing of Peptide Analogs.

To determine the role of individual residues in activation of the T cell clone and, based on this, the importance of overall characteristics such as helical potential or helical amphipathicity, we have systematically replaced each residue in the minimal activating peptide, LQAAPALDKL, with amino acids with varying functional groups. These are listed in Table 1. Several classes of substitution were used: residues were individually deleted from the sequence, their side chains were omitted by substitution with glycine (a similar approach has used substitution with alanine (15)) and they were individually replaced by residues with charged, sterically similar or sterically different side chains. The peptide mutants were then tested for activation of clone 2F10 in place of the original peptide LQAAPALDKL.

TABLE 1
SUBSTITUTIONS OF INDIVIDUAL RESIDUES IN PEPTIDE ANALOGS

CLASS:	I	II	III	IV	V	
			Sterically			
	Omit	Omit	Similar	Sterically	Charged	
Residue	Residue	Side Chain	Residue	Different	+	-
L	omit L	G	I,N,A,V	F,P	K	E
Q	omit Q	G	N,E,L	F	K	E
A	omit A	G	V,L	F,P	K	E
P	omit P	G		A,N,V	K	E
D	omit D	G	E,N,L	A,F,P	K	E
K	omit K	G	R,M,Q	P,D	K	E

A Variety of Analogs Stimulate 2F10.

The results of stimulation of clone 2F10 by peptide analogs are summarized in Table 2. In a number of cases, all or nearly all substitutions inactivate the response. These residues should thus be critical for some step involved in the events leading to T cell activation. When substitutions are allowed, the residue should be less critical for T cell activation. Deletion analogs were synthesized as the most severe test for the contribution of a residue. Deletion of an amino acid will not only remove the possibility of interactions involving its side chain, it may affect the spacing between segments of the peptide on either side of the deletion. All but one of these analogs are inactive.
A second set of analogs involves replacement by glycine, which should delete side chain interactions without altering spacing of other amino acids in the peptide chain. Only ala-3, ala-6 and leu-10 can be replaced by glycine with retention of significant activity. All other substitutions for these alanines involve larger side chains. If one presumes that these short peptides interact directly with HLA and the T cell receptor without prior processing steps, this

TABLE 2
Activation of T Cell Clone 2F10 by Peptide Analogs of LQAAPALDKL

Analog class		Relative Activity of Analogs									
		L	Q	A	A	P	A	L	D	K	L
I	omit residue	0.1	0	0	0	0	0	0.2	0.3	0.9	7.9
II	G	0.2	1.5	68	0.8	0	24	0.1	0.7	0.3	29
III, IV	I	13						0.5			58
	N	2.7	102			0.1		0.1	24	Q:0.5	43
	A	12.9				0.1		0.1	0.4	D:0.2	75
	V	93		0.5	0.1	0	0.8	0.3			50
	F	134	55	0.2	0.1	0	0.3	1.1	0.2		48
	P	82		0.1	0		0.2	0.4	0.5	0.6	55
	L			0.2	0		0.3		0.5	M:0.3	
V	K	0	42	0.3	0	0	0.2	0.2	0.2	R:28	78
	E	110		0.1	0.1	0.2	0.1	0.4	0.2	0.2	43

Data is presented as a percent of the control response of 2F10 to a saturating concentration (0.83 uM) of the peptide LQAAPALDKL, which gives a response of 100% (121000 cpm above background). Each analog was tested at the same concentration. 3H-thymidine incorporation was used to assay 2F10 proliferation (7). Each epitope residue is listed on the top and the newly substituted residues are listed along the left. Peptides were synthesized in the amide form using 4-methylbenzhydrylamine resin and the tea-bag methodology of Houghten (16), and analyzed as reported (11).

suggests that when bound they are in a sterically confined environment, and perhaps not merely on the outer surface of HLA. Using similar reasoning with the phe substitutions for leu-1, gln-2 and leu-10 suggests these residues are not sterically as restrained as the above alanines. Substitution with other amino acids reveals more active analogs. Leu-1 and leu-10 can be substituted by several amino acids each, and may thus be important for broadening the response of the T cell to a variety of different peptides. Ala-4, pro-5 and leu-7 cannot be substituted without destroying 2F10 activa-

tion, and ala-3, ala-6, asp-8 and lys-9 tolerate only a
single replacement. Thus residues 3-9 seem to be quite critical for T cell activation.

Several analogs give similar or greater activation than
the original peptide, including leu-1 substituted by val,
phe, pro and glu, gln-2 substituted by asn and phe, ala-3
substituted by gly, and leu-10 substituted by ile, ala, val,
phe, pro and lys. This is similar to results reported for a
cytotoxic T cell clone responding to analogs of an influenza
virus nucleoprotein (17) although the most active peptide
was a deletion mutant.

The activity of multiple analogs has implications for
immune system function and design of synthetic vaccines.
First, the finding of peptide analogs with enhanced activity
suggests that synthetic peptide vaccines can be engineered
for increased activity by single mutations and probably by
other alterations of the peptides. The observation here that
a single defined HLA-T cell receptor pair can respond to
multiple peptides also suggests an additional factor to explain the breadth of response of the cellular immune system
to foreign proteins. Previously, explanations included
heterogeneity in the types of the HLA proteins (although
this may be many orders of magnitude less than the heterogeneity found in antibodies) and in T cell receptors. The
observed response to multiple analogs may be consistent with
the recently determined crystal structure of the class I HLA
protein HLA-A2 (18-19), in which the peptide binding site
may consist of a deep trough long enough for binding of
individual peptides in multiple positions, or for more than
one type of peptide to find complementary residues for tight
binding.

Antigen presenting cell-T cell pairs with defined HLA
protein and T cell receptors may respond to more peptides
than discovered by this approach. The clone 2F10 also proliferates in response to the self peptide EQARAAVDTY, which
represents the third hypervariable region of the HLA-DR2
beta 3 chain (9). This response would not have been predicted based on the above results, since all ala-4 and pro-5
analogs, replacement of leu-7 with val, and all uncharged
lys-9 replacements were inactive.

Does T Cell Activation Correlate with Peptide Amphipathicity?

Peptides activating helper and cytotoxic T cells have
been hypothesized to be capable of forming amphipathic alpha
helices (20-23). Seven of ten residues in the epitope studied

here, LQAAPALDKL, are strong helix formers and only proline is a strong helix-breaker (24). Although peptides this short are unlikely to form a stable secondary structure in solution, a particular secondary structure may be stabilized or induced by intermolecular interactions when the peptide is bound to HLA proteins, or bound at any other important step in antigen presentation. The hydrophobic moment is one way to quantitate the degree of amphipathicity of a peptide assuming it adopts a particular secondary structure. Figure 1 shows a hydrophobic moment plot for the M. leprae 65 kDa protein, assuming it is entirely alpha-helical. Similar plots of 3-10 helix or beta sheet hydrophobic moments were relatively featureless. If peptides represented by the peaks are helical, they represent amphipathic alpha helices; the peptide LQAAPALDKL thus represents a predicted amphipathic helix. A number of other known T cell epitopes derived from this protein's sequence are located for comparison; some but not all represent predicted amphipathic helices.

To determine the correlation between the amphipathicity of a (potentially) helical peptide and activation of the T cell clone 2F10, the calculated alpha helical hydrophobic moment (26) is compared to the activity of selected active peptide analogs (Table 3). Overall, there is no strict correlation in this system. Peptides with similar predicted alpha helical moments vary significantly in activity (compare peptides 8-9, 3-5, 4-16 or 13-17). Peptides with enhanced alpha moments can have lower activity than control peptides (c.f. peptides 3 and 13). When similar calculations were performed on 11-residue stretches from 58 different known T cell activating peptides (23, 28), a wide range of maximum values resulted, ranging from 0.022 to 0.515. Using this method, a hypothetical helix designed to be very amphipathic, such as IIRRIRRIIRR, would have a calculated moment of 0.794, while another non-amphipathic hypothetical helix such as an alanine 11-mer would have a helical moment of 0.005. Thus peptides presumed to be helical seem capable of activating T cells over a wide range of amphipathicity.

Does T Cell Activation Correlate with Helical Potential ?

The peptide LQAAPALDKL is predicted to be alpha helical (Table 3) using Chou-Fasman parameters, as are all of the variants shown. However, the predicted helix may be bent due to the presence of a proline in the middle, as in the case of melittin (29). Twenty two of 58 other known T cell epitopes contain proline; nine of these contain proline in the interior of the sequence, and one of these has been shown to be non-helical in solution (30). Here, a number of analogs containing two prolines or a proline and glycine,

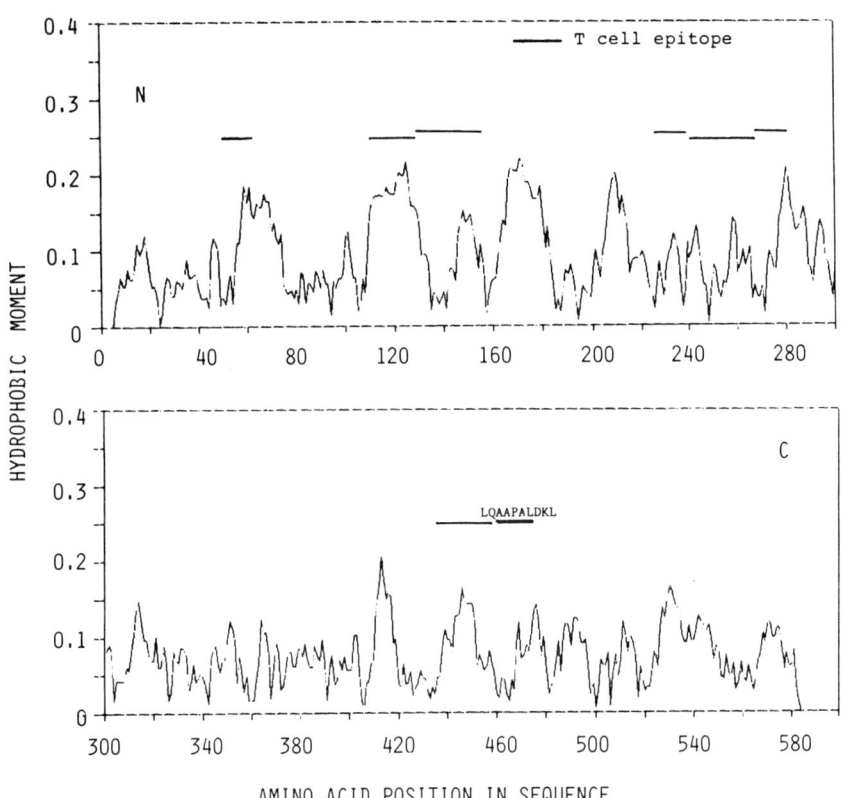

FIGURE 1. Hydrophobic moment plot of the M. leprae 65 kDa protein. The hydrophobic moment for running 11 amino acid segments throughout the sequence is assigned to the sixth residue of the window and is shown assuming the entire protein is helical. Amphipathic alpha helices would correspond to peaks in the plot. Known T cell activating peptides are shown as bars above the plot, including the peptide LQAAPALDKL. Calculations were done using the program MSEQ (25) based on the published sequence of the protein (34).

TABLE 3
COMPARISON OF PEPTIDE ANALOG ACTIVITY WITH CALCULATED ALPHA
HELICAL HYDROPHOBIC MOMENTS AND AVERAGE HELICAL POTENTIAL[a]

	Peptide Sequence	Percent Activity	Alpha Moment	Avg. Helical Potential
1	LQAAPALDKL	100	0.246	1.17
2	I---------	13	0.237	1.16
3	F---------	134	0.242	1.17
4	P---------	82	0.278	1.11
5	E---------	110	0.313	1.20
6	-N--------	102	0.243	1.13
7	-E--------	21	0.241	1.21
8	-F--------	55	0.175	1.18
9	-K--------	42	0.275	1.18
10	--G-------	68	0.240	1.09
11	-----G----	24	0.247	1.09
12	-------N--	24	0.245	1.14
13	--------R-	28	0.303	1.16
14	---------V	50	0.247	1.16
15	---------P	55	0.223	1.11
16	---------K	78	0.212	1.17
17	---------E	43	0.212	1.20
18	EQARAAVDTY	15	0.245	1.15

[a] Hydrophobic moments were calculated using the program MSEQ (25), using Eisenberg's unscaled consensus hydrophobicity parameters (27), with the addition of an N terminal gly to each sequence to make an 11-mer peptide for the calculations. The average helical potential is the average Chou-Fasman helical probability per residue, calculated using published values (24).

both strong helix breakers, are active (c.f. peptides 4, 10, 11 and 15). Although the range of calculated helical potential is limited, there is no strict correlation of helical potential with analog activity. Among the active analogs, the peptide with the highest helical potential, peptide 7, is one-third as active as peptide 10, which has the lowest helical potential. If the active conformation were helical, substitutions tending to cancel the helical dipole (31)

might be expected to enhance activity by stabilizing a
helical conformation in solution. Thus an analog adding a
positive charge near the C terminus of the peptide, or a
negative charge near the N terminus, should have enhanced
activity. Some analogs fit this pattern (peptide 5) while
others do not (peptides 7, 13 and 16). Keeping in mind the
difficulties of secondary structure prediction (32), when
other T cell activating peptides (23, 28) were examined
using the program MSEQ, 32 of 58 were predicted to be non-
helical using the Chou-Fasman method. Thus from these examples we find no necessary correlation between predicted
secondary structure and the ability of a peptide to acti-
vate T cells. The data seem more consistent with a more
extended conformation of the active peptide, with numerous
residues contacting a critical cellular receptor for T cell
activation, such as HLA-DR2.

Testing Other Predictive Schemes.

Rothbard and Taylor (28) have suggested that in cases
where the minimal-sized helper or cytotoxic T cell epitope
has been determined, the epitopes contain residues in the
pattern (charged or glycine, hydrophobic, hydrophobic,
polar or glycine) or (charged or glycine, hydrophobic,
hydrophobic, hydrophobic or proline, polar or glycine).
None of the peptides which they studied included HLA-DR2
restricted epitopes. None of the active peptides described
here fit these patterns. Thus their bound form on HLA-DR2
may differ from that of other known T cell epitopes. The a
priori prediction from primary structure of peptides useful
in stimulating helper or cytotoxic T cells may thus not be
straightforward, and may depend on the HLA protein type of
the cells involved.

Conclusions.

The use of peptide mutants to study molecular properties associated with T cell activation has been very useful.
Focussing on the role of peptides, the above results are
consistent with a more complex picture of T cell activation
than previously proposed. Using the analogs mentioned here,
there seems to be no strict correlation with activity of
either calculated peptide helical potential or helical
amphipathicity. These features are present in some but certainly not all of the known T cell epitopes. The fact that
we find a core of critical residues for T cell activation

suggests that specific bound interactions throughout much of the peptide chain, which is possibly in a more extended form, may dominate the effect on overall activation of T cells. Since hypervariable region residues of HLA proteins may be important in binding these peptides (33) it is not surprising that interactions may differ for each HLA protein type (28).

The finding that a number of peptide analogs retain similar or even higher activity not only suggests an additional possible explanation for the breadth of the cellular immune response to foreign peptides, but also that peptides may be further engineered for enhanced stimulation of T cells. For example it should be interesting to test "double mutant" peptides and peptides with other modifications for effects on T cell activation.

REFERENCES

1. Livingstone AM, Fathman CG (1987). The structure of T-cell epitopes. Ann. Rev. Immunol. 5: 477.
2. Babbitt BP, Allen PM, Matsueda G, Haber E and Unanue, ER (1985). Binding of immunogenic peptides to Ia histocompatibility molecules. Nature 317: 359.
3. Germain RN (1986). The ins and outs of antigen processing and presentation. Nature 322: 687.
4. Watts TH, McConnell HM (1987). Biophysical aspects of antigen recognition by T cells. Ann. Rev. Immunol. 5: 461.
5. Bloom BR (1986). Learning from leprosy: a perspective on immunology and the third world. J. Immunol. 137: 1.
6. van Eden W, deVries RRP, D'Amaro J, Schreuder G, Lecker DL and van Rood JJ (1982). HLA-DR associated genetic control of the type of leprosy in a population from Surinam. Hum. Immunol. 4: 343.
7. van Eden W, deVries RRP, Mehra NK, Vaidya M, D'Amaro J and van Rood JJ (1980). J. Inf. Dis. 141: 693.
8. deVries RRP, Serjeantson SW and Layrisse Z (1984). Leprosy. In Histocompatibility Testing 1984. Albert, E.D. et al, eds. Berlin: Springer Verlag, p. 362.
9. Anderson DC, van Schooten W, Barry ME, Jansson A, Buchanan T and deVries RRP (1988), submitted.
10. Shinnick TM, Vodkin MH and Williams JC (1988). The mycobacterium tuberculosis 65-kilodalton antigen is a heat shock protein which corresponds to common antigen

and to the excherichia coli groEL protein. Inf. Immun. 56: 446.
11. Anderson DC, Barry ME and Buchanan T. Exact definition of species-specific and cross-reactive epitopes of the 65 kDa protein of Mycobacterium leprae using synthetic peptides (1988). J. Immunol., in press.
12. deVries RRP, Ottenhoff THM, Shuguang L, and Young RA (1986). HLA class II restricted helper and suppressor clones reactive with Mycobacterium leprae. Lepr. Rev. 57 (suppl. 2): 113.
13. Emmrich F, Thole J, van Embden J, and Kaufman S (1986). A recombinant 64 kD protein of Mycobacterium bovis BCG specifically stimulates human T4 clones reactive to mycobacterial antigens. J. Exp. Med. 163: 1024.
14. Lamb JR, Ivanyi J, Rees ADM, Rothbard JB, Howland K, Young RA and Young DB (1987). Mapping of T cell epitopes using recombinant antigens and synthetic peptides. EMBO J. 6: 1245.
15. Allen PM, Matsueda GR, Evans RJ, Dunbar Jr. JB, Marshall GR and Unanue ER (1987). Identification of the T-cell and Ia contact residues of a T-cell antigenic epitope. Nature 327: 713.
16. Houghten, RA (1985). Proc. Nat. Acad. Sci. USA 82:5131.
17. Bodmer HC, Pemberton RM, Rothbard JB and Askonas BA (1988). Enhanced recognition of a modified peptide antigen by cytotoxic T cells specific for influenza nucleoprotein. Cell 52: 253.
18. Bjorkman PJ, Saper MA, Samraoui B, Bennett W, Strominger JL and Wiley DC (1987). Structure of the human class I histocompatibility antigen, HLA-A2. Nature 329: 506.
19. Bjorkman PJ, Saper MA, Samraoui B, Bennett W, Strominger JL and Wiley DC (1987). The foreign antigen binding site and T cell recognition regions of class I histocompatibility antigens. Nature 329: 512.
20. DeLisi C and Berzofsky JA (1985). T-cell antigenic sites tend to be amphipathic structures. Proc. Nat. Acad. Sci. USA 82: 7048.
21. Berkower I, Buckenmeyer GK and Berzofsky JA (1986). Molecular mapping of a histocompatibility-restricted immunodominant T cell epitope with synthetic and natural peptides: implications for T cell antigenic structure. J. Immunol. 136: 2498.
22. Berzofsky JA, Cornette J, Margalit H, Berkower I, Cease K and deLisi C (1986). Molecular features of class II MHC-restricted T-cell recognition of protein

and peptide antigens: the importance of amphipathic structures. Curr. Top. Micro. Immunol. 130: 14.
23. Spouge JL, Guy HR, Cornette JL, Margalit H, Cease K, Berzofsky JA and deLisi C (1987). Strong conformational propensities enhance T cell antigenicity. J. Immunol. 138: 204.
24. Chou PY, Fasman GD (1978). Empirical predictions of protein conformation. Ann. Rev. Biochem. 47: 251.
25. Black SD, Glorioso JC (1986). MSEQ: a microcomputer-based approach to the analysis, display and prediction of protein structure. BioTechniques 4: 448.
26. Eisenberg D, Schwarz E, Komaromy M and Wall R (1984). Analysis of membrane and surface protein sequences with the hydrophobic moment plot. J. Mol. Bio. 179: 125.
27. Eisenberg D, Weiss RM, Terwilliger TC and Wilcox W (1982). Faraday Symp. Chem. Soc. 17: 109.
28. Rothbard JB, Taylor WR (1988). A sequence pattern common to T cell epitopes. EMBO J.7: 93.
29. Terwilliger TC, Eisenberg D (1982). J. Biol. Chem. 257: 6010.
30. Heber-Katz E, Hollosi M, Dietzschold B, Hudecz F and Fasman GD (1985). The T cell response to the glycoprotein D of the herpes simplex virus: the significance of antigen conformation. J. Immunol. 135: 1385.
31. Shoemaker KR, Kim PS, York EJ, Stewart JM and Baldwin RL (1987). Tests of the helix dipole model for stabilization of alpha-helices. Nature 326: 563.
32. Kabsch W, Sander C (1983). How good are predictions of protein secondary structure ? FEBS Lett. 155: 179.
33. Guillet J-G, Lai M, Briner T, Buus S, Sette A, Grey HM, Smith JA and Gefter ML (1987). Immunological self, non-self discrimination. Science 235: 865.
34. Mehra V, Sweetser D and Young RA (1986). Efficient mapping of protein antigenic determinants. Proc. Nat. Acad. Sci. USA 83: 7013.

THE USE OF SYNTHETIC PEPTIDES FOR THE ANALYSIS OF THE ENVELOPE OF HEPATITIS B VIRUS (HBV)[1]

Colin Howard, Helen Stirk, Alan Buckley, Sheila Brown and Michael Steward

London School of Hygiene and Tropical Medicine
London WC1E 7HT, U.K.

ABSTRACT The outer coat of HBV contains HBsAg determinants coded by the S and pre-S regions of the genome. We have developed a model of HBsAg structure using predictive algorithms. The presence of 4 membrane spanning helicies is predicted, together with an external loop between residues 111 and 156 where the major a, or protective, determinants are located. To test this model, a panel of synthetic peptide analogues to potentially exposed regions were prepared together with a peptide analogue of an internal loop structure. Reactivities of these peptides with human antibodies were then measured either by ELISA or by the assessment of affinity. Peptides mimicking external regions reacted with antibodies induced by native HBV. High levels of antibody capable of binding to a peptide mimicking carboxyl terminus residues 217 to 226 were detected. An analogue of the possible contact residues for antibody within the major a determinants was performed using animal antisera to related hepadnaviruses having amino acid replacemnts in the region 124 to 139. Affinity measurements showed common residues may be important for anti-HBs induction.

[1]This work was supported by the Wellcome Trust and the Medical Research Council.

INTRODUCTION

The outer coat of the hepatitis B virus (HBV) contains HBsAg determinants coded by the S and preS regions of the genome. Approaches to define the B-cell determinants on preS and S proteins have concentrated largely on immunizing laboratory animals with synthetic peptides predicted as representing antigenic domains on the HBV particle. Several laboratories have defined such structures largely as a result of immunizing animals with peptide-carrier conjugates and seeking reactivity with either intact HBV particles or HBsAg 22nm. particles. Hepatitis B immune globulin (HBIg) preparations contain antibodies capable of neutralizing HBV infectivity or attenuating the extent of infection or both.

The affinity of an antibody for its antigen is an important parameter of antigen binding; high-affinity antibody has been shown to be superior to low-affinity antibody in virus neutralization, and we have previously shown high affinity binding of a cyclical 139-147 peptide analogue of the S protein (1,2). In the present study, we have expanded our analysis of human anti-HBs antibodies by use of a wider panel of peptide analogues of S, pre S and other HBV antigens and compared the results with the predicted location of these determinants on the native molecule. An analysis of the external loop region of the S protein between residues 124 and 147 has been undertaken using sera against the S protein of other hepadnaviruses in order to identify the contact residues in this region. A preliminary comparison of HBV vaccines prepared from plasma and by recombinant techniques has also been undertaken to compare the immunogenicity of selected S determinants.

MATERIALS AND METHODS

Peptides.

Peptides were synthesized by either tBoc- or FMOC- solid phase methods. The sequences of indiviudal peptides were as follows:

```
preS (1-13)     M G G W S S K P R K G M G Y
     (20-34)    N P L G F F P D H Q L D P A F Y
     (92-107)   P P P A S T N R Q S G R Q P T P Y
     (126-140)  A F H Q T L Q D P R V R G L Y

   S (1-10)     M E N I T S G F L G Y
     (35-50)    W W T S L N F L G G S P V C L G Y
     (47-52)    V C L G Q N
     (124-137)  C T T P A Q G N S M F P S C
     (139-147)  C T K P T D G N C
     (217-226)  P I F F C L W V Y I

   C (40-49)    E A L E S P E H C S Y
     (75-84)    N L E D P A S R D L Y
     (132-147)  Y R P P N A P I L S T L P G T T G
```

All peptides were purified by either Sephadex G-25 gel chromatography (eluent: 10% acetic acid) or Sephadex Sp-25 ion exchange chromatography (eluent: pyridine-formic acid pH 2.8 - 5.0 buffer system). The purity of each peptide was assessed by HPLC and the total amino acid content confirmed after acid hydrolysis.

Antibodies.

Human anti-HBs immunoglobulin (HBIg) was obtained from the Blood Products Laboratory, Elstree, UK. Rabbit anti-HBs antibodies were produced by three consecutive immunizations of either plasma-derived or recombinant-derived hepatitis B vaccine (Merck, Sharp and Dohme, Rahway, USA). Each animal was immunized intramuscularly three times with 10ug at 7 to 10 day intervals and animals bled 9 to 10 days after the third dose.

Assays.

Measurements of antibody affinity were carried out as described by Brown et al (3) using a double isotope radioimmunoassay. Each peptide antigen was radiolabelled with ^{125}Iodine prior to use. ELISA procedures were carried out by direct binding of peptide antigen to the solid-phase (30ug/ml peptide in 0.05M carbonate/bicarbonate buffer, pH 9, 50ul per well). Unabsorbed sites were blocked using 0.5% gelatin in PBS. Antisera were titrated in PBS and 0.05% Tween 20 and bound peptide:antibody complexes revealed using the appropriate anti-species antibody conjugates to horse radish peroxidase. O-phenylenediamine was used as a substrate and O.D. values read at 490-492 nm.

RESULTS

Predicted Secondary Structure of Hepatitis B Envelope Proteins.

The major S protein of HBV consists of 226 amino acids and bears B-cell determinants (HBsAg) important for the induction of a protective immune response in man. Analysis of average hydrophilicity indicates that this is an essentially hydrophobic molecule with few areas of hydrophilicity which may represent putative antigenic sites. One approach to the identification of important antigenic epitopes involves the measurement of the affinity of antibodies elicited to the native HBsAg molecule for synthetic peptide analogues of this region (4). This work has shown that a peptide analogue of residues 139-147 cyclised by formation of a disulphide bond between terminal cysteine residues more clearly resembles the conformation of an important a determinant on the virus surface (1). Other regions are also immunogenic and antigenic, as shown by ELISA titration of human antisera to plasma-derived HBsAg particles (containing both preS and S specificities) and rabbit antibodies to recombinant-derived HBsAg particles (S specificities only).

An approach for the future approach is the use of computer algorithms to predict more accurately the secondary structure of HBsAg and compare the results with other molecules whose three-dimensional structure is known from crystallography studies. Figure 1 illustrates the secondary structure elements of the major S protein as generated using Chou and Fasman parameters. The presence of eight helicies is predicted, two of which are on the external surface and one is internal. Helical structures A,B,C, and D are shown as in close association with lipid and separated by beta-sheet structures. These of the four regions that span the lipid bilayer are thought to associate to form a hydrophilic pore through the membrane. A single, complex carbohydrate side chain is attached at residue 146 in the glycosylated form of the S protein.

In contrast to the S protein, our preliminary data suggest that the HBV preS proteins, expressed contiguous with the S peptide, are devoid of alpha helices and are situated external to the lipid bilayer of the HBV. The high content of charged residues and the hydrophilic properties of these sequences also indicated that preS specificities are exposed and form determinants for immune recognition. In contrast to the S proteins, the complete absence of cysteine residues indicates these regions are less constrained in secondary structure. Carbohydrate side chains are at positions 15 and 123; folding of preS1 sequences may present glycosylation at position 123 on preS2 and mark preS2-specific determinants.

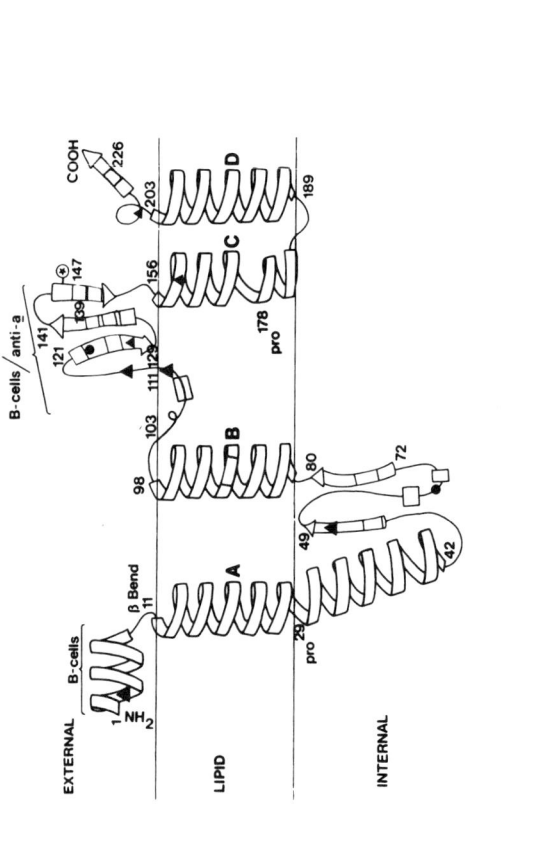

FIGURE 1. Predicted secondary structure of HBV S proteins. Helices A,B,C and D are shown in close association with lipid and separated by beta-sheet structures (broad arrows). A single, complex N-linked carbohydrate side chain is attached at residue 146 on some molecules.

Fine Analysis of anti-HBs antibodies using ELISA.

A total of 13 peptides have been examined by ELISA to determine if human anti-HBs immune globulin and rabbit anti-HBs antibodies recognize these sequences (Table 1). Titration end-point analysis showed that the highest titre of antibodies in immune globulin was against residues 139-147 and 217-226, two regions predicted as being external to the HBV coat surface. Only a low titre of antibody was observed to residues 1-10 at the amino terminus of S, a region predicted with low probability as being alpha-helical in structure. Among the preS peptides, highest antibody levels were observed against residues 126-140, a highly immunodominant region within preS2 (4, S.-H Chen, unpublished observations). No reactivity against preS peptides was observed in the case of the rabbit antisera raised against a recombinant source of HBsAg devoid of preS sequences. It is noteworthy that the titre of human antibodies to the S peptides 139-147 and 217-226 were substantially higher than that against the native molecule, suggesting antibody to these regions has a higher affinity compared to the equivalent native sequences.

Similar competitive inhibition curves were obtained for the S peptide analogues using plasma-derived HBsAg particles as bound antigen in the ELISA assay, suggesting that the determinants mimicked by these peptides are similarly exposed on both preparations.

Analysis of the Group a Antigen of HBsAg.

The major a determinants of HBsAg particles induce a protective antibody response. We have shown previously that S peptides representing residues 124-137 and 139-147 respectively bind a major proportion of anti-a antibodies present in convalescent patients and immunized individuals (5).

TABLE 1
TITRATION END-POINTS OF HUMAN AND RABBIT ANTI-HBS ANTIBODIES BY ELISA USING BOUND SYNTHETIC PEPTIDE ANALOGUES

Peptide	Antibodies Human[1]	Rabbit[2]
pre S (1 - 13)	40[3]	<10
(20 - 34)	160	<10
(92 - 107)	320	<10
(126 - 140)	1280	<10
S (1 - 10)	40	40
(35 - 50)	640	160
(47 - 52)	640	160
(124 - 137)	40	20
(139 - 147)	5120	160
(217 - 226)	2560	80
C (40 - 49)	640	<10
(75 - 84)	640	<10
(132 - 147)	640	<10
native HBsAg	1280	320
native HBcAg	2560	10

[1] Human immunoglobulin (HBIg)

[2] Rabbit anti-HBs serum

[3] Reciprocal of end-point titre

There is extensive sequence variation in the region 124 to 137 both between HBV serotypes and closely related hepadnaviruses, eg. the Woodchuck hepatitis virus (WHV). Despite there being only 5 of 14 residues in common between HRV and WHV in this region, anti-WHV antibodies bind the HBV sequence equally well (6). In contrast, 7 of 9 residues are common between residues 139 to 147. However, approximately a 1 log drop in affinity constant was observed in the binding of anti-WHV antibodies to the HBV-related sequence. These changes are T - L (position 140) and D - A (position 144). A modified peptide sequence (K - V, position 141; D - A, position 144) was used to confirm the importance of these two residues (figure 3).

DISCUSSION

Hepatitis B immune globulin (HBIg) contains antibodies capable of either neutralizing HBV infectivity, or attenuating the extent of infection, or both. The presence of antibodies to the S gene product is protective, particularly if directed to the a determinants (7). However, there are indications from chimpanzee protection studies that antibodies directed to other viral-specific proteins, for example the preS region are also protective (8). We show in this study that HBIg contains a broad spectrum of reactivity with synthetic peptide analogues of both S and preS sequences, and these structures correlate with proposed secondary structure elements on the surface of the HBV envelope. These antibodies were measured by ELISA and confirm previous affinity measurement data which indicated the importance of determinants between residues 139 to 147 (1 -3). In addition, the immunogenic nature of the carboxyl end of the S protein is indicated by these studies. The significance of this region in eliciting protective immunity is at present unknown.

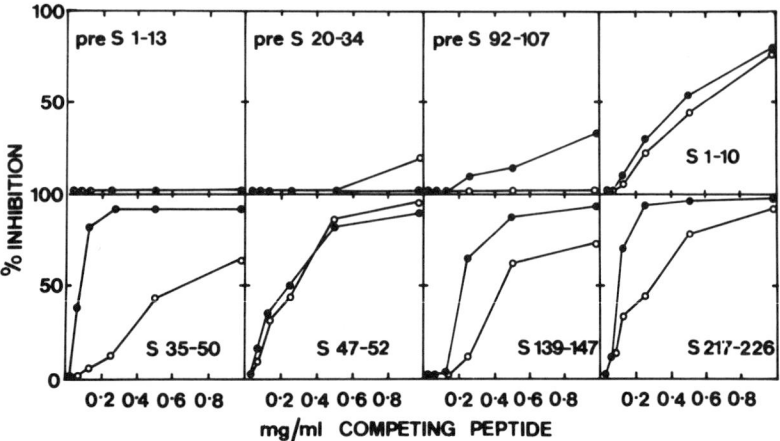

FIGURE 2. Competitive inhibition by preS and S peptides of rabbit (o-o) and human (●-●) antibodies to HBsAg particles.

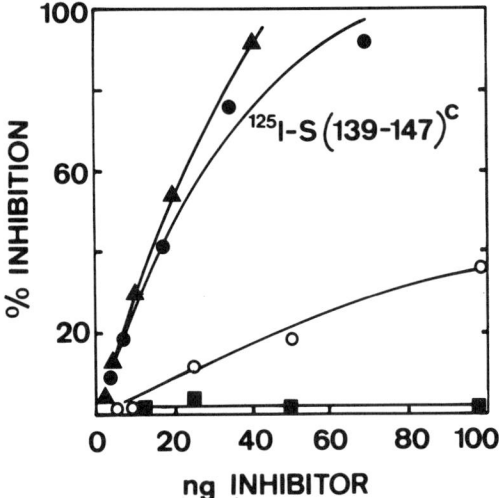

FIGURE 3. Inhibition of human anti-HBs antibody binding to ^{125}I S139-147 peptide in cyclized form by native HBsAg(▲-▲), homologous peptide (●-●), homologous peptide with substitutions at residues 141 & 144(o-o), and preS126-140(■-■).

This study shows that antibodies to both pre-S1 and pre-S2 sequences are to be found in patients long after recovery from acute infection. Antibodies recognizing pre-S1 peptides 26-34 and 92-107 appear during acute hepatitis B (unpublished observations), suggesting that these specificities are not very effective in limiting spread of infection and aiding recovery. There remains the possibility, however, that passive administration of pre-S antibodies can still prevent of modify infection if present sufficiently early in the course of infection and at high titre.

Several residues in the region of the a determinant in the region of 139-147 appear essential for reactivity. Previous studies have shown the importance of the lysine residue at position 141 (9) and in our studies substitutions at either positions 140 or 141 decrease the binding of antibody to this peptide.

REFERENCES

1. Brown SE, Howard CR, Zuckerman AJ, Steward MW (1984). Affinity of antibody responses in man to hepatitis B vaccine determined with synthetic peptides. Lancet ii:184.
2. Brown SE, Stanley C, Howard CR, Zuckerman AJ, Steward MW (1986). Antibody responses to re-combinant and plasma derived hepatitis B vaccines. Brit. med. J. 292:159.
3. Brown SE, Howard CR, Zuckerman AJ, Steward MW (1984). Determination of the affinity of antibodies to hepatitis B surface antigen in human sera. J. Immunol. Methods 72:41.
4. Milich DR, Thornton,GB, Neurath AR, Kent SB, Michel M-L, Chisari F (1985). Enhanced immunogenicity of the pre-S region of hepatitis B surface antigen. Science 228:1195.
5. Brown SE, Howard CR, Zuckerman AJ, Steward MW (1984). The use of synthetic peptides for analysing the specificity and affinity of anti- HBs antibodies in human sera. Ann. Sclavo 2:177.

6. Howard CR, Stirk, HJ, Brown SE, Steward MW (1987). Towards the development of synthetic hepatitis B vaccines. In Zuckerman AJ (ed): "Viral Hepatitis and Liver Disease", New York: Alan R. Liss, in press.
7. Szmuness W, Stevens CE, Harley EJ, Zang EA, Alter HJ, Taylor PE, DeVera A, Chen GTS, Kellner A (1982). Hepatitis B vaccine in medical staff of hemodialysis units - efficacy and subtype cross-protection. New Eng. J. Med. 307:833.
8. Itoh Y, Takai E, Ohnuda H, Kitajima K, Tsuda F, Machida A, Mishiro S, Nakamura T, Miyakawa Y, Mayumi M (1986).A synthetic peptide vaccine involving the product of the pre-S(2) region of hepatitis B virus DNA: protective efficacy in chimpanzees. Proc. Natl. Acad. Sci, USA, 83:9174.

ANALYSIS OF TOPOGRAPHIC ANTIGENIC DETERMINANTS ON HUMAN CHORIOGONADOTROPIN USING SYNTHETIC PEPTIDES

Jean-Michel BIDART, Frédéric TROALEN, Claude BOHUON and Dominique H. BELLET

Département de Biologie Clinique, Institut Gustave-Roussy 94805 Villejuif- France

ABSTRACT Two discontinuous antigenic determinants associated to either the α/β dimer or the free β subunit of human choriogonadotropin were investigated at the primary structure level using synthetic peptides and monoclonal anti-peptide antibodies. This approach provides new insight into the quaternary structure of the glycoprotein hormones, which remains an unsolved problem.

INTRODUCTION

Human choriogonadotropin (hCG) belongs to the glycoprotein hormone family which includes the pituitary hormones lutropin (hLH), follitropin, and thyrotropin (1). Each hormone consists of two polypeptide chains, namely the α and β subunits, held in association by noncovalent forces. Within a species, the α sequence is essentially identical for each hormone, whereas the hormone-specific β subunit possesses regions that differ in sequence. Both subunits are glycosylated at specific residues and are extensively cross-linked internally by disulfide bonds. The three-dimensional structure of either the subunits or the intact dimers remains yet unsolved. Gordon and Ward (2) provided an update on functionnal group studies which have been used to probe both the structural and topographical features of the hormones interacting with the lutropin receptor. Recently bovine TSHα has been crystallized (3) but structural data are not available. Using the technique of computer modeling, different groups attempted to model the structure of the α and β subunits of hCG (4,5). Thus, it is obvious that more work needs to be done to approach the three-dimensional structure of the glycoprotein hormones.

Immunochemical methods have been successfully applied to the determination of local conformation of several proteins (6,7). The study of antibody-binding sites provides a unique way of probing surface structure of proteins as the recognized epitopes are externally located externally on the surface of the molecule (8). Epitopes on proteins have been classified as continuous and discontinuous sites (9). Continuous antigenic determinants comprised amino acid residues that are close to each other on the primary structure, whereas discontinuous or topographical sites consist of residues far apart in the polypeptide sequence but brought together on the surface by the folding of the protein. The ability of synthetic peptides representative of different parts of a protein to elicit antibodies that cross-react with the native molecule constitutes also a new approach in the determination of the structure of protein (7,10).

In this article, attemps to study the topographical antigenic determinants recognized by monoclonal antibododies raised against the native hCG and its β subunit are described. Several approaches were developped including: 1) The synthesis of numerous peptides mimicking various portions of the α and β subunits, 2) The preparation of a chemically modified β subunit and of various α subunits from different species, and 3) The production of monoclonal antibodies to either native molecules or synthetic peptides conjugated to carrier proteins. Two antigenic determinants were particularly studied : one is located on the free β subunit (not present on the α/β dimer) and is defined by the monoclonal antibody FBT11. The second epitope, recognized by the monoclonal antibody C8, is specific for the hCG molecule, i.e. only present on the α/β dimer.

METHODS

Synthetic peptides and peptide-carrier conjugates

The peptides were synthesized by conventional solid phase peptide synthesis in an Applied Biosystems Model 430A apparatus (11) (Table 1a and 1b).

TABLE 1a
MAIN SYNTHETIC PEPTIDES MIMICKING hCGα PORTIONS

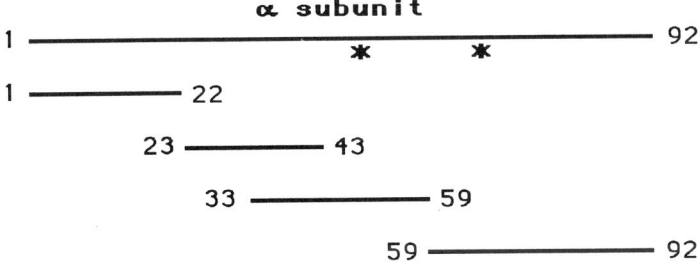

(*) Carbohydrate moieties

The purity and identity of each peptide were checked by: 1) amino acid analysis, 2) high-performance liquid chromatography, 3) fast-atom bombardment mass spectrometry, and 4) peptide sequencing. Twenty-five synthetic peptides mimicking different portions of the α and β subunit of hCG were prepared and the main peptides are listed under tables 1a and 1b. The peptide-carrier conjugates were obtained with tetanus toxoid (TT) and glutaraldehyde as the coupling agent.

TABLE 1b
MAIN SYNTHETIC PEPTIDES MIMICKING hCGβ PORTIONS

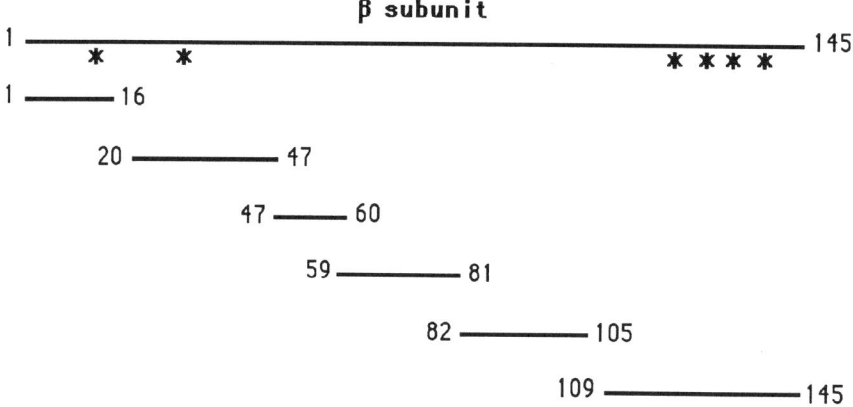

(*) : Carbohydrate moities

Preparation of the chemically modified β subunit, of the various α subunits and of monoclonal antibodies.

hCG β minus the carboxyl-terminal 33-residue peptide (hCG β core) was obtained by the cleavage of the Asp^{112}-Pro^{113} bond using a 30 min incubation in 0.03 N HCl at 110°C (12). The resulting peptides were purified by Sephacryl S-200 chromatography and by reverse phase HPLC. Glycoprotein α subunits were prepared as previously reported (13). Purity and identity of the polypeptides were assessed by sodium dodecyl sulfate-polyacrylamide gel electrophoresis and amino acid analysis.

Production of monoclonal antibodies against native hCG, βhCG or the synthetic peptide analogous to the 109-145 carboxyl-terminal part of βhCG (CTP) and conjugated to tetanus toxoid, cell fusions screening assays and purification of the monoclonal antibodies were as previously described (14,15). The characteristics of the monoclonal antibodies depicted in the present report are shown in table 2.

TABLE 2
CHARACTERISTICS OF THE MONOCLONAL ANTIBODIES

Monoclonal Antibody	Immunogen	Epitope Specificity	Molecules Recognized
C8	hCG	α/β	hCG
HT 13	hCG	α, α/β	hCG α, hCG hLH α, hLH hFSH α, hFSH hTSH α, hTSH
FBT 10	hCG β	β, α/β	hCG β, hCG hLH β, hLH
FBT 11	hCG β	β	hCG β
FB 12	CTP-TT	β(110-116)	hCG β, hCG

Current methods for epitope identification

A variety of immunoassays have been first employed to examine the fine specificities of the monoclonal antibodies. Competitive inhibition assays were carried out by incubating a defined dilution of monoclonal antibody with the radiolabeled antigen (either ^{125}I-hCG-β or ^{125}I-hCG) in the presence of increasing concentrations of either unlabeled competitive hormones, intact or modified subunits or synthetic peptides (16). Dose-response curves generated by the radioimmunoassays (RIAs) provided a half-maximal inhibitory dose for each molecule tested. The specificities of the monoclonal anti-peptide antibodies were also tested using inhibition experiments in enzyme-linked immunosorbent assays (ELISAs) (15,16). Two-site "sandwich" monoclonal immunoradiometric assays (M-IRMAs) were established to approach the localization of the antigenic determinants recognized by monoclonal antibodies C8 and FBT11. They were developed as "simultaneous" (one-step) or "forward" (two steps) M-IRMAs as described previously (17). Briefly, these assays are based on an antibody-coated solid phase support which is used to "capture" antibody-reactive molecules and a ^{125}I-labeled monoclonal antibody to detect molecules bound on the antibody-coated beads.

We analyzed the recombination of native α and β subunits to test 1) The availability of the antigenic determinant recognized by the specific α/β antibody C8 on various hybrids molecules and 2) The influence of the binding of antibody FBT11 to its epitope on the formation of α/β dimer. Reassociation experiments between α and β subunits were performed as depicted by Garnier et al (18). The finals products were assayed for their binding activity by either an M-IRMA using C8 antibody-coated phase support and ^{125}I-labeled FBT10 or a radiohormone receptor assay.

A third approach was to checked the accessibility of the antigenic determinants on the hCG-receptor complex. To this end, purified monoclonal antibodies were iodinated to a specific activity of 12-15 μCi/μg and utilized to detect the epitopes which remained accessible on hCG after its binding to the CG/LH receptor of porcine testicular homogenates. Experiments included a first incubation of hCG (500 IU/ml) in the presence of the homogenates(2mg/ml) for 1 h at 37°C, followed by extensive washing. Radiolabeled antibodies were then added for a second 18 h incubation at 4°C. After extensive washing, the precipitates were counted.

RESULTS

Characterization of an antigenic determinant specifically associated to the hCG dimer.

This epitope is defined by a monoclonal anti-hCG antibody designated as C8 which binds only to hCG and does not cross-react with either the free α and β subunits as assessed by conventional RIAs. Further, the antibody C8 does not bind to homologous hormones such as hLH, hFSH or hTSH suggesting that specific residues of the β subunit are involved in the epitope. We investigated the use of synthetic peptides in the delineation of the boundaries of the antibody-binding site. Figure 1 shows the competitive inhibition assays with various peptides of the β subunit. It is striking that the synthetic peptide 109-122 was the only one capable of inhibiting the binding of ^{125}I-hCG to antibody C8. Different subpeptides corresponding to this portion were prepared by recurrent synthesis and revealed that subpeptides shorter than the 113-122 portion were unable to compete. The essential role of the 111-113 portion in participating to the epitope recognized by C8 was subtantiated by the study of the simultaneous binding of both C8 and an anti-peptide antibody directed to the hCGβ 110-116 portion (19,20).

Two-site immunoradiometric assay of hCG based on the combination of anti-protein C8 and anti-peptide FB12 antibodies revealed that the two antibodies do not bind simultaneously to the molecule. Such results suggest that the corresponding epitopes are in close proximity to each other and induce a steric hindrance.

Using competitive inhibition assays with peptides spanning various regions of the α subunit, we attempted to delineate the antigenic portions recognized by antibody C8 on the α subunit but no inhibition was observed with any of the six peptides tested. Recombination experiments were performed by incubating the hCGβ subunit with α subunits of different species (human, equine, porcine) and the resulting hybrids were tested by an M-IRMA based on C8 as capture antibody and ^{125}I-FBT10 as the indicator probe of the hCG-β structure. Results shown in Figure 2 indicate that C8 is capable of binding to the various hybrids composed of the hCGβ subunit and homologous α subunits. Further, we were able to recombine the hCGα subunit with the hCGβ core (missing the 113-145 portion). This deleted hybrid was recognized by an m-IRMA based on the antibodies C8 and FBT10 suggesting the crucial role of the 111-113 sequence in the antibody-binding site of C8. Finally, we determined that the antigenic region is partially accessible on the hCG-receptor complex (21).

Characterization of an antigenic determinant specifically associated to the free beta subunit of hCG.

We have produced a monoclonal antibody (FBT11) that binds to the free hCGβ subunit but not to hCG, hLH and hLHβ (20). The corresponding antigenic determinant was localized to the hCGβ core (region

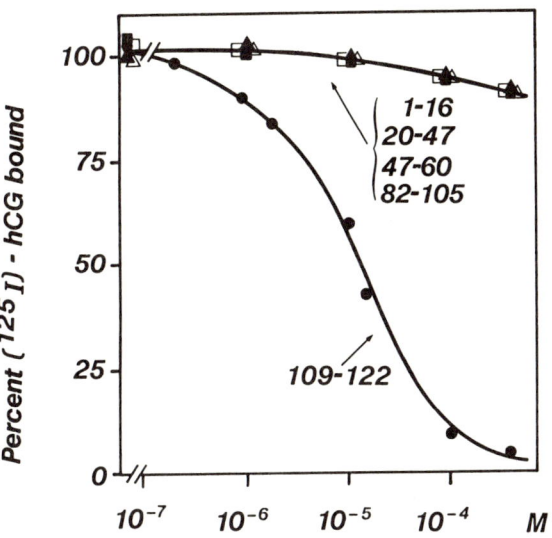

Figure 1. Inhibition of C8 binding by synthetic peptides (reprinted from reference 21).

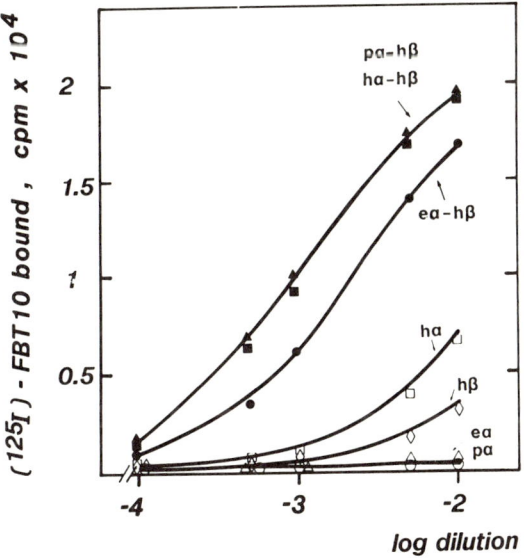

Figure 2. Recombination experiments of hCGβ with the α subunits from different species (reprinted from reference 21).

1-112) and a synthetic peptide mimicking the (1-7) NH_2-terminal portion was found to be active in competitive inhibition assays (Figure 3). We also found that FBT11 inhibits the recombination of hCGβ with the α subunit suggesting that the 1-7 portion of hCGβ is not accessible for antibody binding on hCG (22). In contrast, the monoclonal antibody FBT10 recognizing an epitope present on both the free and combined β subunit was not inhibited by the synthetic peptides. Moreover its combining site appeared to be located on a portion of the hCGβ present at the surface of the hCG-receptor complex.

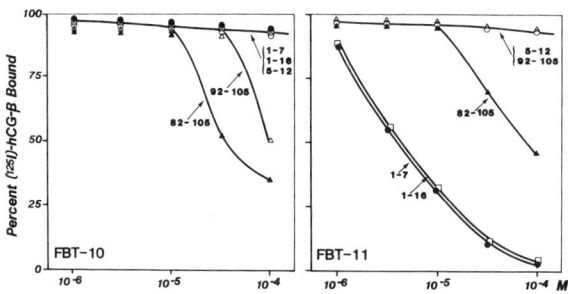

Figure 3. Inhibition of FBT10 and FBT11 binding by synthetic peptides (reprinted from reference 22).

DISCUSSION

Linear peptides representing short segments of polypeptide chains have been successfully utilized to identify epitopes recognized by antibodies directed against proteins (23,24). However, there are limitations to this approach; antibodies directed to protein in its native conformation preferentially recognize that form, and a smaller fraction of them may recognize a linear array of the primary structure in the epitope (9). Even though certain antibodies react with synthetic peptides, they exhibit a higher binding affinity for the native protein than for the cross-reacting peptides. In spite of these limitations, synthetic peptides are potent tools to study interactions between macromolecules and, in the case of gonadotropins, they have been successfully used in structure-activity studies of FSH (25) and hCG (26).

Hitherto, the location on the primary structure of antigenic determinants on hCG and its subunits has not been elucidated although polyclonal and monoclonal antibodies are widely described in the litterature (27). Using a library of monoclonal antibodies directed to hCG, its α or β subunit, we attempted to localize some topographic epitopes. Using a combination of different approaches, two portions of residue participating to an epitope specifically associated to the hCG dimer were delineated on the primary structure of both α and β subunits: The portion defined on the β subunit included the $Asp^{111}-Asp^{112}-Pro^{113}$ residues whereas residues of αhCG appeared to be located on a highly conserved portion of the α subunit. Moreover we were able to delineate partially an antigenic region present only on the free hCGβ subunit: This region included the N-terminal portion of the molecule. All together, our results constitute a new approach in the evaluation of the three-dimensional structure of the gonadotropins.

Such informations are necessary for a full understanding of how the hormones interact with their receptors. In addition, the precise location of antigenic determinants on hCG is particularly important for the production of antibodies possessing protective on neutralizing activities against the hormone. Future directions in the immunological approaches to contraception require the identification of hCG-specific epitopes and the construction of appropriate synthetic vaccines based on a rational design.

ACKNOWLEDGEMENTS

The authors are grateful to the NIADKK, NIH and Dr. R. Canfield for their generous supply of hCG and its subunits, to Drs. GR Bousfield and DN Ward for their continuous help, and to Mrs C. Bombled and C. Potentini for their excellent technical assistance. Research in this laboratory has been continuously supported by Grants from the Association pour la Recherche sur le Cancer (ARC), Villejuif, France. JM. Bidart expresses appreciation to V. Martel who prepared and type-writed the present manuscript.

REFERENCES

1. Pierce JG, Parsons TF (1981). Glycoprotein hormones: structure and function. Ann Rev Biochem 50:465.
2. Gordon WL, Ward DN (1985). Structural aspects of luteinizing hormone actions. In Ascoli M (Ed): "Luteinizing hormone action and receptors", Boca Raton: CRC Press, p 173.
3. Mc Pherson A, Keszelale S, Axelrod H, Day J, Williams R, Robinson L, Mc Grath M, Cascia D (1986). An experiment regarding crystallization of soluble proteins in the presence of β-Octyl glucoside. J Biol Chem 261:1969.
4. Ryan RJ, Keutmann HT, Charlesworth MC, Mc Cormick DJ, Milius RP, Calvo FO, Vutyavanich T (1987). Structure-function relationships of gonadotropins. Recent Prog Horm Res 43:383.
5. Salesse R, Bidart JM, Troalen F, Garnier J, Biou V, Gibrat JF, Bellet D, Genty N, Bohuon C

(1988). Mapping of subunit interaction and receptor binding regions of LH and hCG with synthetic peptides. Manuscript submitted to publication.
6. Benjamin DC, Berzofsky JA, East IJ, Gurd FRN, Hannum C, Leach SJ, Margoliash E, Michale JG, Miller A, Prager EM, Reichlin M, Sercars EE, Smith-Gill J, Tood PE, Wilson AC (1984). The antigenic structure of proteins: a reappraisal Ann Rev Immunol 2:67.
7. Lerner RA (1984). Antibodies of predetermined specificity in biology and medicine. Adv Immunol 36:1.
8. Amit AG, Mariuzza RA, Phillips SEV, Poljak RJ (1985). Three dimensional structure of an antigen-antibody complex at 6A resolution. Nature 313:156.
9. Jemmerson R, Paterson Y (1986). Mapping antigenic sites on proteins: implications for the design of synthetic vaccines. BioTechniques 4:18.
10. Geysen HM, Barteling SJ, Meloen RH (1985). Small peptides induce antibodies with a sequence and structural requirement for binding antigen comparable to antibodies raised against the native protein. Proc Natl Acad Sci USA 82:178
11. Merrifield RB (1963). Solid-phase peptide synthesis. 1. The synthesis of a tetrapeptide. J Am Chem Soc 85:2149.
12. Bousfield GR, Ward DN (1986). Biologic activity of eLH and hCG following removal of the C-terminal glycopeptide. Endocrinology 118:530.
13. Bousfield GR, Ward DN (1984). Purification of lutropin and follitropin in high yield from horse pituitary glands. J Biol Chem 259:1911.
14. Bellet D, Bidart JM, Jolivet M, Tartar A, Caillaud JM, Ozturk M, Strugo MC, Audibert F,

Gras-Masse H, Assicot M, Bohuon C (1984). A monoclonal antibody against a synthetic peptide is specific for the free native human chorionic gonadotropin βsubunit. Endocrinology 115:330.
15. Bidart JM, Ozturk M, Bellet D, Jolivet M, Gras-Masse H, Troalen F, Bohuon C, Wands JR (1985). Identification of epitopes associated with hCG and the βhCG carboxyl terminus by monoclonal antibodies produced against a synthetic peptide. J Immunol 134:457.
16. Bellet D, Ozturk M, Wands JR, Bidart JM, Assicot M, Troalen F, Bohuon C (1984b). Monoclonal antibodies antibodies directed toward unique epitopes on human chorionic gonadotropin: a new approach to the development of sensitive and specific radioimmunoassays. In Bizollon CA (Ed): "Monoclonal antibodies and new trends in immunoassays", Amsterdam: Elsevier, p43.
17. Ozturk M, Bellet D, Manil L, Hennen G, Frydman R, Wands J (1987). Physiological studies of human chorionic gonadotropin (hCG) αhCG, and βhCG as measured by specific monoclonal immunoradiometric assays. Endocrinology 120:549.
18. Garnier J, Salesse R, Pernollet JC (1974). Reversible folding of human chorionic gonadotropin at acid pH or upon recombination of the alpha and beta subunits. FEBS letter 45:106.
19. Bidart JM, Bellet DH, ALberici GF, Van Besien F, Bohuon C (1987). The immune response to a synthetic peptide analogous to the 109-145 βhCG carboxyl-terminus is directed against two major and two minor regions. Molec Immun 24:339.
20. Bellet D, Ozturk M, Bidart JM, Bohuon C, Wands JR (1986). Sensitive and specific assay for

human chorionic gonadotropin (hCG) based on anti-peptide and anti-hCG monoclonal antibodies: construction and clinical implications. J Clin Endocrinol Metab 63:1319.
21. Bidart JM, Troalen F, Bohuon CJ, Hennen G, Bellet DH (1987). Immunochemical mapping of a specific domain on human choriogonadotropin using anti-protein and anti-peptide monoclonal antibodies. J Biol Chem 262:15483.
22. Bidart JM, Troalen F, Salesse R, Bousfield GR, Bohuon CJ, Bellet DH (1987). Topographic antigenic determinants recognized by monoclonal antibodies on human choriogonadotropin beta-subunit. J Biol Chem 262:8551.
23. Atassi MZ (1984). Antigenic structures of proteins. Their determination has revealed important aspects of immune recognition and generated strategies for synthetic mimicking of protein binding sites. Eur J Biochem 145:1.
24. Paterson Y (1985). Delineation and conformational analysis of two synthetic peptide models of antigenic sites on rodent cytochrome C. Biochemistry 24:1048.
25. Schneyer AL, Sluss PM, Huston JS, Ridge RJ, Reichert LE (1988). Identification of a receptor binding region on the β subunit of human follicle-stimulating hormone. Biochemistry 27:666.
26. Charlesworth MC, McCormick DJ, Madden β, Ryan RJ (1987). Inhibition of human choriotropin binding to receptor by human choriotropin α α peptides. A comprehensive approach. J Biol Chem 262:13409.
27. Erlich PH, Moyle WR and Canfield R.E. (1985). Monoclonal antibodies to gonadotropin subunits. In Birbaumer L and O'Malley BW (Eds): "Methods in Enzymology, Peptide hormone" New York: Academic Press, p 638.

ANALYSIS OF THE FUNCTIONAL AND ANTIGENIC DOMAINS OF THE HERPES SIMPLEX VIRUS MAJOR REGULATORY PROTEIN α4.

Jeff Hubenthal-Voss[#], Richard A. Houghten[$], and Bernard Roizman[#]

[#]The Marjorie B. Kovler Viral Oncology Laboratories, The University of Chicago, Chicago IL 60637, [$]The Department of Molecular Biology, The Scrips Clinic and Research Foundation, San Diego CA 92037

INTRODUCTION

The expression of herpes simplex virus 1 (HSV-1) β and γ genes during productive infection requires functional α4 (1, 2, 3, 4, 5, 6, 7, 8, 9, 10, 11). The α4 protein is predicted to contain 1298 amino acids (12); its apparent molecular weight deduced from its electrophorectic mobility ranges from 163,000 to 171,000, depending on the nature and extent of post translational processing (13). In infected cells α4 is phosphorylated (14) on both serine and threonine residues (15) and can be poly ADP ribosylated in vitro (16). In denaturing one dimensional gels it forms at least 3 bands differing in electrophoretic mobility (13). In two dimensional separation systems, it forms at least 20 spots (17). In its native state it exists as a homodimer (18). Mapping studies of ts mutations in the α4 gene have suggested that the functional domains of the gene extend over a large portion of its coding domain (2, 1). More recent studies of deletion mutants have shown that the functions related to the transition from α genes to genes expressed later in infection map in the N terminal two thirds of the gene (19, 20).
 Recently, the α4 protein has been shown to associate stably and specifically with the promoter/regulatory domains of each kinetic class of HSV gene (21, 22, 23). The initial analyses of the functional domains of the α4 gene centered on the observation that in DNA band shift assays 2 of 10 monoclonal antibodies tested blocked the binding of the protein to its DNA binding site (22). An initial goal of these studies was to map the epitopes recognized by these antibodies and to determine if α4 proteins which contained these epitopes also contained the DNA binding activity associated with the

intact α4 protein. The 10 monoclonal antibodies mentioned above were tested for their capacity to bind truncated α4 proteins and these assays were ultimately extended to synthetic oligopeptide antigens. The results showed that all 10 monoclonal antibodies react with epitopes contained in the N terminal 288 amino acids of the protein and that the epitope of one of the blocking monoclonal antibodies maps near a cluster of oligopeptides that enhance the binding of α4 to DNA. In addition, one oligopeptide (No.19) inhibited the DNA binding activity of α4 and two, oligopeptides No.19 and 27 have an intrinsic DNA binding activity. The results presented here also establish that α4 protein is competent to bind to DNA in the absence of other viral proteins, and that the C-terminal 450 amino acids of the α4 protein are not required for DNA binding activity.

RESULTS

10 monoclonal antibodies to the α4 protein react with the N terminal domain of the α4 protein. In the initial series of experiments 5 plasmids, one with an intact α4 gene and four with deletions in this gene were constructed as shown diagrammatically in Figure 1. The results of western blot analysis of lysates BHK cells transfected with three of these constructs and then reacted with each member of a panel of 9 monoclonal antibodies is shown in figure 1. One monoclonal antibody (H1091) was tested separately (data not shown). The significant finding was that each of the truncated α4 genes expressed polypeptides which reacted with all of the monoclonal antibodies. The smallest truncated gene cloned as pRB3824 formed several bands possibly reflecting either premature termination of translation or degradation. The results indicate that the epitopes recognized by all of the monoclonal antibodies are located within the N- terminal 288 amino acids of the protein.

Mapping of the epitopes recognized by the monoclonal antibodies to the α4 protein. To map the epitopes recognized by the monoclonal antibodies, 29 oligopeptides covering the 288 amino acid α4 polypeptide specified by the truncated gene contained in pRB3824 were synthesized. Each oligopeptide was prepared as described by Houghten (24), contained 15 amino acids and overlapped its adjacent peptides by 5 amino acids, as shown in Figure 3B and 3C. The reaction of each monoclonal antibody with the 29 oligopeptides was assayed by enzyme linked immunosorbant assay (ELISA). The results of these
(text continued on page 242)

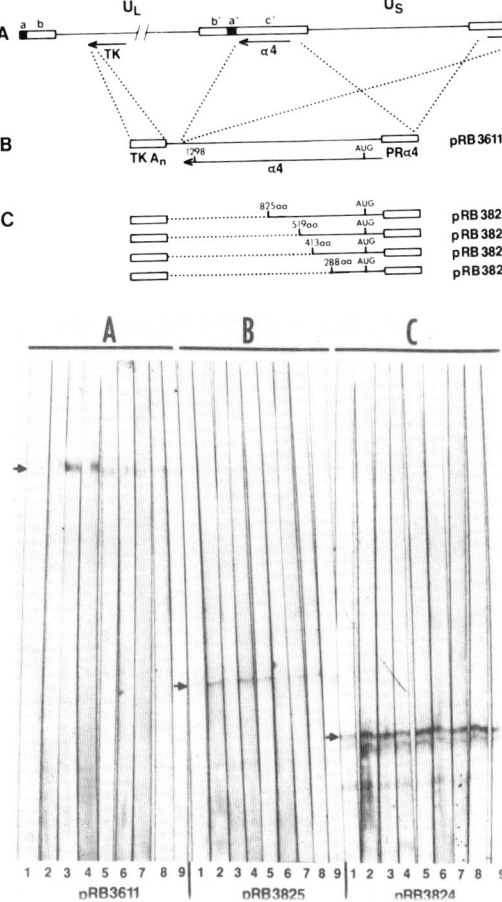

Figure 1. Sequence map and location of deletion end points used in this study and the reactivity of full length and truncated α4 proteins with monoclonal antibodies. Top panel: Section A illustrates the location of both copies of the α4 gene and the single copy of the TK gene in the prototype arrangement of the HSV-1 genome. Section B is a map of pRB3611 which contains the entire α4 gene from HSV-1 (F) extending 3' from nucleotide -330 relative to the transcription initiation site to approximately 130 base pairs (bp) downstream from its polyadenylation site. The plasmid also contains the polyadenylation sequences of the TK gene, engineered to contain stop codons in all possible reading frames and inserted in the correct transcriptional orientation 3' to the α4 gene. Section C summarizes the contents of the deletion plasmids used in this study. The end point of each deletion is given in amino acids (aa) from the initiating methionine of the α4 gene at the end of the solid bar on each map. Deletion derivatives of pRB3611 were constructed as described elsewhere (25). Each deletion results in the fusion of an internal portion of the α4 gene to a stop codon followed by a polyadenylation signal. Bottom panel: Panels A, B, and C contain wetern blots of nuclear proteins isolated from cells transfected with the plasmids pRB3611, pRB3825 and pRB3824 respectively. DNA transfections were done as described by elsewhere (25). In each case nuclear proteins were separated
(legend continued on following page)

by PAGE and transfered to nitrocellulose as described elsewhere (26, 21, 22). The nitrocellulose sheets were cut into strips and reacted individually with monoclonal antibodies designated 1-9 at the bottom of each strip. The designations 1-9 refer to monoclonal antibodies H640, H953, H948, H942, H949, H950, H924, and H944 respectively.

---

assays, shown in Figures 2 and 3, were as follows: (i) Only 1 of 10 monoclonal antibodies tested, H640, failed to react with any of the oligopeptides. (ii) Two monoclonal antibodies, H950 and H953 reacted with oligopeptide No. 3. (iii) Monoclonal antibodies H943, H944, and H948 each reacted with oligopeptide No.13. (iv) H924 reacted with both peptides 5 and 6, suggesting that the epitope contains the shared amino acids. (v) Monoclonal antibody H942 reacted with oligopeptide No. 25. (vi) Hydropathic analyses (26) of the N-terminal stretch of 288 amino acids (Figure 3) reveals that oligopeptides containing epitopes are above below average hydropathicity and range in hydropathic index from between about -3 to about -1.

The C-terminus of α4 is dispensable for DNA binding activity. In this series of experiments, nuclear extracts from transfected cells were reacted for 30 minutes with 40,000 cpm (0.2 ng) of 5' end labeled α0 promoter probe and subjected to electrophoresis on a 4% polyacrylamide gel as described elsewhere (21, 22, 25). Previous studies have shown that the α4 protein specifically forms complexes with a specific sequence contained in this probe DNA (21). The results of the DNA binding assay are shown in Figure 4. Lanes 1 and 2 each contain an α4 protein-DNA complex (arrow) formed by the full length (from pRB3611) or truncated (from pRB3827) α4 proteins respectively. Lanes 3 and 4, contain the products of reactions that included nuclear extracts derived from cells transfected with plasmids pRB3826 or pRB3825. These do not contain complexes that were not present in extracts made from untransfected cells (lane 5).

Synthetic peptides can affect the DNA binding activity of α4. The purpose of this series of experiments was to determine whether the synthetic peptides affect the binding of α4 to its cognate sites in α0 promoter fragment. The labeled DNA probe used in this assay consisted of the α4 binding site in the promoter-regulatory domain of the α0 gene described in Kristie and Roizman (21). The standard conditions for all DNA binding reactions that included synthetic oligopeptides are given in the legend to Figure 5. Nuclear extracts of infected Hela

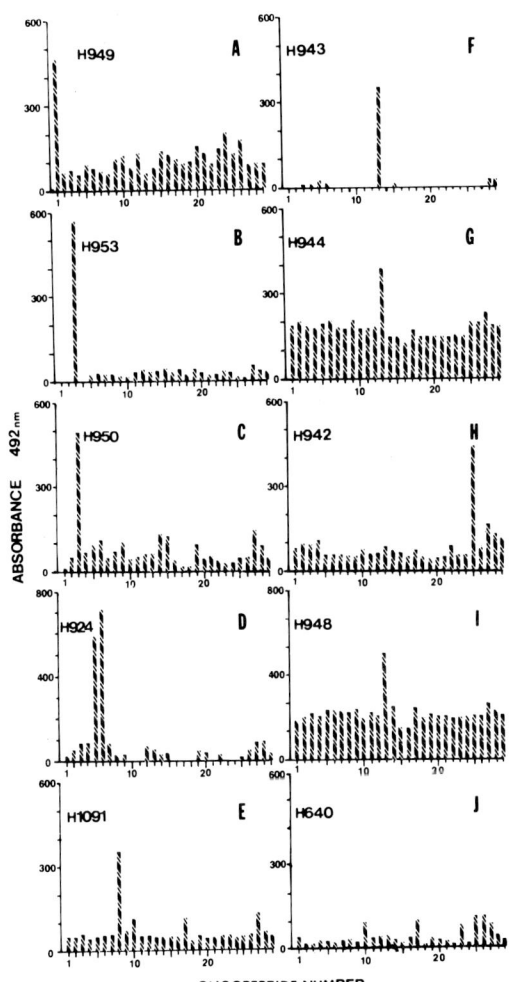

Figure 2.
Reactivity of monoclonal antibodies with synthetic polypeptides in ELISA assays. Panels A-J show the results of ELISA assays of the oligopeptides shown in figure 3 with a panel of 10 monoclonal antibodies known to react with the α4 protein. Each panel shows the reactivity of a monoclonal antibody with each of the oligopeptides, as measured in absorbance units at 492 nm, in an immunoperoxidase reaction using o-phenylenediamine as a substrate.

cells served as a source of the α4 protein. The results were as follows: (i) Initial experiments showed that only 5 oligopeptides, Nos. 17, 19, 26, 27, 28, affected the binding of α4 protein to the α0 DNA probe or bound directly to the DNA in reaction mixtures containing 5 mM concentrations of individual oligopetides (data not shown). (ii) Oligopeptide No. 17 reproducibly appeared to precipitate the labeled DNA probe into complexes that failed to enter the polyacrylamide gel (data not shown). (iii) Oligopeptide No. 19 blocked the binding of α4 to the labeled α0 DNA probe (Figure 5A lanes 6-9). The inhibition was oligopeptide concentration dependent

(text continued on page 245)

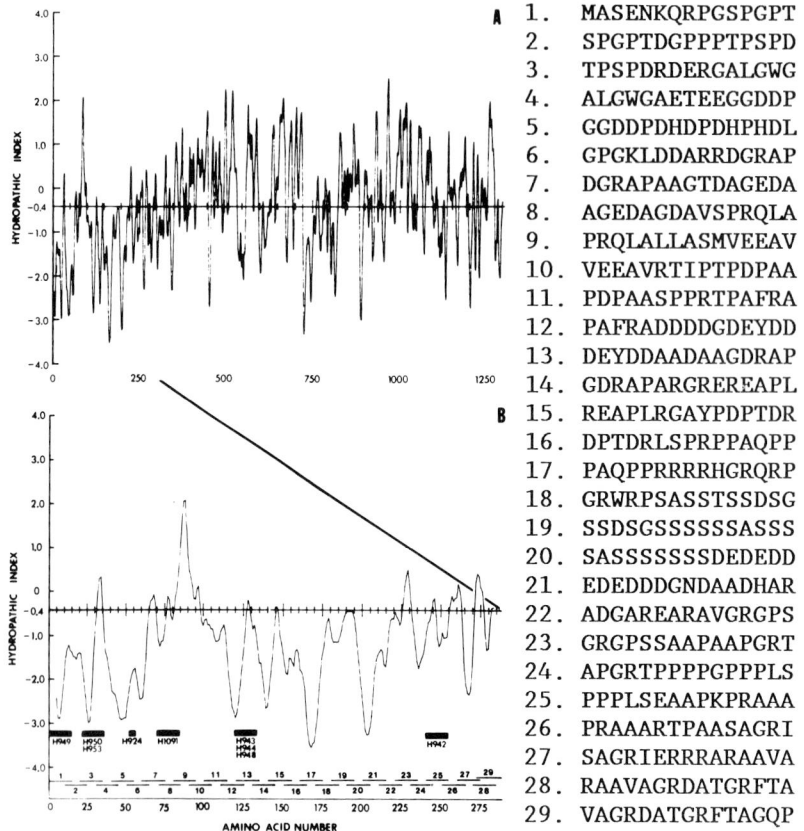

1. MASENKQRPGSPGPT
2. SPGPTDGPPPTPSPD
3. TPSPDRDERGALGWG
4. ALGWGAETEEGGDDP
5. GGDDPDHDPDHPHDL
6. GPGKLDDARRDGRAP
7. DGRAPAAGTDAGEDA
8. AGEDAGDAVSPRQLA
9. PRQLALLASMVEEAV
10. VEEAVRTIPTPDPAA
11. PDPAASPPRTPAFRA
12. PAFRADDDDGDEYDD
13. DEYDDAADAAGDRAP
14. GDRAPARGREREAPL
15. REAPLRGAYPDPTDR
16. DPTDRLSPRPPAQPP
17. PAQPPRRRRHGRQRP
18. GRWRPSASSTSSDSG
19. SSDSGSSSSSSASSS
20. SASSSSSSDEDEDD
21. EDEDDDGNDAADHAR
22. ADGAREARAVGRGPS
23. GRGPSSAAPAAPGRT
24. APGRTPPPPGPPPLS
25. PPPLSEAAPKPRAAA
26. PRAAARTPAASAGRI
27. SAGRIERRRARAAVA
28. RAAVAGRDATGRFTA
29. VAGRDATGRFTAGQP

Figure 3. Hydropathic analysis of the predicted α4 amino acid sequence and of the positions of the antigenic determinants recognized by the monoclonal antibodies used in this study. Panel A contains a plot of the hydropathic character of the entire α4 amino acid sequence as determined by the method of Kyte and Doolittle (26). Panel B contains an expansion of Panel A with particular reference to the N-terminal portion of the α4 protein. Both panels A and B show the average hydropathicity values of incremental 7 amino acid segments of the α4 protein plotted as a function of their position in the α4 coding sequence. Panel B also contains the postions of the sythetic oligopeptides used in this study (thin bars) and their corresponding numerical designations. The amino acid compositions of the peptides used in this study and their numerical designations are given in figure 3. Also shown in panel B and represented by thick bars are the locations of the

positioned over the oligopeptide to which a monoclonal antibody bound and below each bar is a list of the antibody or antibodies that react with that oligopeptide. Panel C contains a list of the oligopeptides used in this study in single letter code.

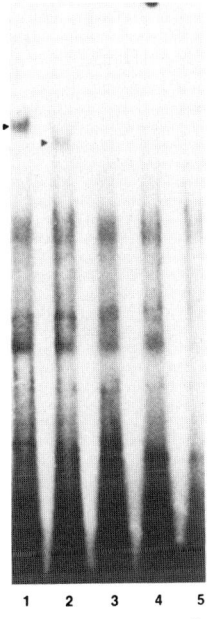

Figure 4. Autoradiographic images of [^{32}P]-labeled DNA protein complexes electrophoretically separated in non denaturing polyacrylamide gels. Nuclear extracts from BHK cells transfected with plasmids containing full length and C terminal truncated α4 genes were reacted with a DNA fragment containing the α4 protein recognition site of the promoter sequence of the α0 gene. Lanes 1-5: electrophoretically separated DNA-protein complexes formed by nuclear extracts prepared from BHK cells transfected with the plasmids pRB3611, pRB3827, pRB3826, pRB3825 and mock transfected cells, respectively. Arrows in lanes 1 and 2 indicate the positions of the α4 protein DNA complexes.

and reduced the binding of α4 protein to the probe DNA by approximately 50% at the lowest concentration tested.
(iv) The partially overlapping oligopeptides Nos. 26, 27 and 28 enhanced the binding of α4 to the labeled α0 DNA probe (Figure 5B). The enhancement was most pronounced in the case of oligopeptide No. 27 and least pronounced in the case of oligopeptide No. 28. In each instance, the extent of binding enhancement was oligopeptide concentration dependent.

Peptides Nos. 19 and 27 bind to the labeled α0 DNA probe. The ability of the oligopeptides 19 and 27 to either inhibit or enhance the binding of the α4 protein to the α0 DNA probe raised the possibility that the synthetic oligopeptides were able to bind to the DNA probe. To test this hypothesis, each of the oligopeptides, at a concentration of 5 mM was mixed with 0.1 ng of the labeled α0 DNA probe in binding reaction buffer containing increasing amounts of sonicated salmon sperm DNA. The mixtures were allowed to react for 30 min at room

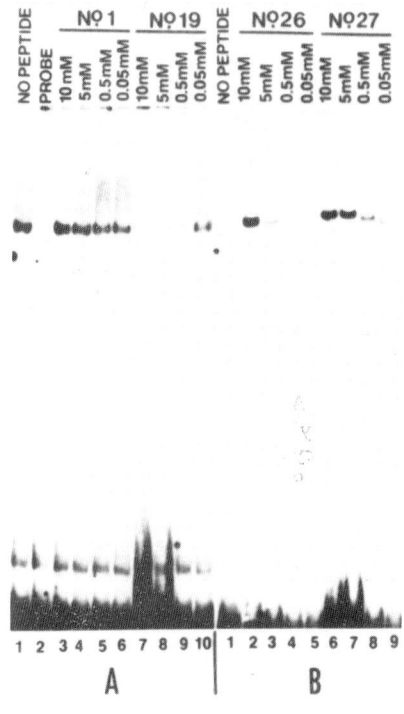

Figure 5. Autoradiographic images of electrophoretically separated α4 protein-DNA complexes formed in the presence of oligopeptides No 1, 19, 26, 27, and 28. Nuclear extracts (0.5 μg) from infected cells containing the α4 protein were reacted first with the oligopeptides at concentrations indicated in the figure for 15 min. at 25 °C and then with a DNA probe for an additional 20 min. Each reaction also included 4μg of unlabeled salmon sperm competitor DNA. Panel A, lane 1: α4 protein-DNA complex formed in the absence of oligopeptides; Lane 2: labeled DNA probe only. Lanes 3-6: α4 protein DNA complexes formed in the presence of oligopeptide No. 1. This oligopeptide like most oligopeptides, was found to be inert in the DNA binding reaction. Lanes 7-10: α4 protein-DNA complexes formed in the presence of various concentrations of oligopeptide No. 19. Panel B, lane 1: α4 protein-DNA complexes formed in the absence of oligopeptides; lanes 2-5 and 6-9 α4 protein DNA complexes formed in the presence of various concentrations of oligopeptides No.26 and No.27 respectively. The amount of oligopeptide included in the DNA binding reaction is given at the top of the corresponding lane.

temperature and then electrophoretically separated in an 8% polyacrylamide gel. The results (Figure 6, left panel) showed that oligopeptides Nos. 19 and 27 were capable of retarding the electrophoretic mobility of the α0 probe DNA. Salmon sperm DNA competed with the probe DNA for the peptides at concentrations ranging from 5 μg for oligopeptide No. 19 to 5 ng for oligopeptide No. 27.

The specificity of the effects of oligopeptides Nos. 19 and 27 on the binding of proteins to DNA. To determine whether the effects of oligopeptides Nos. 19 and 27 on the binding of the α4 protein were specific, we tested the effects of these peptides on the binding of the host proteins αH1 and αH2-αH3 to their binding sites on the 48bp fragment (48α27R) from the regulatory domain of the α27 gene (27). These proteins have been shown to bind to the cis-acting site of the α gene trans-inducing factor, a structural component shown to induce the α genes of HSV-1 (27). Furthermore, both αH1 and αH2-αH3

Figure 6. Autoradiographic images of electrophoretically separated oligopeptide-DNA complexes formed by the reacting of synthetic oligopeptides No. 19 and No. 27 with a DNA. Lanes 2-4 and lanes 5-6 contain peptide No. 19 and No. 27, respectively, and increasing μg amounts of competitor salmon sperm DNA as indicated at the top of each lane. Lane number 1 contains only the α0 promoter DNA probe and indicates the position where unbound DNA migrates.

proteins can bind concurrently to DNA forming a DNA-protein complex (αH1 + αH2-αH3) which migrates more slowly than either αH1 or αH2-αH3 protein-DNA complexes (28). As illustrated in Figure 7, oligopeptide No. 19 inhibited the binding of αH1 and αH2-αH3 proteins to DNA. Oligopeptide No. 27 had no effect on the binding of αH1 but did enhance binding of the αH2-αH3 protein and the concurrent binding of both αH1 and αH2-αH3 proteins to the DNA fragment. Therefore it appears that the enhancement effect of oligopeptide No. 27 is not limited to the α4 protein and yet does not increase the affinity of all proteins for DNA whereas the oligopeptide No. 19 inhibited the formation of all complexes tested.

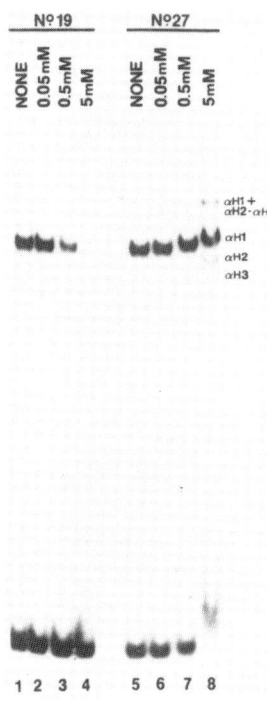

Figure 7. Autoradiographic images of electrophoretically separated αH1 host protein - HSV-1 DNA complexes formed in the presence of increasing amounts of the synthetic oligopeptides No. 19 or 27. The reactions were composed exactly as described in the ledgend to Figure 5 except that 1 μg of nuclear extract prepared from uninfected Hela cells was used. Lanes 1-4 and 5-8 contain the electrophoretically separated products of DNA binding reactions to which various amounts of oligopeptides Nos. 19 and 27, respectively, were added. The amount of oligopeptide added to the reaction mixture is indicated on the top of each lane.

DISCUSSION

The epitopes recognized by the monoclonal antibodies against the α4 protein bind to the N terminal 288 amino acid domain of the protein. A significant and unexpected feature of the results presented in this paper is that all 10 of the monoclonal antibodies tested reacted with the N terminal domain of the protein. Of the 10 monoclonal antibodies, one failed to react with the synthetic oligopeptides and the epitope that it recognizes may be discontinuous. The remaining 9 reacted with the oligopeptides and identified the location of a minimum of 6 epitopes. Two oligopeptides reacted with more than one monoclonal antibody suggesting that either the peptide contains more than one epitope or that the monoclonal antibodies are clonally related. The former may be the case for monoclonal antibodies H950 and H953 inasmuch as H950 inhibits the binding of α4 protein to its DNA binding site whereas H953 does not. The observation that all of the antibodies tested react with the native α4 molecule in the DNA band shift assays suggests that the epitopes are on the surface of the protein in its native conformation.

Accordingly, the N terminal domain is predicted to be, on the average, more hydrophilic than the remainder of the protein (Figure 3A).

Of the 10 monoclonal antibodies tested, two, i.e. H950 and H942, were shown to inhibit the binding of the α4 protein to its DNA and a major objective of this study was to determine whether the epitopes recognized by these monoclonals were distinct from the epitopes recognized by monoclonal antibodies which lack this property. The epitope recognized by H950 is in close proximity to those recognized by monoclonal antibodies that do not affect the binding of α4 protein to its DNA binding site, and none of the oligopeptides from this region of the α4 protein bind to DNA or interfere with the binding of α4 to the DNA probe. Conversely, the epitope recognized by H942 is located in close proximity to two oligopeptides, No. 19 and 27 which competed with or enhanced the binding of α4 to DNA, respectively, and which also bound to DNA. In this instance, H942 may block the binding reaction by sterically hindering the interaction of α4 with its target or the mutual interaction of different domains of the α4 protein required for binding to DNA.

The role of the amino acid sequences contained in the oligopeptides Nos. 19 and 26-28 in the binding of α4 protein to its binding site. A striking feature of the results was that the oligopeptide No. 19 interfered with the binding of α4 protein to its DNA binding site whereas the oligopeptides Nos. 26-28 enhanced the binding. In as much as the oligopeptide No.19 inhibited the binding of αH1 and αH2-αH3 host proteins to DNA, the inhibitory effect of this oligopeptide on the DNA binding activity of α4 protein is probably not specific.

The enhancement effect of the oligopeptides Nos. 26-28 is less readily explainable. The unresolved questions are the specificity of binding and the sequence responsible for the effect of the oligopeptide No. 27. The former issue arises from the observation that oligopeptide No. 27 enhances α4 and αH2-αH3 but not αH1 protein binding to DNA whereas the latter arises from the observation that oligopeptides Nos. 26, 27, and 28 as a group do not share the same amino acid sequence (there is none between No. 26 and No. 28) and yet both have a similar effect on the binding of α4 protein to its DNA binding site. Further studies on variants of the amino acid sequences contained in this region of the intact α4 protein may be necessary to resolve these questions and to determine whether the sequence plays a determinant role in the DNA binding properties of the α4 protein.

It is of interest to note that analyses of the amino acid

sequences of oligopeptides Nos. 19 and 27 by the method of Chou and Fasman (29) predicts that there is a high probability that these sequences would adopt ordered structures in the α4 protein. Oligopeptide No. 19 is predicted to be involved in a succession of reverse turns with a turn probability product that is greater that 1.1×10^{-4} over the entire sequence, the minimum turn probability product for predicted reverse turns being 0.75×10^{-4} (29). This type of structure is atypical of DNA binding motifs but the sequence of oligopeptide 19 contains a predominance of serine residues which are capable of forming hydrogen bonds with guanine residues or the phosphate backbone of the DNA helix. Oligopeptide 19 represents a subset of a larger serine rich domain of α4 in which 24 of 29 amino acids are serine. Consistent with our observation that peptide No. 19 has intrinsic affinity for DNA is the recent observation by DeLuca and Schaffer (20) that α4 proteins which contain a deletion of the serine rich domain do not bind to DNA *in vitro*. Oligopeptide No. 27 is predicted to adopt a 9 amino acid α helical structure with a tetrapeptide α helical propensity average of at least 1.11 over the length of the predicted helix, the mininum value for a predicted α helical structure being 1.03 (29). The structure of this helix is interesting in that it positions on one side of the helix three of the four hydrophilic arginine residues and on the other side of the helix the three hydrophobic alanine residues of the peptide. The predicted structure of this oligopeptide therefore exhibits a number of motifs that are present in the DNA binding domains of other proteins such as the site specific DNA binding *cro* protein from bacteriophage λ and the nonspecific DNA binding protamines (30). It should mentioned that sequences which could give rise to a DNA binding "zinc finger" motif (31) seem to be absent from the α4 protein.

The functional and DNA binding domains of the α4 protein. Previous studies have shown that the truncated α4 gene specifying an 825 amino acid product was able to induce HSV-1 genes in a transient expression system (20). In this study we demonstrated that the truncated polypeptide is able to bind to viral DNA. Analyses of the phenotype of the truncated 825 amino acid gene are incomplete and the precise functions of the α4 protein encoded in the 3' domain of the gene have not been identified. The observation that the sequence of the varicella-zoster virus analogue of the α4 gene (32) contains several conserved domains of the α4 gene in the 3' domain deleted from the truncated gene argues strongly for functions encoded in that domain. In this regard, it has been reported that a recombinant virus which expresses a truncated α4 protein which appears to be very nearly the same size as the the 825 amino acid polypeptide reported here and elsewhere (25, 20) is essentially unable to induce the synthesis of viral DNA and must be propagated on a complementing cell line (20).

These results presented here indicate that the full length and truncated α4 proteins derived from transfection of pRB3611 or pRB3827 are competent to bind to the α0 probe and demonstrated conclusively that no other viral factors need be present for α4 protein to bind to DNA. These observations demonstrate that the amino acids essential for the binding of α4 protein to its cognate site in the α0 promoter are contained in the first 825 amino acids of the protein and suggest, but do not, prove that the DNA binding domain lies between amino acid residue 519 and residue 825. The inference that this region contains a sequence element that contributes significantly to the DNA binding activity of α4 protein stems from the inability of the 519 amino acid polypeptide to bind to DNA. In addition to the interpretation that the DNA binding site of α4 is contained entirely between residues 519 and 825 there are at least 3 formal hypothese to explain the failure of the 519 amino acid polypeptide to bind to DNA. First, a single sequential polypeptide chain may not be competent to bind DNA. The DNA binding domain of α4 in this case would derive from the juxtaposition of two discontinuous segments of amino acid sequence, only one of which would be contained between residues 519 and 825. Hence, the DNA binding domain would be discontinuous and would result from "long range" intramolecular interaction between one or more seaquences contained between amino acid residues 519 to 825 and amino acods sequences located elsewhere. This interpretation is consistent with the antagonistic effects of monoclonal antibodies H950 and H942 on α4 DNA binding activity. Second, the conformation of the 519 amino acid polypeptide could be so altered from that of the 825 amino acid polypeptide as to preclude or grossly reduce the efficiency of its binding activity. Finally, it is conceivable that only multimeric forms of the α4 protein bind to DNA. In this instance elimination of polypeptide sequences which facilitate multimerisation would result in the loss of DNA binding activity. Recently, chimeric proteins have been used to define the functional polypeptide segments of a number of *trans*-activating DNA binding proteins (33, 34, 35). The recovery of chimeric proteins containing defined α4 polypeptide sequences bearing the sequence specificity of intact α4 protein would be of tremendous value in distinguishing between these alternatives.

ACKNOWLEDGMENTS

The studies were aided by grants from the National Cancer Institute (CA08494 and CA19264), the National Institute for Allergy and Infectious Diseases (AI124009), and the American Cancer Society (MV2W) to the University of Chicago, from the National Institute for Allergy and Infectious Diseases (AI23592) and the National Institute for Dental Research (DE08275) to University of California, and from The Heart and Lung Institute (P01HL31950) to Scripps Clinic and Research Foundation. J.H-V. was a predoctoral trainee, National Institute of General Medical Sciences (GM0797).

LITERATURE CITED

1. Preston, VG. (1981). Fine-structure mapping of herpes simplex virus type 1 temperature-sensitive mutations within the short repeat region of the genome. J. Virol. **39**:150-161.
2. Dixon, RAF, and P Schaffer (1980). Fine-structure mapping and functional analysis of temperature sensitive mutants in the gene encoding the herpes simplex virus type 1 immediate early protein VP175. J. Virol. **36** 189-203.
3. Preston, CM (1979). Control of herpes simplex virus 1 mRNA synthesis in cells infected with wildtype virus or the temperature-sensitive mutant tsK. J. Virol 29:275-284.
4. Pereira, L, M Wolff, M Fenwick, and B Roizman (1977). Regulation of herpesvirus syntheis. V. Properties of α polypeptides specified by HSV-1 and HSV-2. Virology. 77:733-749.
5. DeLuca, NA, MA Courtney and PA Schaffer (1984). Temperature-sensitive mutants in herpes simplex virus 1 ICP4 permissive for early gene expression. J. Virol. 52:767-776.
6. DeLuca, NA, and PA Schaffer (1985). Activation of immediate-early, early and late promoters by temperature sensitive and wild-type forms of herpes simplex virus type 1 protein ICP4. Mol. Cell. Biol. 5:1997-2008.
7. Everett, RD (1984). Trans activation of transcription by herpes virus products: Requirements for two HSV-1 immediate-early polypeptides for maximum activity. EMBO J. 3:3135-3141.
8. Gelman, IH and S Silverstein (1985). Identification of immediate early genes from herpes simplex virus that transactivate the virus thymidine kinase gene. Proc. Natl. Acad. Sci. USA **82**:5265-5269.
9. Quinlan, MP and DM Knipe (1985). Stimulation of expression of a herpes simplex virus DNA-binding-protein by two viral functions. Mol. Cell Biol. 5:957-963.
10. DeLuca, NA, AM McCarthy, and PA Schaffer (1985). Isolation and characterization of deletion mutants of herpes simplex virus type 1 in the gene encoding immediate-early regulatory protein ICP4. J. Virol. 56:558-570.
11. O'Hare, P, and GS Hayward (1985). Evidence for a direct role for both the 175,000- and 110,000-molecular-weight immediate early proteins of herpes simplex virus in the trans-activation of delayed-early promoters. J. Virol. 53:751-760.

12. McGeoch, DJ, A Dolan, S Donald and HK Bauer-Dieter (1986). Complete DNA sequence of the short repeat region in the genome of herpes simplex virus type 1. Nucl. Acid Res. 14:1727-1745.
13. Morse, LS, L Pereira, B Roizman, and PA Schaffer (1978). Anatomy of HSV DNA. XI. Mapping of viral genes by analysis of polypeptides and functions specifies by HSV-1 X HSV-2 recombinants. J. Virol. 26:389-410.
14. Wilcox, KW, A Kohn, E Sklyanskaya and B Roizman (1980). Herpes simplex virus phosphoproteins. I. Phosphate cycles on and off some viral polypeptides and can alter their affinity for DNA. J. Virol. 33-167-182.
15. Faber, SW and KW Wilcox (1986). Characterixation of a herpes simplex virus regulatory protein: aggregation and phosphorylation of a temperature sensitive variant of ICP4. Arch. Virol. 91:297-312.
16. Preston, CM and EL Notarianni (1983). Poly(ADP-ribosyl)ation of a Herpes Simplex Virus immediate early polypeptide. Virology 131:492-501.
17. Ackermann, M, DK Braun, L Pereira, and B Roizman (1984). Characterization of α proteins 0, 4, and 27 with monoclonal antbodies. J. Virol. 52:108-116.
18. Metzler, DW and K Wilcox (1985). Isolation of herpes simplex virus regulatory protein ICP4 as a homodimeric complex. J. Virol. 55:329-337.
19. Mavromara-Nazos, P, S Silver, J Hubenthal-Voss, JLC McKnight, and B Roizman (1986). Regulation of herpes simplex virus 1 genes: αsequence requirements for transient induction of indicator genes regulated by β or late γ_2 promoters. Virology. 149:152-164.
20. Deluca, NA, and PA Schaffer (1988). Physical and Functional domains of the Herpes Simplex Virus Transcriptional Regulatory Protein ICP4. 62:732-743.
21. Kristie, TM and B Roizman (1986). α_4, the major regulatory protein of herpes simplex virus 1, is stably and specifically associated with promoter-regulatory domains of α genes and selected other viral genes. Proc. Natl. Acad. Sci. USA 83:3218-3222.
22. Kristie, TM and B Roizman (1986). DNA-binding site of major regulatory protein α_4 specifically associated with promoter-regulatory domains of α genes of herpes simplex virus type 1. Proc. Natl. Acad. Sci. USA 83:4000-4007.
23. Michael, N, D Spector, P Mavromara-Nazos, T Kristie, and B Roizman (1988). The DNa binding properties of the Major Regulatory protein α_4 of Herpes Simplex Virus. Science. In press, March 1988

24. Houghten, RA (1985). General method for the rapid solid-phase synthesis of large numbers of peptides: Specificity of antigen-antibody interaction at the level of individual amino acids. Prod. Natl. Acad. Sci. USA 82:5131-5135.
25. Hubenthal-Voss, J, RA Houghton, L Pereira, and B Roizman (1988). Mapping of Functional and Antigenic Domains of the α_4 protein of Herpes Simplex Virus 1. 62:454-462.
26. Kyte, J and RF Doolittle (1982). A simple method for displaying the hydropathic character of a protein. J. Mol. Biol. 157:105-132.
27. Kristie, TM and B Roizman (1987). Host cell proteins bind to the cis-acting site required for virion-mediated induction of herpes simplex virus 1 α genes. Proc. Natl. Acad. Sci. USA 84:71-75.
28. Kristie, TM and B Roizman (1988). Differentiation and DNA contact points of Host Proteins That Bind at the cis-site for Virion-Mediated Induction of α of Herpes Simplexs Virus 1. J.Virol. 62:1145:1157.
29. Chou, PY, and GD Fasman (1978). Empirical predictions of protein conformation. Ann. Rev. Biochem. 47:251-276.
30. Pabo, CO, and RT Sauer (1985). Protein-DNA recognition. Ann. Rev. Biochem. 53:293-3212.
31. Miller, J, AD McLachlan, and A Klug (1985). Repetitive zinc-binding domains in the protein transcription factor IIIA from Xenopus oocytes. EMBO J. 4:1069-1614.
32. Davison, AJ, and JE Scott (1986). The complete sequence of varicella-zoster virus. J. Gen. Virol. 67:1759-1816.

33. Brent, R and M Ptashne (1985). A eukaryotic transcriptional activator bearing the DNA specificity of a prokaryotic repressor. Cell 43:729-736.
34. Hope, IA and K Struhl (1986). Functional dissection of a eukaryotic transcriptioal activator protein, GCN4 of yeast. Cell 46:885-894.
35. Struhl, K (1987). The DNA-Binding domain of the jun oncoprotein and the yeast GCN4 transcriptional activator proteins are fuctionally homologous. Cell 50:841-846.

V. BIOACTIVE CONFORMATIONS OF PEPTIDE HORMONES

MIMICS OF SECONDARY STRUCTURAL ELEMENTS OF PEPTIDES AND PROTEINS[1]

W.F. Huffman, J.F. Callahan, E.E. Codd,
D.S. Eggleston, C. Lemieux,[2] K.A. Newlander
P.W. Schiller,[2] D.T. Takata and R.F. Walker

Departments of Peptide Chemistry, Physical and Structural Chemistry, and Reproductive and Developmental Toxicology, Smith Kline & French Laboratories, P.O. Box 1539
King of Prussia, PA 19406-0939

ABSTRACT A novel γ-turn mimic (<u>1</u>) has been incorporated into cyclic enkephalin analogs in an attempt to define the role of reverse turns in the biologically-active conformation of these molecules. The lack of receptor affinity observed with these analogs (<u>3a,b</u>) makes it imperative to understand the structural and conformational restrictions imposed by the mimic. X-ray crystallographic data indicates that <u>1</u> is a good mimic of an equatorial γ-turn but not an axial γ-turn.

INTRODUCTION

Reverse turns, an important component of protein secondary structure, are believed to occur frequently on the outer surfaces of proteins and therefore have the potential to define important features of epitopes or receptor pharmcophores (1). In proteins, the most common

---

[1]The work of P.W.S. and C.L. was supported by operating grants from the Medical Research Council of Canada (MT-5655), the Quebec Heart Foundation and the National Institute on Drug Abuse (1RD1 DA-04443-01).
[2]Laboratory of Chemical Biology and Peptide Research, Clinical Research Institute of Montreal, 110 Pine Avenue West, Montreal, Que., Canada, H2W1R7

β-turn   γ-turn

form of reverse turn observed in X-ray crystallographic studies is the ß-turn; the occurrence of a γ-turn is rare (2). However in smaller peptides, it is frequently hypothesized that γ-turns may represent important features of secondary structure as determined by theoretical modeling studies (3), NMR structure determinations (4) and X-ray crystallography (5).

Verification of predicted secondary structural elements is a difficult but nonetheless critical step if one is to define chemically the function of reverse turns in the biologically-active conformations of peptides and proteins. Conformationally-restricted mimics of reverse turns limit the possible conformational array and thereby allow one to potentially validate the presence of reverse turns in pharmacophores. To be successful, a mimic must restrict the peptide backbone to a known set of conformations in the region of the proposed turn and allow for the incorporation of sidechains which are important to receptor recognition and activation. Recently, we described the design and synthesis of the first examples of a γ-turn mimic (6). In this mimic, represented by structure <u>1</u>, the normal amide bond between the i and i + 1 residues in a γ-turn is replaced with a <u>trans</u> olefin and the oxygen and hydrogen atoms of the hydrogen bond (C=<u>O</u>   <u>H</u>-N) are each replaced with a methylene unit.

γ-turn   <u>1</u>

Our initial peptide target for turn mimics was the pentapeptide leucine enkephalin (Tyr-Gly-Gly-Phe-Leu). In addition to their obvious synthetic ease, the enkephalins are an attractive model in which to study turn mimics since they are representative of a growing number of biologically important small peptides which are 1) highly flexible, 2) possess multiple low energy conformations, some of which are characterized by the presence of reverse turns (7), and 3) appear to interact via different pharmacophores at a variety of receptor subtypes. Recently, molecular modeling studies (8) and NMR-based structure determination (8,9) of cyclic enkephalin analogs such as 2 (10) have resulted in the proposal that a γ-turn about the phenylalanine residue is present in low energy structures. Since cyclic structures, of necessity, form some type of reverse turn, a γ-turn may be an important structural feature of the pharmacophore of certain cyclic enkephalins. We attempted to test this idea by incorporation of a γ-turn mimic in cyclic enkephalin 2.

RESULTS

The preparation of the desired cyclic enkephalin analogs 3a,b is depicted in Scheme 1. The key intermediate in this synthesis was the γ-turn mimic γ(Gly-D,L-Phe-Leu), compound 4, which was described previously in our preparation of conformationally constrained linear enkephalins (6). Compound 4 was a mixture of diastereomers with roughly equal amounts of the R and S isomers about the benzyl sidechain of the

SCHEME 1
SYNTHESIS OF CYCLIC ENKEPHALINS CONTAINING
γ-TURN MIMICS

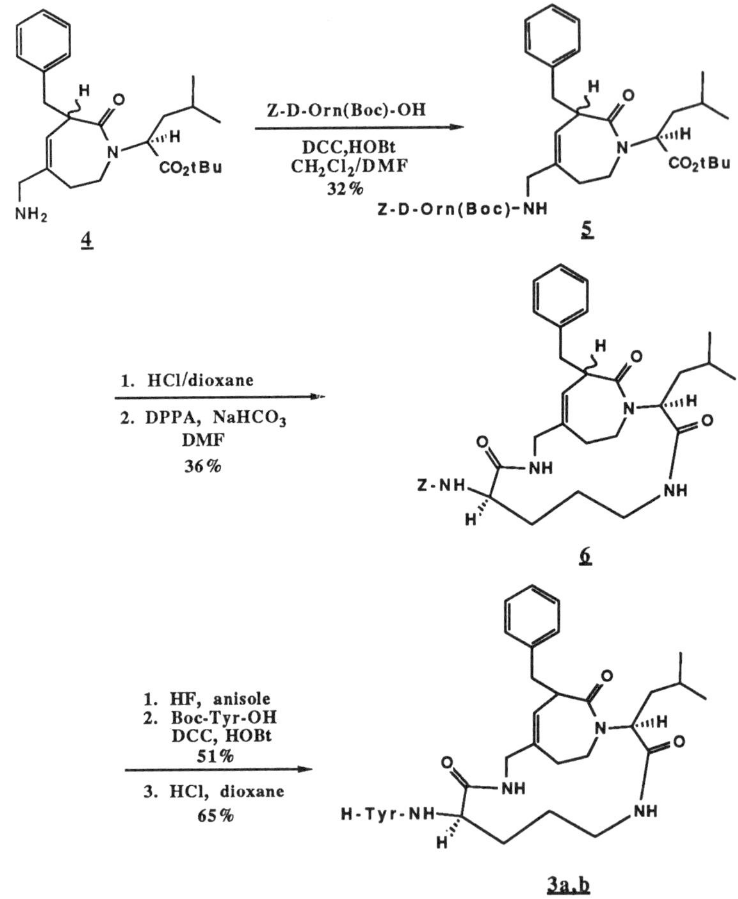

seven-membered ring. The R and S isomers could be separated by flash chromatography at the stage of the bicyclic intermediate 6. Both diastereomers were carried on separately to afford stereochemically pure 3a and 3b. The exact stereochemistry about the benzyl residue in each final product has not been determined.

As indicated in Table 1, incorporation of a γ-turn mimic about Phe[4] in the cyclic enkephalin analog 2 resulted in analogs (3a,b) which display substantially reduced opioid activity. This is reflected by loss of affinity for the mu and delta receptors as well as very low level activity in the classical mouse vas deferens and guinea pig ileum assays (11).

TABLE 1
BIOLOGICAL ACTIVITY OF CYCLIC ENKEPHALINS

Analog	IC_{50} (nM)		K_i (nM)	
	GPI	MVD	DAGO[a]	DSLET[b]
2	48	475		
3a	19,500	>11,400	5,040	10,700
3b	>24,600	>11,400	4,400	6,790
Leucine enkephalin	246	11.4	9.4	2.5

[a]Competitive binding versus the mu-selective ligand [^{3}H]DAGO in rat brain tissue.
[b]Competitive binding versus the delta-selective ligand [^{3}H]DSLET in rat brain tissue.

DISCUSSION

The lack of biological activity observed with the cyclic enkephalin analogs 3a,b is similar to the results obtained when γ-turn mimics were incorporated into linear enkephalin structures (6). In each of these cases it is tempting to interpret the results as implying that γ-turns are not structural features of the pharmacophores of linear or cyclic enkephalins at the mu or delta opiate receptors. However, before such a definitive conclusion is drawn, one must appreciate that a key element in the use of any type of conformational mimic is an understanding of the chemical and conformational features of the mimic itself. It is safe to say that every mimic introduces or removes elements of structure which, when compared to the native situation, have the

distinct possibility of influencing the shape of the
ligand and/or the interaction between ligand and
receptor. In the case of the γ-turn mimic 1, there are
three main areas of difference from the native structure
which need to be understood in order to properly interpret
the results obtained when using this type of mimic in
cyclic or linear enkephalin analogs. These areas include
1) replacement of a peptide bond with a trans olefin, 2)
replacement of a hydrogen bond with two methylene units,
and 3) the overall success of the mimicry of a γ-turn by
the cycloheptene lactam 1.

Experimental data from SAR studies with linear or
cyclic enkephalin analogs is potentially helpful in
addressing the stereo-electronic impact of a trans olefin
or an ethylene bridge. There has been no reported case of
a trans olefin replacement for the Gly^3-Phe^4 peptide
bond present in cyclic enkephalin 2. As a result, it is
not possible to predict how the loss of the CO and NH in
the peptide bond might affect receptor affinity or
activation. However, in linear enkephalins, replacement
of this amide bond with a thiomethylene unit [ψ(CH_2S)]
resulted in an analog with substantially reduced activity
(12). Similar results were obtained (13) when a
retro-inverso modification (CO-NH to NH-CO) was made at
the Gly^3-Phe^4 amide bond in a cyclic enkephalin analog
closely related to 2. N-Methylation of Phe^4 in leucine
enkephalin affords an analog which retains 10% of the
biological activity (14). Taken together, these trends
could be interpreted to indicate that the Phe^4-NH is not
a key binding element, while the Gly^3 carbonyl may be an
important feature in the pharmacophore (13). In the
context of understanding turn mimic 1, this might mean
that the carbonyl either interacts directly with the
receptor or plays a role in the stabilization of a
γ-turn or both.

Equally difficult to assess are the potential effects
of the ethylene bridge in 1 which was designed to
approximate the hydrogen bond. In leucine enkephalin,
substitution of N-MeLeu for Leu^5 results in a molecule
which retains almost full biological activity (14). No
equivalent substitution has been reported for cyclic
analogs. However, to the extent that the pharmacophore of
cyclic enkephalins may be similar to that of linear
enkephalins, this might argue against the presence of a
γ-turn in the pharmacophore. From modeling studies, it
is clear that the steric bulk of the ethylene bridge can

exert an intramolecular effect on the conformational preference of the torsional angles of the backbone and sidechains on either side of the mimic. Depending on the proximity of the receptor to various regions of the ligand, it is also easy to envisage the possibility that the ethylene bridge might result in an unfavorable steric interaction between the ligand and receptor.

Of utmost importance for any mimic is an evaluation of the effectiveness of the mimicry. A comparison between torsional angles for the i + 1 residue of idealized γ-turns and the same angles for γ-turn mimics la,b is presented in Table 2. Mimics la,b are partial representations of the structures of two different mimics as derived by X-ray crystallography. The torsional angles assigned to the idealized γ-turns are the expected values when the sidechain of the i + 1 residue is in the equatorial orientation. An important result which emerges from the X-ray structures of la,b is that the γ-turn mimics of the cycloheptene lactam variety appear to be mimics of only equatorial γ-turns. Their ability to imitate an axial γ-turn appears to be dramatically reduced due to a severe steric interaction between the sidechain in the i + 1 position and the ethylene bridge portion of the ring. This could be of particular relevance in the cyclic enkephalin studies since both an axial (8) and equatorial γ-turn (9) about Phe^4 have been proposed. Clearly, if the pharmacophore for cyclic enkephalin 2 involves an axial γ-turn, then it is very likely that mimic 3 will lock the molecule in the wrong conformation.

These structural analyses, aimed at gaining an understanding of the chemical and conformational features of the novel γ-turn mimic 1, lead to several conclusions which are applicable to all attempts at conformational mimicry. The ability to make definitive conclusions from data involving structural or conformational mimics is directly related to the number of ways in which the overall structure or pharmacophore interaction can be altered by incorporation of the mimic. Even in the best case, where insertion of a mimic provides an analog which either retains or displays improved activity, one must be careful not to overinterpret the results. In the context of chemical design, negative results are always difficult to interpret. The data presented with respect to γ-turn mimics incorporated in cyclic enkephalins is a graphic illustration of this principle.

TABLE 2
COMPARISON OF γ-TURN MIMICS TO γ-TURNS

Analog	Chirality of R_1	Torsional Angles (°)	
		Φ	Ψ
Theoretical(1)			
γ-turn	D	70 to 85	−60 to −70
inverse γ-turn	L	−70 to −85	60 to 70
Mimics			
1a	"L"[a]	−53	73
1b	"D"[a]	56	−67

[a]Stereochemical definition based on conversion of mimic back into classical amino acid nomencalature.

Rationally-designed molecular mimicry is at a very early stage of development. Although our current ability to interpret the results of this approach is still limited, the use of chemically-defined conformational mimics is essential if we are to gain a truly molecular understanding of the function of reverse turns in the biologically-active conformations of peptides and proteins.

REFERENCES

1. Rose G, Gierasch LM, Smith JA (1985). Turns in Peptides and Proteins. In Anfinsen CB, Edsall JT, Richards FM (eds): "Advances in Protein Chemistry, Vol. 37," Orlando: Academic Press, Inc., p 1.
2. Sapse A-M, Mallah-Levy L, Daniels SB, Erickson, BW (1987). The γ Turn: Ab Initio Calculations on Proline and N-Acetylproline Amide. J Amer Chem Soc 109:3526.
3. Hagler AT, Osguthorpe DJ, Dauber-Osguthorpe P, Hempel JC (1985). Dynamics and conformational energetics of a peptide hormone:vasopressin. Science 227:1309.
4. Narutis VP, Kopple KD (1983). Amide proton exchange rates in a cyclic dodecapeptide of defined conformation. pH and conformation dependence. Biochem 22:6233.
5. Loosli HR, Kessler H, Oschkinat H, Weber HP, Petcher TJ, Widmer A (1985). 76. Peptide conformations. Part 31: The conformation of cyclosporin A in the crystal and solution. Hel Chim Acta 68:682.
6. Huffman WF, Callahan JF, Eggleston DS, Newlander KN, Takata DT, Codd EE, Walker RF, Schiller PW, Lemieux C, Wire WS, Burks TF (1988). Reverse turn mimics. In Marshall GM (ed): "Peptides: Chemistry and Biology: Proceedings of the Tenth American Peptide Symposium," Leiden: ESCOM, p. 105.
7. Paine GH, Scheraga HA (1987). Prediction of the native conformation of a polypeptide by a statistical mechanical procedure. III. Probable and average conformations of enkephalin. Biopolymers 26:1125.
8. Mammi NJ, Hassan M, Goodman M (1985). Conformational analysis of a cyclic enkephalin analogue by ^{1}H NMR and computer simulations. J Amer Chem Soc 107:4008.
9. Kessler H, Holzemann G, Zechel C (1985). Peptide conformations. 33. Conformational analysis of cyclic enkephalin analogs of the type Tyr-cyclo-(-N$^\omega$-Xxx-Gly-Phe-Leu). Int J Peptide Protein Res 25:267.
10. Dimaio J, Nguyen TM-D, Lemieux C, Schiller PW (1982). Synthesis and pharmacological characterization in vitro of cyclic enkephalin analogues: effect of conformational constraints on opiate receptor selectivity. J Med Chem 25:1432.
11. Leslie FM (1987). Methods used for the study of opioid receptors. Pharmcol Rev 39:197.
12. Spatola AF, Saneii H, Edwards JV, Bettag AL, Anwer MK, Rowell P, Browne B, Lahti R, Von Voigtlander P (1986). Structure-activity relationships of enkephalins

containing serially replaced thiomethylene amide bond surrogates. Life Sci 38:1243.
13. Richman SJ, Goodman M, Nguyen TM-D, Schiller PW (1985). Synthesis and biological activity of linear and cyclic enkephalins modified at the Gly^3-Phe^4 amide bond. Int J Peptide Protein Res 25:648.
14. Hansen PE, Morgan BE (1984). Structure-activity relationships in enkephalin peptides. In Udenfriend S, Meienhofer J (eds) "The Peptides: analysis, synthesis, biology, Vol 6," Orlando: Academic Press, Inc, p. 269.

DEVELOPMENT OF A BIOACTIVE MODEL OF ATRIAL NATRIURETIC FACTOR

R.F. Nutt, S.F. Brady, T.A. Lyle, T.M. Ciccarone,
C.D. Colton, W.J. Paleveda, T.M. Williams,
G.M. Smith, R.J. Winquist, and D.F. Veber

Merck Sharp & Dohme Research Laboratories
West Point, Pennsylvania 19486

Abstract: Analogs of atrial natriuretic factors (ANFs) containing carboxy-terminal amides and D-amino acids were synthesized and evaluated as inhibitors of methoxamine-induced contraction in rabbit aorta tissue. The three ANF peptides ANF(3-28)-OH, Arg^3-Arg-Ser-Ser-Cys-Phe-Gly-Gly-Arg-Ile-Asp-Arg-Ile-Gly-Ala-Gln-Ser-Gly-Leu-Gly-Cys-Asn-Ser-Phe-Arg-Tyr^{28}-OH (I), ANF(5-26)-NH$_2$ (II), and ANF(6-26)-NH$_2$ (III) exhibited high potencies and were used as reference compounds for other substitutions. Chiral amino acids were replaced individually by their respective D-enantiomers and glycines were replaced by D-alanine. Highly potent analogs (≥ 0.33 relative to the reference) were obtained with D-amino acids in positions Ser^6, Gln^{18}, Asn^{24}, Ser^{25}, Phe^{26} and in four out of the five glycine positions. Pronounced loss of potency (≤ 0.1) was observed with D-residues at Phe^8, Ile^{12}, Asp^{13}, Arg^{14} and Leu^{21}. Combined with information from other conformationally constrained analogs, a bioactive model is proposed.

An important problem in medicinal chemistry involves the design of nonpeptide ligands for peptide receptors. Although no general solution to this problem exists, advances toward this end have included the design of conformationally constrained peptide analogs. In the case of somatostatin, this approach has yielded simplified

highly rigid cyclic hexapeptides.(1) In contrast, a
nonpeptidal 3-substituted benzodiazepine antagonist of CCK
has been developed based on the natural product, asperlicin, itself discovered in a receptor screening program.(2)
It has been noted that the proposed interacting surface of
the somatostatin cyclic hexapeptides has dimensions of 10 Å
X 15 Å.(3) The molecular dimensions of the benzodiazepine
CCK antagonist also has long dimensions of 10 Å X 15 Å in
the crystal. Tifluadom, a 2-substituted benzodiazepine is
an opiate agonist which interacts at the enkephalin receptor.(4) Both enkephalin and CCK have been constrained in
small cyclic peptide structures in a manner similar to
somatostatin.(5) It has therefore been proposed that it is
reasonable to expect that nonpeptide structures are feasible for peptides which have an interacting surface of about
10 Å X 15 Å or smaller.(3) In contrast, X-ray crystal
structure data on insulin and glucagon suggest the likelihood of a much larger interacting surface and corresponding
lower probability of a nonpeptide or small molecule mimetic. We present herein structure-activity data on analogs
of atrial natriuretic factor (a 28-peptide) which allow
derivation of a bioactive molecular conformation which can
present all of the known receptor intracting groups within
a surface less than 10 Å X 15 Å.

Atrial natriuretic factors (ANFs) are a group of
structurally related peptides which have been isolated from
atrial cardiocytes and represent carboxy-terminal fragments
of a common precursor protein (see reference 6 for review
and references cited therein). The 28-peptide, Ser^1-Leu-
Arg-Arg-Ser-Ser-Cys-Phe-Gly-Gly-Arg-Ile-Asp-Arg-Ile-
Gly-Ala-Gln-Ser-Gly-Leu-Gly-Cys-Asn-Ser-Phe-Arg-Tyr^{28}-OH
(A), has been shown to be the circulating form of the
hormone (7). ANFs elicit a multitude of biological activities, including potent diuretic and natriuretic effects,
smooth muscle relaxation, blood pressure reduction, and
inhibition of aldosterone and renin secretion. These
biological effects are of short duration with half-lives of
$\leq$3 min (8). As in many other peptide hormones, the short
duration of action can in part be ascribed to rapid degradation by metabolic enzymes. For ANFs, facile cleavage of
the Arg^4-Ser^5 bond in the circulating hormone to give the
less active 24-peptide has been proposed (7). A carboxy
terminal dipeptidase has been isolated from atrial tissue

which readily converts ANF(5-27)-OH to the much less potent ANF(5-25)-OH (9). In addition, facile inactivation of atrial extracts has been observed after exposure to the proteolytic enzymes kallikrein, trypsin, carboxy peptidase-B, A-chymotrypsin and elastase (10).

In order to find structural modifications which might lead to ANF analogs with enhanced metabolic stability, a number of carboxy-terminal amides and analogs singly substituted with D-amino acids were synthesized and examined for retention of biological activities as measured by inhibition of smooth muscle contraction. In addition to the potential for metabolic stabilization, the different conformational preferences when D-amino acids replace L-amino acids can be helpful in elucidating important conformational features needed for bioactivity.

MATERIALS AND METHODS

ANF analogs were synthesized using a strategy similar to the one used for the preparation of multigram quantities of ANF(3-28) OH (I) (11). The approach entailed assembly of protected peptide-resins by a combination of single amino acid (S) and fragment (F) couplings on solid support with sidechain protecting groups as indicated in Figure 1. For analogs in the ANF(3-28)-OH series, chloromethylated resin (1 mm Cl/g ®, Lab Systems) was used for attachment of Boc-Tyr(2,6-Cl_2-Bzl)-OH (12), whereas the carboxy-terminal amides in the series ANF(5-26)-NH_2 (II) and ANF(6-26)-NH_2 (III) were prepared using p-methylbenzhydrylamine resin (1% crosslinked, 0.57 mmole amine/g ®, USB). Single amino acid couplings were carried out using <u>in situ</u> activation with DCCI in CH_2Cl_2 or preformed symmetrical anhydrides in DMF. If retreatment was required, DCCI/HOBt activation was used. Arg, Gln, and Asn were routinely activated by DCCI/HOBt. Whenever possible, the indicated fragments (Figure 1) were used as NH_2-terminal Boc derivatives and COOH-terminal acids. Fragments were used in 50% excess and activation was carried out using DCCI/HOBt for the tri- and tetrapeptides (F-3,F-4) and DCCI/HOSu for the N-terminal penta-, hexa-, hepta-, and octapeptides (F-5,-6,-7,-8). The indicated fragments were prepared by stepwise solid phase synthesis followed by transesterification with NEt_3/MeOH (10-20%, 24-48 hours) and saponification with NEt_3/THF/H_2O (1:2:2, 6-16 hours). Purification of fragments to $\geq 97\%$

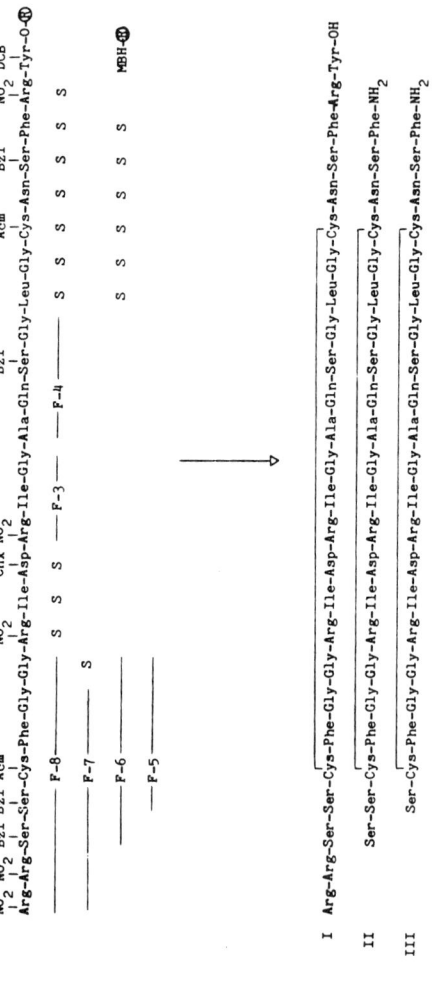

Figure 1

Synthetic scheme for compounds I, II and III using a combination of standard solid phase peptide synthesis and fragment couplings on solid support. Residues marked S were added as single protected amino acids and those marked F were added as independently synthesized fragments (see reference 11 for details).

purity was accomplished by crystallization of esters
(MeOH-EtOAc, MeOH) or silica gel chromatography of acids
(EtOAc-Pyr-HOAc-H_2O). Cleavage of the 26-peptide from the
solid support was accomplished by treating a mixture of
peptide-resin (2 g), m-cresol (4 ml), and Met (1 g) with HF
(30 ml) at $0°$ for 75 minutes. After evaporation of HF and
precipitation with Et_2O, the bis-Cys(Acm) containing linear
intermediate was leached from the resin with 50% HOAc and
purified by gel chromatography on Sephadex G-25F using 2N
HOAc. Deprotection of cysteines and cyclization to the
disulfide was carried out using iodine (15-20 equiv.) in
HOAc-H_2O (8:2). After reduction of excess I_2 with Zn dust,
small amounts of polymer were removed by gel filtration on
Sephadex G-50F using 50% HOAc. Final purification of
analogs was carried out by preparative HPLC (Vydac C_4,
15-20 μ, 5-65% acetonitrile gradient in 0.1% TFA-H_2O).
Conversion of TFA salts into HOAc salts was accomplished by
ion exchange using Bio Rad AG3-X4A (acetate form; 6:1
resin:peptide). All analogs were $\geq$97% pure (HPLC); gave
the expected amino acid ratios after 70 hours acid hydrolysis, contained $\leq$1 mole % free SH (13) and exhibited NMR
spectra consistent with structure and degree of purity.
Each analog was found to be free of the parent L-diastero-
mer which could be separated by HPLC.

    <u>Biological Assays.</u> The rabbit and rat aorta assays
were carried out as described previously (14). Aorta rings
were contracted to 80% of maximal effect by methoxamine
(10^{-6}M). IC_{50} values for the ANF analogs were determined
using 3 tissue samples for each data point.

RESULTS AND DISCUSSION

    The two C-terminal amide ANF analogs ANF(5-26)-NH_2
(II) and ANF(6-26)-NH_2 (III) exhibited high relative
potencies compared to ANF(3-28)-OH (I) when evaluated as
inhibitors of methoxamine induced contraction in rabbit and
rat aorta tissues (Table 1). The relatively high potencies
of the two truncated amide analogs are noteworthy, in that
they do not contain the Arg^{27} residue considered by other
investigators (15) to be a "primary contributor to activity". Subsequent studies (21) have shown that the important arginine can be contributed from either the amine or
carboxy terminus. The different absolute potency of I in
rabbit or rat aorta tissues could represent a species-

Table 1. Inhibition of Methoxamine Contraction of Aorta Tissue

ANF Analog	Rabbit		Rat	
	IC_{50} (M)	Rel. Pot.	IC_{50} (M)	Rel. Pot.
I ANF(3-28)OH	6.7×10^{-10}	1	1.35×10^{-9}	1
II ANF(5-26)NH_2	2.88×10^{-9}	0.23	1.05×10^{-9}	1.29
III ANF(6-26)-NH_2	1.98×10^{-9}	0.32	1.08×10^{-9}	1.26

dependent selectivity or may be due to differences in degradative enzyme profiles in the two tissues. Similar species-related differences in potencies have been reported previously (15,16). Relaxation of rabbit aorta tissue appears to correlate better with other species such as dog and monkey than does corresponding data from rat tissue (17).

The potencies of analogs with single D-amino acid substitutions for relaxation of rabbit aorta are presented in Table 2. In all cases full dose response curves and full effects were obtained. Only modifications of potency were seen, indicating that none of the analogs are antagonists.

Replacement of the exocyclic amino acids Ser^6 (16), Asn^{24}, Ser^{25}, and Phe^{26} (5, 6, 7) with the respective D-enantiomers resulted in retention of a high relative potency. The retention of high potency with D-Ser^6 and D-Ser^{25} (16 and 6) substitutions offer the possibility of blocking metabolism by exopeptidases. The D-Ser analog 6, in particular, might stabilize ANFs towards enzymatic degradation by the dipeptide carboxypeptidase mentioned earlier (9).

Additional points for both conformational and metabolic stabilization were identified by substitution of D-alanines for the five glycine residues in the ring portion of ANF_2. Replacement of Gly^9 (8), Gly^{16} (3), Gly^{20} (13), and Gly^{22} (15) resulted in retention of potency of at least 60%. The potency of 8 is in agreement with literature reports for the D-Ala^9 containing analog in the ANF(7-28)-OH series (18). The low activity of 9 combined with the high potency of 8 draws analogy to the results

Table 2. Biological Activities of ANF Analogs as Inhibitors of Methoxamine Induced Contraction in Rabbit Aorta Tissue.

Cmpd	Structure	IC_{50} nM	95% (Conf. Limits)	Rel. Pot.
I	Arg-Arg-Ser-Cys-Phe-Gly-Gly-Arg-Ile-Asp-Arg-Ile-Gly-Ala-Gln-Ser-Gly-Leu-Gly-Cys-Asn-Ser-Phe-Arg-Tyr-OH	0.676	(0.58, 0.78)	1
1	———D-Arg———	2.6	(1.3, 5.4)	0.26
2	———D-Ile———	4.8	(4.3, 5.4)	0.14
3	———D-Ala———	1.0	(0.75, 1.3)	0.68
4	———D-Cys———	3.1	(1.3, 7.1)	0.22
5	———D-Asn———	2.0	(0.81, 4.9)	0.34
6	———D-Ser———	0.77	(0.45, 1.3)	0.88
7	———D-Phe———	1.6	(0.86, 3.0)	0.42
II	Ser-Ser-Cys-Phe-Gly-Gly-Arg-Ile-Asp-Arg-Ile-Gly-Ala-Gln-Gly-Leu-Gly-Cys-Asn-Ser-Phe-NH$_2$	2.88	(2.0, 4.2)	1
8	———D-Ala———	3.1	(1.2, 8.1)	0.96
9	———D-Ala———	20	(12, 33)	0.14
10	———D-Ile———	146	(76, 281)	0.02
11	———D-Asp———	196	(123, 311)	0.01
12	———D-Arg———	141	(30, 681)	0.02
13	———D-Ala———	1.75	(0.56, 5.5)	1.65
14	———D-Leu———	33	(15, 75)	0.09
15	———D-Ala———	2.6	(1.5, 45)	1.1
III	Ser-Cys-Phe-Gly-Gly-Arg-Ile-Asp-Arg-Ile-Gly-Ala-Gln-Ser-Gly-Leu-Gly-Cys-Asn-Ser-Phe-NH$_2$	1.98	(0.7, 5.6)	1
16	D-Ser	5.4	(2.1, 14)	0.37
17	———D-Cys———	6.94	(4.9, 9.8)	0.29
18	———D-Phe———	81.8	(16, 417)	0.02
19	———D-Ala———	10.5	(5.6, 20)	0.19
20	———D-Gln———	2.33	(1.4, 4)	0.85
21	———D-Ser———	16.1	(7.5, 35)	0.12

seen for D-Ala replacements in the Gly-Gly sequence of enkephalins and suggests the possibility of a similar conformation in this region of ANF.

Pronounced loss in potency was observed by substituting Phe^8, Leu^{21}, Ile^{12}, Asp^{13}, and Arg^{14} with their D-enantiomers (18, 14, 10, 11, 12). The potency-reducing effect of the D-Asp modification has been described previously in another analog series (18). Reasons for loss of activity in D-residue substituted analogs can be two-fold. First, the specific sidechain may play an important role in eliciting bioactivity so that in the D-enantiomer the sidechain is sterically inaccessible to the active site. Alternatively, the specific sidechain may not play an important role, but a change in chirality can have an important conformational effect on neighboring residues which are crucial for effectively eliciting bioactivity. Differentiating these effects in the analogs mentioned above is aided by the synthesis of L-alanine replacement analogs described in reference 21. These studies define the hydrophobic amino acids at positions 8, 12, 15 and 21 as being particularly important for receptor binding.

The cyclic feature of ANF is an important structural constraint for efficiently attaining the bioactive conformation. That this feature, however, is not an absolute requirement for eliciting bioactivity is shown by the full activity but low potency of the linear analogs Boc-Cys-(Acm)^{7,23}ANF(3-28) acid and amide (19), and ANF(7-27) (20). Analogs 4 and 17 represent cyclic structures with one of the cysteines in the D-configuration. The relative potencies of 0.22 and 0.29, respectively, demonstrate that steric proximities of important elements in the termini and/or ring are attained but not as efficiently as with an L,L-cystine cyclizing unit.

Thus, analogs of ANF have been synthesized which offer the possibility of blocking metabolism by both exo-and endopeptidases. These involve the introduction of D-amino acids in the amino and carboxy termini as well as carboxy terminal amidation. The spread of acceptable D-substitutions throughout the ring also has favorable potential for overall metabolic stabilization. In addition a repeating pattern of allowed D-replacements in residues 21-27 suggests a regular 3-dimensional structure in the bioactive form of this molecule.

In addition to the D-alanine replacements for each glycine residue, L-alanine replacements have also been placed at each of these positions (21). The combinations of these results for glycines 9 and 10 are summarized in Table 3 and taken together suggest a turn structure in the bioactive conformation. Thus, the analog having D-alanine in position 9 retains essentially full potency, while the L-alanine-9 analog (22) suffers about a ten fold loss in potency. The reverse is true in the 10 position, with the L-alanine analog (23) retaining full potency and the D-alanine analog showing reduced potency. At position 16 high potency is seen for both the L- and D-alanine replacements for glycine (Table 4). In addition, the α-branched, α-amino-isobutyric acid (Aib) as a replacement for glycine in ANF(3-28) gives an analog (25) of almost full potency (74%). These results are most consistent with a helix (3_{10}- or α-type) (22). Furthermore, the high potency seen on α-branching at position 13 strengthens the idea of a helical conformation in this region. The α-amino isobutyric acid replacement for aspartic acid-13 has a potency about the same as that of the analog having the unbranched aminobutyric acid at this same position (10% and 20% of the unmodified parent respectively) (23).

In order to create a working model of the ring portion of ANF in the Merck molecular modeling system the turn conformation at residues 9 and 10 was coupled with a helical structure from residues 12 through 18. The structure was manipulated to allow cyclization through formation of the disulfide required for biological activity (19, 20). In model building studies, the importance of the hydrophobic side chains of residues 8 and 15 was kept in mind and an attempt was made to bring these into some proximity. Once a structure having nearly reasonable bond lengths was obtained in this crude model building process, the energy minimization program, AMBER, (24) was applied to yield the structure shown in figure 2. This model has proven to be a useful working model for the bioactive conformation of the ring portion of ANF. The structure retained a turn segment around residue 9 and a helical conformation (3_{10} type)in the sequence 12-18. A turn (inverse γ) is also centered at glycine-20. A particularly interesting feature of the model is the clustering of 4 important hydrophobic side chains of residues 8, 12, 15 and 21 all shown to be high contributors to the potency of ANF (21). These residues lie within a surface of 10 Å x 10 Å.

Table 3. Biological Activities of ANF Analogs as Inhibitors of Methoxamine Induced Contraction in Rabbit Aorta Tissue.

Cmpd	Structure	IC_{50} nM	95% (Conf.Limits)	Rel.Po
II	Ser-Ser-Cys-Phe-Gly-Gly-Arg-Ile-Asp-Arg-Ile-Gly-Ala-Gln-Ser-Gly-Leu-Gly-Cys-Asn-Ser-Phe-NH_2	2.88	(2.0, 4.2)	1
8	----------D-Ala--	3.1	(1.2, 8.1)	0.93
9	--------------------D-Ala--	20	(12, 33)	0.24
22	--------------------Ala---	12		0.14
23	--------------------Ala---	2.5		1.15

Table 4.

Inhibitor of methoxamine contraction of rabbit aorta.

		Rel. Pot. (95% C.L.) Rabbit Aorta
I	ANF(3-28)	1
3	D-Ala^{16} ANF(3-28)	0.64 (0.29,1.4)
24	Ala^{16} ANF(3-28)	2.2 (10,4.8)
25	AIB^{16} ANF(3-28)	0.74 (0.3,7)

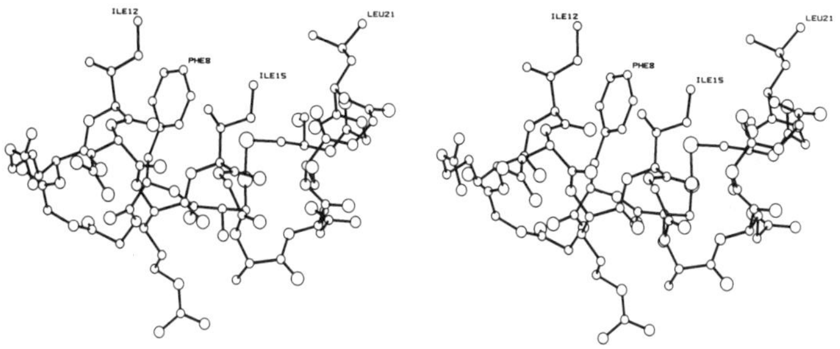

Figure 2

Stereo view of proposed bioactive model of the ring portion of ANF.

In addition to the hydrophobic amino acids within the ring structure of ANF, it has been shown that a single charged side chain of an arginine which can be supplied from either the amine or carboxyl end of ANF is important for potency.[21] Although a specific placement for this charge group cannot be made, it can easily come close enough to the hydrophobic cluster that all of the important binding elements of ANF can be contained within a surface of 10 Å x 15 Å.

This model building process does not give a uniquely defined bioactive conformation for ANF. It is rather a working model to be tested by the design of further constrained analogs. Nonetheless, the conclusion that all of the important known binding elements of ANF can be located in close proximity in an energetically reasonable structure suggests that a small molecule ligand for the ANF receptor is a reasonable goal of either a medicinal chemistry or natural products screening program. Such a conclusion is a useful step in the drug discovery process.

REFERENCES

1. Veber DF, Freidinger RM, Perlow DP, Paleveda WJ, Jr, Holly FW, Strachan RG, Nutt RF, Arison BH, Homnick C, Randall WC, Glitzer MS, Saperstein R, Hirschmann RF (1981). Nature, 292:55-58.

2. Evans BE, Bock MG, Rittle KE, DiPardo RM, Whitter WL, Veber DF, Anderson PS, Freidinger RM (1986). Proc. Nat. Acad. Sci. 83:4918-4922.
3. Veber DF, Freidinger RM (1988). Pharmacology: Proc. 10th International Congress of Pharmacology, MJ Rand and C Raper (Eds) Exerpta Medica, Amsterdam (1987) pp 215-218.
4. Romer D, Buscher HH, Hill RC, Maurer R, Petcher TJ, Zeugner H, Benson W, Finner E, Milkowski W, Thies PW (1982). Nature, 298:759.
5. Freidinger RM, Anderson PS, Bock MG, Chang RSL, DiPardo RM, Evans BE, Garsky VM, Lotti VJ, Rittle KE, Veber DF, Whitter WL (1988). Proc. 10th Am. Pept. Symp. Peptides: Chemistry and Biology, GR Marshal, Ed., Escom, Leiden pp 97-100.
6. Nutt RF, Veber DF (1987) Endocrinol and Metab. Clin. of North Amer. 16:19-41.
7. Schwartz D, Geller DM, Manning PT, Siegel NR, Fok KF, Smith CE, Needleman P (1985). Science 229:397-400.
8. Espiner, EA, Crozier IG, Nicholls MG, Cuneo R, Yandle TG, Ikram H (1985). The Lancet 2:398-9.
9. Harris RB, Wilson IB (1985). Peptides, 6(13):3936.
10. Briggs JP, Marin-Grez M, Steipe B, Schubert G, Schnermann J (1984). Amer. J. Physiol. 247:F480-484.
11. a) Nutt RF, Brady SF, Lyle TA, Ciccarone TM, Colton CD, Paleveda WJ, Veber DF, Winquist RF (1986). Proceedings of the XXXIVth Colloquium-Protides of the Biological Fluids, Brussels, Belgium, Pergamon Press, Oxford London, pp 55-58.
b) Lyle TA, Brady SF, Ciccarone TM, Colton CD, Paleveda WJ, Veber DF, Nutt RF (1987). J. Org. Chem., 52:3752.
12. Gisin BF (1973). Helv. Chim. Acta, 56:1476-1482.
13. Habeeb A.F.S.A. (1972). Methods in Enzym. 25:457-464.
14. Winquist RJ, Faison EP, Nutt RF (1984). Eur. J. Pharmocol., 102:169-173.
15. Needleman P, Adams SP, Cole BR, Currie MG, Geller DM, Michener ML, Saper CB, Schwartz D, Standaert DG (1985). Hypertension, 7:469-82.
16. Rapoport RM, Winquist RJ, Baskin EP, Faison EP, Waldman SA, Murad F (1986). Eur. J. Pharmacol., 120:123-126.
17. Winquist RJ (1987). Blood Vessels, 24:128-131.

18. Chino N, Nishiuchi Y, Noda Y, Watanabe TX, Kimura T, Sakakibara S (1985). Peptides: Structure and Function, Proceed. of Ninth Amer. Pept. Symp., 945-948.
19. Napier MA, Vandlen RL, Albers-Schonberg G, Nutt RF, Brady S, Lyle T, Winquist R, Faison EP, Heinel LA, Blaine EH (1984). Proc. Natl. Acad. Sci. USA, 81:5946-5960.
20. Schiller PW, Maziak L, Nguyen MD, Godin J, Garcia R, DeLean A, Cantin M (1985). Biochem. Biophys. Res. Commun., 131(3):1056-1062.
21. Nutt RF, Ciccarone TM, Brady SF, Colton CD, Paleveda WJ, Lyle TA, Williams TM, Veber DF, Wallace, A, Winquist RJ (1988). Proc. 10th Am. Pept. Symp. Peptides: Chemistry and Biology, GR Marshal, Ed., Escom, Leiden pp 444-446.
22. Toniolo C, Bonora GM, Bavoso A, Benedetti E, DiBlasio B, Pavone V, Pedone C (1983). Biopolymers, 22:205-215.
23. Nutt RF, Ciccarone TM, Brady SF, Colton CD, Williams TM, Winquist RJ, Veber DF (1987). Peptide Chemistry 1987, Proceedings of the 25th Japanese Peptide Symposium, Protein Research Foundation, Osaka (in press).
24. Weiner SJ, Kollman PA, Case DA, Singh UC, Caterina G, Alagona G, Profeta S, Jr., Weiner P (1984). J. Am. Chem. Soc. 106:765-784.

VI. PEPTIDE HORMONES AND GROWTH FACTORS

SYNTHESIS OF INSULIN-LIKE GROWTH FACTOR I THROUGH RECOMBINANT DNA TECHNIQUES AND SELECTIVE CHEMICAL CLEAVAGE AT TRYPTOPHAN

Richard DiMarchi[1], Harlan Long[1], Janet Epp[2], Brigitte Schoner[2], and Rama Belagaje[2]

Departments of [1]Biochemistry and [2]Molecular Biology
Lilly Research Laboratories, Indianapolis, Indiana 46285

ABSTRACT. A new approach to chemical cleavage of proteins has been developed. Through the combined action of DMSO/HCl/TFA, followed immediately by DMS/HCl/TFA treatment, a selective modification of proteins at only tryptophan can be achieved. In combination with recombinant DNA techniques, this method is capable of providing a rapid and efficient synthesis of peptides which are free of tryptophan. The application of this synthetic strategy in the production of Insulin-Like Growth Factor I is described.

Selective peptide bond cleavage with chemical reagents has proven to be of immeasurable value in the determination of amino acid sequence. Methionine is the most utilized amino acid site of cleavage due to the exquisite action of cyanogen bromide (1). The cleavage target of second preference is tryptophan, for which a variety of reagents have been described (2,3). Despite more than twenty-five years of investigation none of the tryptophan reagents can routinely match the cleavage efficiency and selectivity achievable at methionine.

Methods in selective protein cleavage have assumed an increased level of importance with the advent of protein-semisynthesis (4,5), and genetically-engineered protein synthesis (6,7). In these areas, utilization of cyanogen bromide has been near-exclusive. The decided preference for this reagent has been most evident in insulin synthetic studies. A large selection of potential cleavage reagents are applicable to insulin as it is devoid of tryptophan and methionine. Cyanogen bromide has been chosen without exception in each synthetic study (6-9).

The use of cyanogen bromide for synthetic purposes is precluded when methionine is an inherent residue of a peptide. Several peptides, in particular atrial naturietic factor (ANF), growth hormone releasing factor (GRF), and insulin-like growth factor-I (IGF-I) possess a single methionine and no tryptophan. Determination of the physiological significance, and clinical utility of each peptide requires a reliable synthetic source. Currently the most high-yielding production of peptides in this molecular size is through selective cleavage of an E. coli synthesized fusion-protein (6,7,10).

Bacterial synthesis of a fusion-protein in which tryptophan immediately precedes a natural peptide devoid of it provides a potential site for selective cleavage. Chemical cleavage at tryptophan peptide bonds is achieved through oxidative halogenation, and has been extensively reviewed (3). Cleavage yields approaching 60% have been attained through the action of BNPS-skatole (2-(2-nitrophenylsulfenyl)-3-methyl-3-bromoindolenine). More recently selective cleavage by a mixture of DMSO and HBr in acetic acid has been recommended (11). Modification at tryptophan is not selective by either of these methods. Formation of methionine sulfoxide can occur at near quantitative levels. To a lesser extent irreversibly modified amino acids such as methionine sulfone, cysteic acid, and/or brominated tyrosine have been observed.

The degree of side-reactions which occur in the course of tryptophan cleavage have been reported to be minimized with the reagent N-chlorosuccinimide (NCS), (12). Methionine conversion to its sulfoxide was the only other modification originally detected. However in a subsequent study with NCS, appreciable levels of cysteic acid and methionine sulfone were observed (13). In amino acid sequence determination, these side-reactions can be distracting but do not preclude successful utilization of these reagents (14,15). More stringent requirements apply to the synthetic use of these reagents.

IGF-I is identical to somatomedin C and belongs to a family of regulatory hormones which includes insulin and IGF-II (16). Structural similarity amongst these peptides is readily apparent. Each peptide is devoid of tryptophan, but IGF-I alone contains a single methionine. The peptides display a series of common actions, but do differ appreciably in potency. Physiologically insulin principally serves to regulate the immediate fluctuations in circulating metabolites, while IGF-I is envisioned to primarily promote growth and differentiation (17). Therapeutically IGF-I has been suggested to be of potential utility in the treatment of growth deficiencies, Type II diabetes, wound and fracture healing (18). Antagonists of IGF-I may prove of use in the management of lung and breast cancer (19, 20).

The amino acid sequence of IGF-I was determined in 1978 (21). Its supply for biological evaluation has been restricted by its limited availability as isolated from human plasma. Recently through improvements in isolation methodology, approximately 100 µg of IGF-I was reportedly obtained from 900 g of Cohn fraction IV-I (22). The chemical synthesis of IGF-I was a formidable achievement which yielded 6.8 mg of peptide (23). The biosynthesis of natural IGF-I has been desdribed (24). This synthesis being reportedly achieved in genetically altered yeast cells. Bacterial syntheses of IGF-I have been primarily restricted to analogs which were dictated by the chosen chemical cleavage reagents. Cyanogen bromide provided the means for synthesis of a methionine-substituted threonine analog (25). Formic acid promoted cleavage of a particular aspartyl-proline peptide bond was used to yield an NH_2-terminally deleted IGF-I (26). The absence of appropriate cleavage methodology has impeded the biosynthesis and subsequent biological assessment of IGF-I.

Our initial objective was to utilize one of the previously described tryptophan cleavage reagents in the biosynthesis of natural IGF-I. This peptide was initially perceived as a target of appreciable difficulty. The ability to preserve the integrity of the six cysteine residues through the course of oxidative tryptophan cleavage, and subsequent methionine sulfoxide reduction was envisioned to be a crucial requirement for success. Formation of natural IGF-I through proper disulfide pairing of the cysteines serves as a sensitive monitor of structural modification (10).

The genetic manipulation of E. coli to code for the expression of an appropriate IGF-I fusion-protein was similar to that described for the biosynthesis of IGF-II (27). A precursor bearing a single tryptophan immediately adjacent to the natural IGF-I sequence was expressed in the form of cytoplasmic inclusion bodies (28). Physical isolation of these inclusion bodies through differential centrifugation yielded a trp le´-IGF-I fusion protein of approximately 60% purity. Four previously described methods of tryptophan cleavage were assessed with this partially pure fusion-protein. The degree of cleavage was initially determined by NH_2-terminal sequence analysis of the reaction products. Amino acid analysis was used to quantitate specific modifications, in particular those at methionine, tyrosine and cysteine. Quantitation of the IGF-I potentially attainable was determined by reversed-phase chromatographic analysis.

The cleavage yield with each reagent was surprisingly constant (Table 1), and considerably less than has been optimally reported (3). The extremely low cleavage level observed with DMSO-HCl in aqueous acetic acid was anticipated (11). Tyrosine modification appeared to be significant only through treatment with cyanogen bromide in heptafluorobutyric acid. Irreversible modification of cysteine to cysteic acid occurred to a variable degree. The magnitude of this side-reaction could be suppressed to as little as 3% through reduction in reaction temperature, duration of cleavage and/or the reagent excess. Further reduction in cysteic acid formation could not be achieved without an appreciable decrease in the cleavage yield. All reagents provided near-quantitative oxidation of methionine to its sulfoxide with just minimal further reaction to the sulfone. Recovery of methionine from its sulfoxide is potentially possible by several methods (30-33). For synthetic purposes the efficiency and ease of sulfoxide reduction immediately following tryptophan cleavage are of utmost importance.

The most encouraging results were obtained with BNPS-skatole, but even this reagent was deemed undesirable for IGF-I biosynthesis. Only a modest cleavage yield was achieved, and the formation of cysteic acid served to diminish chromatographic performance and the degree of proper disulfide pairing. The reagent is relatively

TABLE 1
CLEAVAGE OF IGF-I FUSION-PROTEIN WITH
VARIOUS TRYPTOPHAN SELECTIVE REAGENTS

Amino Acid [a]	Untreated	BNPS-Skatole [b]	NCS [c]	CNBr [d]	DMSO-HCl [e]	DMSO-HCl HBr [e]
Cysteic Acid	<0.1	0.6	0.7	0.3	0.1	0.8
Homoserine	<0.1	<0.1	<0.1	<0.1	<0.1	<0.1
Methionine	2.0	<0.1	<0.1	<0.1	<0.1	<0.1
Methionine Sulfoxide	<0.1	1.7	1.9	1.6	1.7	1.9
Methionine Sulfone	<0.1	<0.1	<0.1	<0.1	<0.1	0.1
Tyrosine	3.0	2.8	2.8	2.3	2.9	2.8
Cleavage	----	33%	25%	33%	1%	30%

[a] Amino acid analysis was performed following hydrolysis in 4 N methane sulfonic acid (29). The results are expressed in amino acid equivalents in the IGF-I fusion-protein. The cysteine content in untreated fusion-protein is six. Cleavage is estimated by two cycles of automated sequence analysis. The yield of the NH_2-terminal dipeptide Met-Lys is compared to that of the newly generated NH_2-terminal IGF-I sequence of Gly-Pro.
[b] Procedure as described in Reference 2
[c] Procedure as desdribed in Reference 12
[d] Procedure as described in Reference 15
[e] Procedure as described in Reference 11

unstable at room temperature and displays only a limited solubility in aqueous acetic acid (2). These features in combination with the inefficiency and inconvenience of the subsequent methionine sulfoxide reduction rendered BNPS-skatole unattractive. In sum, two principal improvements in current methodology were required to produce IGF-I with reasonable efficiency. The first being elimination of cysteic acid formation without an adverse effect on cleavage yield. The second improvement requires the identification of a highly efficient and cleavage-compatible method of methionine regeneration.

Conceptually tryptophan cleavage with DMSO and HBr in acetic acid is most attractive (11). Cleavage is selective with reagents that are inexpensive and readily available. More importantly there is the potential opportunity for immediate regeneration of methionine. Reduction of methionine sulfoxide is achievable in concentrated hydrochloric acid through the addition of dimethyl sulfide (30, 34). However, due to the deleterious effects of the strong acidic conditions on peptide structure, these approaches to tryptophan cleavage and methionine sulfoxide reduction are rarely used. In principle, if an appropriate solvent were identified, DMSO-promoted tryptophan cleavage could be rapidly followed by DMS-induced methionine regeneration.

Efficient cleavage at tryptophan cannot be achieved with DMSO and 4N hydrochloric acid in acetic acid (11). The action of a more potent halogenating agent such as HBr is required. Unfortunately, its presence also results in an undesirable increase in the level of cysteic acid (Table 1). It was suspected that the inability of DMSO-HCl to provide cleavage resulted from a reduced potency due to water hydration. To reduce the water concentration trifluoroacetic acid was selected to replace acetic acid. It possesses improved solvating properties without the risk of irreversible peptide acylation. The increased acidity of trifluoracetic acid relative to acetic acid can compensate for the use of a markedly reduced concentration of HCl. Synthetic hexapeptides were prepared and treated with DMSO/HCl/TFA, to assess if cleavage without cysteic acid formation could be achieved. One synthetic peptide chosen was FWGPET-NH$_2$ as it models the desired site of cleavage in the IGF-I fusion-protein. An analogous peptide in which cysteine substitutes for tryptophan was prepared and used to assess the degree of cysteic acid formation. The results

obtained with these peptides were most gratifying as a level of cleavage exceeding 50% was achieved with less than 1% formation of cysteic acid (Table 2).
These findings were substantiated by analogous observations made after similar treatment of several natural peptides that contain tryptophan and/or cysteine. With each peptide examined cleavage was selective, and a variable degree of methionine conversion to its sulfoxide was the only other modification of any signficance.

Treatment of methionine sulfoxide with DMS/TFA provided only minimal reduction to methionine while the additional presence of an appropriate concentration of HCl yielded quantitative conversion (Table 3).
The action of DMS is required to achieve complete reduction, although its precise concentration appears to be of lesser importance than that of HCl. The selectivity and efficiency of reduction with DMS/HCl/TFA was displayed through the regeneration of natural GRF from its sulfoxide derivative in 30 min at 95% yield. This method offers a new approach in sulfoxide reduction whose only apparent limitation is the potential coincident destruction of tryptophan. Methionine sulfoxide can serve similarly to DMSO in the oxidation of tryptophan. Where tryptophan has not been purposefully destroyed, its integrity can be largely retained through the course of reduction by the presence of appropriate scavengers. Glucagon possesses a tryptophan two residues apart from a methionine. It was observed to be recoverable in 85% yield from its sulfoxide analog by the addition of an appropriate excess of tryptophan.

The newly developed cleavage reagent was applied to the IGF-I fusion protein. A nearly equal mixture of IGF-I and its methionine sulfoxide analog were formed in a total yield of approximately 40% (Figure 1). Amino acid analysis of the cleavage product revealed cysteic acid formation to have occurred at less than a percent of the initial cysteine/cystine content. No other modifications were apparent. Reduction of the sulfoxide analog of IGF-I immediately following cleavage and solvent evaporation was achieved in 95% yield through DMS/HCl/TFA treatment (Figure 1). The IGF-I was initially purified in the S-sulfonate form by cation-exchange chromatography. Through analogous conditions developed for insulin and proinsulin formation, natural IGF-I was obtained from its S-sulfonate in nearly 50% yield (10). Unlike proinsulin,

TABLE 2
TREATMENT OF FWPGET-NH$_2$ WITH VARIOUS
TRYPTOPHAN CLEAVAGE REAGENTS

Reagent	Cleavage Yield (PGET-NH$_2$)	Oxidation Yield (FC(SO$_3$)PGET-NH$_2$)
BNPS-skatole	78%	3.8%
NCS	62%	15.9%
CNBr/HFBA	<5%	0.9%
DMSO/HCl/HBr/HOAc	43%	72.0%
DMSO/HCl/TFA	55%	0.1%

The conditions are described in Table 1 with the single exception that CNBr-cleavage was for 60 min. The DMSO/HCl/TFA treatment was at a reagent concentration of 1/1/98, on a volume basis. Peptides were treated at a concentration of 1 mg/ml. Quantitation of cleavage and oxidation yields was by reversed-phase chromatography against an appropriate standard.

TABLE 3
CONVERSION OF METHIONINE SULFOXIDE TO
METHIONINE IN DMS/HCl/TFA

DMS (M)	HCl (N)	TFA (%, v/v)	Yield of Methionine %
---	0.1	99	31
0.01	0.1	99	94
0.1	0.1	98	93
1.0	0.1	90	95
5.0	0.1	47	99
1.0	---	91	1
1.0	0.01	90	14
1.0	0.1	90	95

Reduction was conducted at a concentration of 300 µg/ml for 60 min. The HCl was added as an aqueous 12 $\underline{N}$ solution.

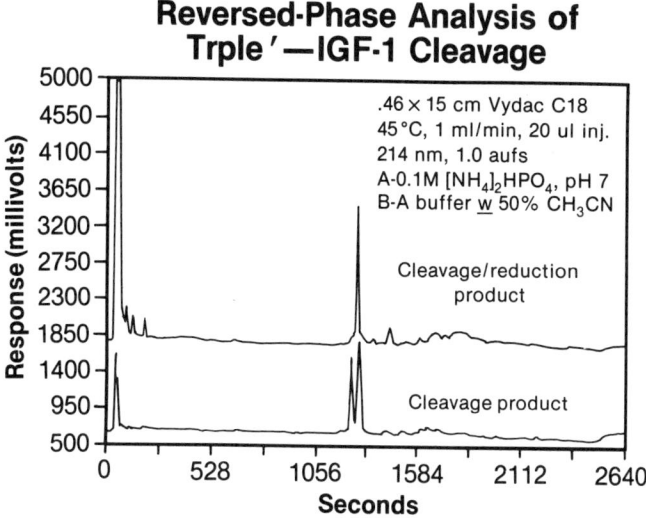

Figure 1

a major single impurity was formed in this conversion at one-half the yield of IGF-I. The remaining IGF-peptides were present as high molecular weight disulfide polymers. The natural IGF-I was purified to near-homogeneity by preparative reversed-phase chromatography.

The exact nature of the major impurity generated in the formation of IGF-I is presently unknown. Initial examination suggests it to be a monomeric disulfide isomer of IGF-I. Subjection of this peptide or IGF-I to the conditions of disulfide interchange from which each was originally obtained yielded analogous results to that initially observed with the IGF-I S-sulfonate. Identical treatment of a Leu^{59}, IGF-I S-sulfonate analog, produced through cyanogen bromide cleavage, yielded a similar chromatographic profile of two peptides. Consequently, the formation of this disulfide isomer does not appear to be a result of the newly developed methodology but instead an inherent feature of the disulfide interchange conditions.

Appreciable quantities of biosynthetic IGF-I can be rapidly produced by this newly developed methodology in tryptophan cleavage and methionine sulfoxide reduction.

The IGF-I which has been prepared by this methodology has been extensively characterized by chemical, physical, and biological analyses. Its amino acid content and sequence are identical to natural IGF-I. Peptide-mapping has verified the proper formation of disulfide bonds. Chromatographic analysis by high-performance methods of differing selectivity have revealed the IGF-I to be no less than 98% pure. In radioimmunoassay the biosynthetic peptide is indistinguishable from natural IGF-I. The biosynthetic IGF-I has proven to be extremely potent in its ability to stimulate protein synthesis and amino acid transport in human fibroblasts (35).

The synthetic methodology described within this report provides a highly efficient alternative to cyanogen bromide cleavage for the biosynthesis of peptides. The efficiency and selectivity achieved in the synthesis of IGF-I closely mimics that which was reported for proinsulin through the action of cyanogen bromide. The application of this cleavage strategy to the amino acid sequence determination of proteins is currently being conducted. The regeneration of methionine following tryptophan cleavage facilitates additional fragmentation through the action of cyanogen bromide.

REFERENCES

1. Gross E and Witkop B (1962). J. Biol. Chem 237:1856.
2. Omenn, GS, Fontana A and Anfinsen CB (1970). J. Biol. Chem 245:1895.
3. Fontana A (1980). In Biochemical and Medical Aspects of Tryptophan Metabolism: Elsevier/North-Holland Biomedical Press, Amsterdam, p 59.
4. Offord RE (1978). In Semisynthetic Peptides and Proteins, Offord RE and DiBello C (eds): Academic Press/London, p 3.
5. Chaiken IM (1981). CRC Critical Reviews Biochem. 11:255.
6. Itakura K, Hirose T, Crea R, Riggs AD, Heyneker HL, Bolivar F and Boyer H. W. (1977). Science 198:1056.
7. Goeddel DV, Kleid DG, Bolivar F, Heyneker HL, Yansura DG, Crea R, Hirose T, Kraszenski A, Itakura K and Riggs AD (1979). Proc. Natl. Acad. Sci. USA 76:106.
8. Busse WD and Carpenter FH (1974). J. Am. Chem. Soc. 96:5947.

9. Sanders DJ and Offord RE (1977). Hoppe-Seyler's J. Physiol. Chem. 358:1469.
10. Chance RE, Hoffmann JA, Kroeff EP, Johnson MG, Schirmer EW, Bromer WW, Ross MJ and Wetzel R (1981). Proc. Am. Pept. Symp. 7th, 721.
11. Savige WE and Fontana A (1977). Methods Enzymol. 47:459.
12. Schechler Y, Patchornik A and Burstein Y (1976). Biochemistry 15:5071.
13. Lischwe, MA and Sung MT (1977). J. Biol. Chem. 252:4976.
14. DiMarchi RD, Wang CC, Hemenway JB and Gurd FRN (1978). Biochemistry 17:1968.
15. Huang HV, Bond MW, Hunkapiller MW and Hood LE (1983). Methods Enzymol 91:318.
16. Froesch ER and Zapf J (1985). Diabetologia 28:485.
17. Zapf J. Froesch ER and Humbel RE (1981). Current Topics in Cellular Regulation 19:257.
18. Underwood LE, D'Ercole AJ, Clemmons DR and Van Wyk JJ (1986). Clinics in Endocrinology and Metabolism 15:59.
19. Minuto F, Del Monte P, Barreca A, Fortini P, Carida G, Catrambone G and Giordano G (1986). Cancer Research 46:985.
20. Dickson RB, McManaway ME and Lippman ME (1986). Science 232:1540.
21. Rinderknecht E and Humbel RE (1978). J. Biol. Chem. 253:2769.
22. Peprides PE, Hintz RL, Bohlen P and Shively JE (1986). Endocrinology 118:2034.
23. Li CH, Yamashiro D, Gospodarowicz D, Kaplan SL and Van Uhet G (1983). Proc. Natl. Acad. Sci. U.S.A. 80:2216.
24. Mullenbach GT, Choo QL, Urdea MS, Barr PJ, Merryweather JP, Brake AS and Valenzuela PV (1983). Fed. Proc. 42:1832.
25. Peters, MA, Lau EP, Snitman DL, Van Wyk JJ, Underwood LE, Russell WE and Svoboda ME (1985). Gene 35:83.
26. Nilsson B, Holmgren E, Josephson S, Gatenbeck S, Philipson L and Uhlen M (1985). Nucleic Acids Research 13:1151.
27. Furman TC, Epp J, Hsiung HM, Hoskins J, Long GL, Mendelsohn LG, Schoner B, Smith DP and Smith MC (1987). Biotechnology 5:1047.
28. Williams DC, Van Frank RM, Muth WL and Burnett JP (1982). Science 215:687.

29. Ozols J. and Gerard C (1977). J. Biol. Chem. 252:5986.
30. Savige WE and Fontana A (1977). Methods Enzymol 47:453.
31. Houghten RA and Li CH (1979). Anal Biochem. 98:86.
32. Tam JP, Heath WF and Merrifield RB (1983). J. Am. Chem. Soc. 105:6442.
33. Fugii N, Kuno S, Otaka A, Funakoshi S, Takagi K and Yajima H (1985). Chem. Pharm. Bull. 33:4587.
34. Schechter Y (1986). J. Biol. Chem. 251:66.
35. Johnson BG, DiMarchi RD, Smith MC and Mendelsohn LG (1988). J. Cell. Biol., Submitted.

ELUCIDATION OF A NOVEL GASTRIN RELEASING PEPTIDE ANTAGONIST BY MINIMAL LIGAND ANALYSIS

David C. Heimbrook, Mark E. Boyer, Victor M. Garsky*, Nancy L. Balishin, David M. Kiefer, Allen Oliff, and Mark W. Riemen

Departments of Cancer Research
and *Medicinal Chemistry
Merck Sharp and Dohme Research Laboratories
West Point, PA 19486

ABSTRACT. Gastrin releasing peptide (GRP) is a peptide hormone containing 27 amino acids which is structurally analogous to the amphibian peptide bombesin. To investigate whether small GRP fragments could bind to the GRP receptor without stimulating mitogenesis, we performed binding inhibition and thymidine uptake assays in Swiss 3T3 fibroblasts. These studies were facilitated by the development of a tritiated GRP-based radioligand, [^{3}H-Phe^{15}] GRP 15-27, which exhibits enhanced chemical stability compared to iodinated GRP derivatives. We examined a series of C-terminal GRP fragments, from the pentapeptide to the octapeptide, with both N-acetyl and free amine moieties at the N-terminus. N-acetylated derivatives were more potent than their primary amine counterparts in both assays. Deletion of N-terminal residues from GRP 20-27 resulted in loss of potency in the binding inhibition assay, with corresponding loss of activity as a mitogen. In contrast, deletion of Met^{27} from the parent structure generated a mitogenic antagonist, N-acetyl GRP 20-26 amide, with an EC$_{50}$ of 1.5 μM. These results suggest that: 1) for the series of N-terminal deletions, binding to the receptor and mitogenic activity are tightly coupled, and 2) the C-terminal residue, Met^{27} amide, plays an essential role in triggering the mitogenic response caused by GRP.

INTRODUCTION

Gastrin releasing peptide (GRP[1]) is a widely distributed peptide hormone containing 27 amino acids which is structurally analagous to the amphibian peptide bombesin (1). Bombesin and GRP mediate a wide variety of biological responses, including stimulation of smooth muscle contraction, amylase secretion (2,3), effects on temperature regulation (4,5), and stimulation of mitogenesis in mouse fibroblast cells (6-8). A growing body of evidence indicates that GRP also stimulates proliferation of small cell lung cancer cells in vitro and in vivo (9-12). GRP may stimulate the growth of small cell lung cancers via an autocrine loop mechanism. Delineation of the binding domain of GRP may facilitate the development of a compound to disrupt this autocrine loop. Previous work indicates that the C-terminal octapeptide, N-acetyl GRP 20-27, exerts full agonist activity in temperature and glucose regulation (5) and gastric secretion (13) in rats. Additional reports have characterized the effect of structural deletions and modifications of bombesin on agonist activity (2,4,5,7), but no systematic study comparing binding inhibition to mitogenic activity has been presented. Since this comparison might identify compounds which bind to the GRP receptor but fail to stimulate mitogenesis, we examined a series of GRP derivatives with N-terminal or C-terminal deletions in receptor binding and mitogenesis assays.

MATERIALS AND METHODS

Peptide ligands were synthesized by standard solid-phase methodology on benzhydrylamine resin using commercially

1. Abbreviations used:
BSA-Bovine serum albumin
BOC-tert-butyloxycarbonyl
DMEM-Dulbecco's modified Eagle medium
EC_{50}-Concentration of ligand required to obtain 50% of maximal response.
GRP-Gastrin releasing peptide
HPLC-High performance liquid chromatography
PBS-Phosphate-buffered saline
TFA-Trifluoroacetic acid

available BOC amino acids containing the normal side chain protecting groups (14,15,16). Cleavage and deprotection of the resin-bound peptide was achieved by treatment with liquid HF containing anisole as a scavenger. Purification of the crude product by reverse-phase HPLC on a Vydac C-18 support yielded a product whose purity was > 98% by analytical HPLC (17). Quantitative amino acid analysis gave the correct ratios for the given sequence, and yielded an effective molecular weight, which was used in all calculations.

[Phe^{15}] GRP 15-27 was prepared by hydrogenolysis of [I-Phe^{15}] GRP 15-27 (18). Ten milligrams of [I-Phe^{15}] GRP 15-27$_2$ was dissolved in 3 ml N,N-dimethylformamide. Ten mg of MgO and 10 mg of 20% Pd on carbon were added, and the reaction vessel was purged of oxygen by three N_2/vacuum cycles with stirring. Hydrogen gas was then admitted to the reaction vessel at 1 atm. The progress of the reaction was monitored by HPLC (Vydac C-18 column, linear gradient from 5% (CH_3CN + 0.1% TFA) / 95% (H_2O + 0.1% TFA) to 60% (CH_3CN + 0.1% TFA)) or by thin layer chromatography (EM Science silica plates, 7:5:3:1 ethyl acetate:pyridine:water:acetic acid). After completion, the catalyst was removed by filtration, and the solvent was removed in vacuo. The residue, a pale yellow oil, was dissolved in 15 ml 1% acetic acid, filtered, and lyophilized.

[^{3}H-Phe^{15}] GRP 15-27 was prepared by New England Nuclear (Boston, MA) using a protocol similar to the one described above, using tritium gas in lieu of hydrogen. The product was purified by preparative HPLC, and the TFA was removed by chromatography on Bio-Rad AG-3 acetate in 1% acetic acid. The final product had a specific activity of 10 Ci/mmol, and was >95 % pure by radiometric HPLC analysis.

^{125}I-[Nle^{27}] GRP 15-27 was prepared by chloramine-T treatment of [Nle^{27}] GRP 15-27 (19). The final product had a specific activity of 10 Ci/mmol. ^{125}I-[Tyr^4] bombesin was prepared from [Tyr^4] bombesin by a similar protocol or was purchased from Amersham (Arlington Heights, IL).

Human [Nle^{27}] GRP 1-27 and human GRP 1-16 were custom synthesized by Bachem (Torrance, CA). [Tyr^4] bombesin and N-acetyl GRP 20-27 were obtained from Peninsula Laboratories (Belmont, CA). All other peptides were synthesized as described above.

2. BOC-o-iodo-Phe was prepared by Dr. S. Varga by an unpublished procedure.

Binding Inhibition Studies

Swiss 3T3 cells were obtained from Dr. K. Brown (Institute of Animal Physiology, Cambridge, U.K.). The cells were grown to confluency in Costar 12-well plates containing DMEM (Gibco) supplemented with 10% fetal bovine serum, 2 mM glutamine and 1% penicillin-streptomycin. The cells were washed twice with binding buffer {1:1 DMEM:Waymouths MB752/1 medium, plus 1 mg/ml BSA (Fraction V, Calbiochem)}. The inhibitor was dissolved in 10 mM HCl, and diluted to the appropriate concentration in binding buffer. The inhibitor was then added to the cells, followed by radioligand at a final concentration of 7 nM. After 30 min incubation at 37°C, the supernatant was removed and the cell monolayer rinsed four times with washing buffer (150 mM NaCl, 20 mM Na_2HPO_4, 5mM KCl, 1.8 mM KH_2PO_4, 1 mg/ml BSA, pH 7.4). The cells were then lysed with 1 ml/well of lysis buffer (1% Triton X-100, 0.1% BSA), and the solution was aspirated into scintillation vials for counting. Each data point was collected in triplicate.

Mitogenesis Studies

Swiss 3T3 cells were grown in monolayer culture in 24-well plates (Costar) in serum-free DMEM for 48 hr, at which time the mitogen and 23 nM ^{3}H-thymidine were added. After an additional 48 hr, the cell monolayer was washed twice with phosphate-buffered saline, and the cells were then removed with 1 ml 10x trypsin containing 5 mM EDTA. The cells were harvested with a Skatron filter apparatus (Skatron, Inc., Sterling, VA), and the filters counted in a scintillation counter. Mitogenic inhibition studies were conducted by a similar protocol, except that varying concentrations of the antagonist were added to the cells at the same time as the mitogen, 7 nM [Nle^{27}] GRP 15-27.

RESULTS

Our first goal in developing a reproducible binding assay was to select a suitable radioligand. ^{125}I-[Tyr^4] bombesin is commercially available, but in our hands this ligand exhibited high non-specific binding and was too unstable for routine use. Similar results were obtained when we prepared ^{125}I-[Tyr^4] bombesin in our own laboratory. Based on the hypothesis that the variabilty in using this radioligand may result from oxidation of the C-terminal methionine (16), we synthesized a GRP fragment with a norleucine substitution at the C-terminus. This peptide, [Nle^{27}] GRP 15-27, does not contain methionine, and was readily iodinated using chloramine-T (19). Binding of this ligand to the GRP receptor of Swiss 3T3 cells was saturable and >90% specific at concentrations below 50 nM. Scatchard analysis indicated the dissociation constant of this radioligand was 4 nM, with approximately 400,000 receptors/cell (20). The ligand was used in a competitive binding assay with porcine GRP 1-27, [Nle^{27}] GRP 15-27, acetyl GRP 20-27, and porcine GRP 1-16; fragments containing the C-terminal region of GRP inhibited binding of the radioligand with EC$_{50}$'s of approximately 5 nM, while the N-terminal GRP fragment showed no inhibition of binding at concentrations as high as 1 μM (Table 1). These results are in accord with previous reports using radiolabeled GRP ligands containing the C-terminal methionine (7,12,13) and demonstrate that isosteric substitution of Nle for Met at position 27 of this GRP fragment results in little or no change in specificity of the ligand for its receptor in Swiss 3T3 cells.

While the iodinated Nle^{27}-radioligand showed increased stability and ease of preparation compared to the radioligand derived from the native sequence, significant loss of specificity was noted after storage at -70 C. We therefore investigated the possibility of preparing a tritiated radioligand, which might offer improved stability, as well as a reduced radiation hazard. This ligand was prepared by reduction of [o-iodo Phe^{15}] GRP 15-27 with ^{3}H$_2$ over a Pd/C catalyst (18). Binding of this ligand to Swiss 3T3 cells was saturable and > 90% specific at concentrations below 25 nM. Scatchard analysis yielded a dissociation constant of 2 nM. Competitive binding inhibition curves were generated vs human [Nle^{27}] GRP 1-27, N-acetyl GRP 20-27, and human GRP 1-16. As with the iodinated ligand, the EC$_{50}$'s of peptides containing the C-terminal portion of GRP occured at a 1:1 ratio of radioligand : antagonist (Table 1). GRP 1-16 showed no inhibition of binding at concentrations up to 100 μM. In addition, the radioligand was assayed in a mitogenic stimulation assay, and showed activity similar to human [Nle^{27}] GRP 1-27. The tritiated GRP derivative also exhibited significantly improved chemical stability compared to the iodine-labeled compounds. No degradation in binding affinity or specificity was observed upon storage of the radioligand at -70°C for three months. Because of its exceptional stability, high specific activity, and appropriate specificity for the GRP receptor, the tritiated ligand was selected for use in routine binding assays.

The N-acetylated C-terminal octapeptide of GRP is the smallest GRP fragment reported to yield a full agonist response (5,13). We prepared a number of shorter C-terminal peptides to identify the smallest derivative capable of binding to the receptor. The results of the binding inhibition studies are shown in Table 2. Removal of the N-acetyl group caused a small but significant increase in the EC_{50} of the C-terminal octapeptide. Deletion of His^{20} dramatically reduced the affinity of the peptide for the GRP receptor. The affinity of the N-acetylated heptapeptide was reduced by a factor of 10^2, while the primary amine derivative exhibited 10^3-fold lower affinity, compared to their octapeptide counterparts. Deletion of His^{20} and Trp^{21}

TABLE 1. CHARACTERIZATION OF RADIOLIGANDS BY COMPETITIVE BINDING INHIBITION

Competing Ligand	EC_{50} (nM) vs:	
	^{125}I-[Nle^{27}] GRP 15-27	[^3H-Phe^{15}] GRP 15-27
Porcine GRP 1-27	3	N.D.
Human [Nle^{27}] GRP 1-27	N.D.	10
[Nle^{27}] GRP 15-27	5	9
N-Acetyl GRP 20-27	4	9
Porcine GRP 1-16	>1,000	>100,000

N.D. - Not Determined

TABLE 2. POTENCY OF GRP DELETION PEPTIDES

Sequence[a]									EC_{50} (nM)	
20	21	22	23	24	24	26	27		Binding Inhibition	Mitogenic Stimulation
Acetyl His	Trp	Ala	Val	Gly	His	Leu	Met NH_2		9	3
His	Trp	Ala	Val	Gly	His	Leu	Met NH_2		40	30
Acetyl Trp	Ala	Val	Gly	His	Leu	Met NH_2			900	500
Trp	Ala	Val	Gly	His	Leu	Met NH_2			70,000	5,000
Acetyl Ala	Val	Gly	His	Leu	Met NH_2				90,000	20,000
Ala	Val	Gly	His	Leu	Met NH_2				>100,000	>100,000
Acetyl Val	Gly	His	Leu	Met NH_2					>100,000	70,000
Val	Gly	His	Leu	Met NH_2					>100,000	>100,000
Acetyl His	Trp	Ala	Val	Gly	His	Leu	NH_2		1,500	>100,000

[a] The sequence numbers correspond to the sequence of GRP.

yielded a hexapeptide which exhibited significant inhibition of binding of the radioligand at a concentration of 100 μM in the acetylated form, but no inhibition as the primary amine. The C-terminal pentapeptide, lacking His[20], Trp[21], and Ala[22] compared to the octapeptide, showed slight (20 %) inhibition of binding at 100 μM as the acetylated derivative, and did not inhibit binding of the radioligand as the primary amine at concentrations up to 100 μM (data not shown).

Mitogenesis studies were conducted to determine if these GRP fragments also stimulated tritiated thymidine uptake in Swiss 3T3 cells. The results of these experiments are presented in Table 2. In general, binding affinity and mitogenic potency increased in parallel for these peptides. N-acetyl GRP 20-27 yielded a half-maximal mitogenic response at approximately 3 nM, as did [Nle[27]] GRP 15-27 and human [Nle[27]] GRP 1-27. The maximal stimulation of tritiated thymidine uptake with each of these peptides was approximately 35-70 fold greater than controls treated with phosphate-buffered saline, and was 25-50% of the maximal response induced by 5% FBS under similar conditions. The EC_{50} for the primary amine octapeptide was 10-fold higher than the acetylated derivative, reflecting its lower affinity for the receptor. N-acetyl GRP 21-27 had an EC_{50} of approximately 500 nM, while the amino heptapeptide exhibited half-maximal stimulation at 5 μM. The C-terminal hexapeptide and pentapeptide also displayed mitogenic potency which correlated with the results of the binding inhibition studies. Despite variations in their potency, all of the N-terminal deletion peptides with a mitogenic EC_{50} of less than 30 μM generated the same maximal mitogenic response at 100 μM as N-acetyl GRP 20-27.

A dramatically different result was observed with the C-terminal deletion peptide, N-acetyl GRP 20-26 amide. This peptide blocked binding of the radioligand with an EC_{50} of 1.5 μM, but did not stimulate mitogenesis in 3T3 cells (Table 2). In contrast, this peptide blocked the mitogenic effect of a challenge dose of 7 nM [Nle[27]] GRP 15-27 with an EC_{50} virtually identical to that obtained in binding inhibition experiments (Figure 1). These results indicate that N-acetyl GRP 20-26 amide is a mitogenic antagonist.

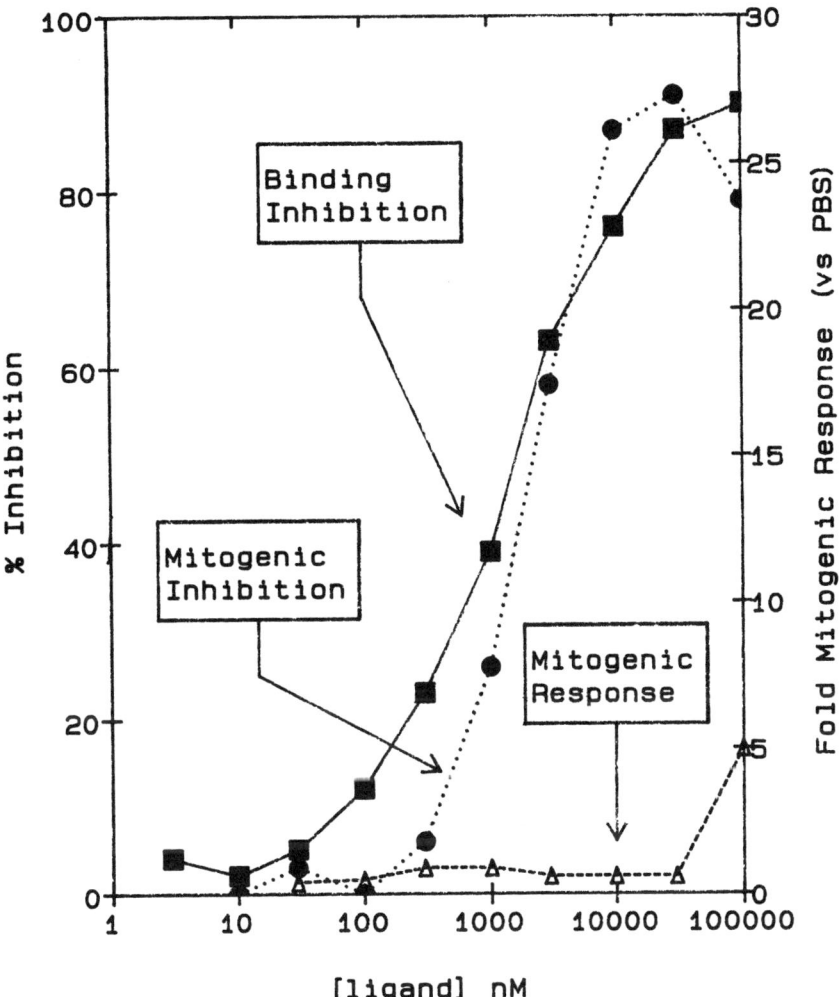

FIGURE 1. N-acetyl GRP 20-26 amide blocks binding of [^{3}H-Phe^{15}] GRP 15-27, but does not stimulate mitogenesis. This peptide also blocks mitogenic stimulation by a challenge dose of 7 nM [Nle^{27}] GRP 15-27. Curves with closed symbols refer to the left Y-axis, curves with open symbols to the right Y-axis.

DISCUSSION

Interest in the bombesin/GRP family of peptides has been stimulated in recent years by reports that they may act as autocrine growth factors in small cell lung cancer cells. A number of GRP antagonists have been identified, based on substance P or bombesin as parent structures (21,22). These antagonists exhibit dissociation constants between 400 nM and 10 µM. Other peptide sequences have been tested for bombesin agonist activity in a variety of biological assays. The smallest peptide reported to exhibit full agonist activity is N-acetyl GRP 20-27 (5,13). Smaller fragments, as well as a number of larger peptides with amino acid substitutions, exhibit reduced agonist activity in temperature regulation and smooth muscle contraction assays (2,4,5). Because competitive binding inhibition data was not obtained in these assay systems, however, it is impossible to determine whether the reduced activity of these peptides is indicative of partial antagonism or the absence of binding of the peptides to the receptor. It is also uncertain how data obtained in these assay systems relates to GRP-stimulated mitogenesis. Therefore, we conducted a systematic comparison of binding inhibition and mitogenic stimulation in a series of GRP C-terminal peptides.

Our first goal was to prepare an improved radioligand for binding assays. Methionine residues are prone to oxidation (15, 16). To avoid this problem, we made the conservative substitution of Nle for Met^{27} in a GRP fragment. The resulting peptide, $[Nle^{27}]$ GRP 15-27, retained full mitogenic activity (data not shown), and was readily iodinated to yield a radioligand which demonstrated saturable and specific binding. Competition experiments with various other GRP derivatives validated the use of the ligand. These results demonstrate that conservative substitutions in the binding region of GRP are capable of generating novel peptides which retain both their affinity for the GRP receptor and mitogenic activity.

While the iodinated $[Nle^{27}]$ GRP 15-27 represented a significant improvement in stability of GRP-related radioligands, it was still subject to decomposition and batch-to-batch variability. The preparation of $[^3H\text{-}Phe^{15}]$ GRP 15-27 addressed both of these difficulties. This ligand showed little or no chemical decomposition at -70 C and exhibited improved binding characteristics, reflected primarily in lower non-specific binding. The tritiated radioligand exhibited less binding to plastic, as well as showing reduced non-specific binding to cells. The improved binding characteristics of the tritiated ligand resulted in a more reproducible assay. In addition, the low-energy beta emissions of the tritiated ligand presented a reduced radiation hazard. The one drawback to the use of [^{3}H-Phe^{15}] GRP 15-27 is that its lower specific activity compared to ^{125}I-labeled peptides reduces the sensitivity of the binding assay. However, this characteristic did not adversely affect our results.

There is evidence that substitutions within the C-terminal region of bombesin and substance P can generate peptides which bind to the GRP receptor without stimulating mitogenesis. These peptides exhibit reduced affinity for the receptor compared to native sequence GRP (21,22). In comparing binding affinity and mitogenic stimulation by deletion peptides containing the native C-terminal sequence of GRP, we find that loss of agonist activity is tightly coupled to loss of binding to the receptor. Nearly proportional increases in the EC$_{50}$'s for binding and mitogenesis were observed over the range of peptides tested. Amino acid deletions towards the C-terminus beyond position 19 of GRP result in a cumulative loss of activity. In addition, it is clear that disruption of the hydrophobicity of these peptides by the presence of a primary amine at the N-terminus dramatically reduces affinity for the receptor, as well as agonist activity (5). Our data confirm that elimination of the positive charge at the N-terminus of shorter GRP fragments by acetylation significantly enhances the activity of the peptide.

In contrast to the N-terminal deletion series, deletion of Met^{27} from N-acetyl GRP 20-27 generates a peptide which binds to the GRP receptor but does not stimulate a mitogenic response. While the binding potency of this compound is 10^3 lower than the parent compound, it is nevertheless remarkable that triggering of the biological response is apparently controlled by a single amino acid, although a similar observation has been reported by Martinez and coworkers with cholecystokinin (23). As of yet, it is unclear system whether the amino acid deletion results in disruption of a biologically active conformation of the peptide, or whether the amino acid specifically interacts with a critical portion of the receptor. The observation that [Nle^{27}] GRP derivatives are potent agonists suggests that the sulfide moiety of the methionine side chain is not responsible for triggering the mitogenic response.

ACKNOWLEDGEMENTS

The authors wish to acknowledge Mr. William Hurni for preparation of iodinated ligands, Ms. Lori Wassel for amino acid analysis, Dr. Avery Rosegay for purification of [^{3}H-Phe^{15}] GRP 15-27, and Ms. Rita Taylor for editorial assistance.

REFERENCES

1. Anastasi, A., Erpsamer, V., and Bucci, M. (1971) Experientia 27 166-167.
2. Broccardo, M., Erspamer, G.F., Melchiorri, P., Negri, L., and De Castiglione, R. (1975) Br. J.Pharmac. 55 221-227.
3. Westendorf, J.M., and Schonbrunn, A. (1983) J. Biol. Chem. 258 7527-7535.
4. Rivier, J.E., and Brown, M.R. (1978) Biochemistry 17 1766-1771.
5. Marki, W., Brown, M. and Rivier, J.E. (1981) Peptides 2 s.2 169-177.
6. Rozengurt, E. and Sinnett-Smith, J. (1983) Proc. Natl. Acad. Sci. USA 80 2936-2940.
7. Zachary, I. and Rozengurt, E. (1985) Proc. Natl. Acad. Sci. USA 82 7616-7620.
8. Rozengurt, E. (1985) in"Molecular Mechanisms of Transmembrane Signalling" (Cohen, P. and Houslay, M.D., eds.) pp. 429-452,Elsevier Science Publishers, B.V., New York.
9. Moody, T.W., Pert, C.B., Gasdar,A.F., Carney, D.N. and Minna, J.D. (1981) Science 214 1246-1248.
10. Moody, T.W., Bertness, B. and Carney, D.N. (1983) Peptides 4 683-686.
11. Cuttitta, F. Carney, D.N., Mulshine, J., Moody, T.W., Fedorko, J., Fischler, A. and Minna, J.D. (1985) Nature 316 833-836.
12. Weber, S., Zuckerman, J.E., Bostwick, D.G., Bensch, K.G., Sikic, B.I., and Raffin, T.A. (1985) J. Clin. Invest. 75 306-309.
13. Tache, Y., Marki, W., Rivier, J., Vale, W., and Brown, M. (1980) Reg. Pept. 1 S112.
14. Merrifield, R.B. (1963) J. Am. Chem. Soc. 85 2149-2154.
15. Stewart, J.M., and Young, J.D. (1969) "Solid Phase Peptide Synthesis", W.H. Freeman, San Francisco.

16. Barany, G., and Merrifield, R.B. (1980) in "The Peptides", vol. 2, (Gross, E., and Meienhofer, J., eds.) pp. 1-285, Academic Press, New York.
17. Rivier, J., McClintock, R., Galyean, R., and Anderson, H. (1984) J. Chromatogr. 288 303-328.
18. Brundish, D.E., and Wade, R. (1973) J. Chem. Soc., Perkin Trans. 1, 2875-2879.
19. Riemen, M.W., Wegrzyn, R.J., Baker, A.E., Hurni, W.M., Bennett, C.D., Olitf, A., and Stein, R.B. (1987) Peptides 8 877-855.
20. Levitzki, A. (1984) "Receptors: A Quantitative Approach" Benjamin/Cummings Publishing Co., Menlo Park, CA.
21. Jensen, R.T., Jones, S.W., Folkers, K., and Gardner, J.D. (1984) Nature 309 61-63.
22. Heinz-Erian, P., Coy, D.H., Tamura, M., Jones, S.W., Gardner, J.D., and Jensen, R.T. (1987) Am. J. Physiol. 252 G439-G442.
23. Lignon, M.F., Galas, M.C., Rodriguez, M., Laur, J., Aumelas, A., and Martinez, J. (1987) J. Biol. Chem. 262 7226-7231.

SYNTHESIS AND BIOLOGICAL EVALUATION OF GROWTH HORMONE RELEASING FACTOR, STRUCTURAL LINEAR AND CYCLIC ANALOGS

Edgar P. Heimer, Mushtaq Ahmad, Theodore Lambros, Timothy McGarty, Ching-Tso Wang, Thomas F. Mowles, Sarah Maines and Arthur M. Felix

Roche Research Center, Hoffmann-La Roche Inc., Nutley, New Jersey 07110

ABSTRACT In the short period of time since the isolation and characterization of human growth hormone releasing factor, GRF(1-44)-NH_2, it has been synthesized and evaluated *in vitro* and *in vivo* in humans and shown to be an effective treatment for growth hormone deficient children. *In vivo* evaluation of GRF and analogs in livestock have also demonstrated its potential application for performance enhancement. Clinical preparations of GRF(1-44)-NH_2 have been carried out on gram-scale by solid phase peptide synthesis and the product fully characterized. An analog program in which the bioactive core, GRF(1-29)-NH_2, was modified led to the development of [desNH$_2$Tyr^1,D-Ala^2,Ala^{15}]-GRF(1-29)-NH_2 which is 10 times more potent than the parent compound *in vivo*. The high potency of this trisubstituted analog is attributed to increased α-helicity and maximization of amphiphilic character (resulting from Gly^{15} replacement by Ala^{15}) and to increased stability to enzymatic degradation by diaminopeptidase (resulting from NH_2-terminal modification). A novel cyclic GRF analog, Cyclo(Asp^8-Lys^{12})-[Ala^{15}]-GRF(1-29)-NH_2, was synthesized and found to retain significant biological activity both *in vitro* and *in vivo*. Molecular dynamics studies reveal that Cyclo^{8,12}[Ala^{15}]-GRF(1-29)-NH_2 is in a preferred bioactive conformation with a nearly ideal α-helical conformation with enhanced rigidity between residues 2-29.

INTRODUCTION

Peptides with growth hormone releasing activity were first isolated from human pancreatic islet tumors, obtained from patients with acromegaly and characterized independently by Vale et al. (1) and

Guillemin et al. (2). Although three homologous growth hormone releasing factor (GRF) peptides were obtained from the tumors [GRF(1-37)-OH, GRF(1-40)-OH and GRF(1-44)-NH_2], the latter has been shown to be the primary structure (3,4) (Figure 1).

H-TYR-ALA-ASP-ALA-ILE-PHE-THR-ASN-SER-TYR-ARG-

LYS-VAL-LEU-GLY-GLN-LEU-SER-ALA-ARG-LYS-LEU-

LEU-GLN-ASP-ILE-MET-SER-ARG-GLN-GLN-GLY-GLU-

SER-ASN-GLN-GLU-ARG-GLY-ALA-ARG-ALA-ARG-LEU-NH_2

FIGURE 1. Primary Sequence of Human Growth Hormone Releasing Factor (Somatocrinin)

Hypothalamic growth hormone releasing peptides have been isolated from several animal species and shown to have significant sequence homology, predominantly from the NH_2-terminus, to human GRF. In spite of the structural differences of GRF among species, it has been shown that human GRF stimulates the specific secretion of growth hormone (GH), both *in vitro* and *in vivo*, in humans as well as a number of other species (5). Although there are differences in the GH response to GRF, all species that have been treated with human GRF have been found to secrete GH. The potential clinical use of GRF for the treatment of growth hormone deficient children (6,7) as well as its application in livestock for performance enhancement and improved feed utilization (8) has been established. Since GRF has been reported to have a short half-life ($t_{1/2}$ 6.8 min) (9), a program was initiated with the goal of preparing analogs with enhanced potency and/or prolonged biological activity for application in the treatment of growth hormone deficiency (human) and for performance enhancement (animal agriculture).

Structure-activity studies have shown that a shortened sequence consisting only of the first 29 amino acid residues in GRF is required for the full *in vitro* intrinsic activity (1) and has led to the synthesis of carboxy-terminus deletion and replacement analogs by several laboratories (10-13). Chou-Fasman empirical probability calculations for GRF predict regions of β-turn (residues 6-10) and α-helix (residues 16-26). Circular dichroism measurements (75% MeOH) are in general agreement with these predictions and conformational analysis of [Nle^{27}]-GRF(1-29)-NH_2 using two-dimensional NMR (14) revealed regions of α-helix between residues 6-13 and 16-29. Our approach to the development of potent GRF analogs focused on the replacement of Gly^{15} by Ala and other hydrophobic residues in order to extend the α-helical region and maximize amphiphilic structure (15). NH_2-terminal replacement analogs were also developed to improve potency and enhance resistance to proteolytic degradation. This approach led to the development of a novel class of cyclic analogs of GRF which retain the bioactive conformation.

METHODS

All solvents used were of the highest purity and trifunctional amino acids were protected as N^αBoc-Lys(2-Cl-Z), N^αBoc-Asp-(OcHex), N^α-Boc-Ser(Bzl), N^α-Boc-Thr(Bzl) and N^α-Boc-Arg(Tos). Couplings were performed by the DCC *in situ* or symmetrical anhydride procedures except for glutamine and asparagine which were coupled as the hydroxybenzotriazole ester. Peptide synthesis was carried out by the solid phase procedure (16) using our established protocol (17).

The medium scale synthesis of GRF(1-44)-NH_2 was performed on the Vega Model 296 synthesizer. The preparation of the GRF(1-29)-NH_2 analogs were carried out either on the Vega Model 1000 synthesizer or by a manual procedure.

Analytical high performance liquid chromatography (hplc) was carried out on an LDC Constametric IIG equipped with a gradient master and Spectromonitor III UV variable wavelength detector. The analytical columns [Waters μBondapak C-18 (0.4 x 30 cm) and Lichrosorb RP-8 (5μ) (0.4 x 25 cm)] were used with a pre-column of Copell ODS pellicular packing (Whatman). Preparative hplc was performed on the Jobin-Yvon Chromatospac Prep 100 (8 x 100 cm) containing C-8 reversed phase resin. Further purification was accomplished with a Nucleosil C-18 (2 x 30 cm) column. Amino acid analyses were performed on the Waters Amino Acid Analyzer. Peptide mapping was achieved by digestion with TPCK-treated trypsin (209 U/mg, 85% protein; Millipore Corp.). The digest was incubated at 25°C for 20 hr, the pH adjusted to 4 with the addition of several drops of AcOH and immediately lyophilized.

Rat cell cultures were prepared by collagenase dispersion of rat pituitary fragments followed by neuraminidase incubation. Cells were plated in a defined medium (BBM-P) containing 2.5% fetal bovine serum as described (18). Cells were washed and plated in 24 well tissue culture plates for 4-5 days at 37°C, 7.5% CO_2 in air. On day 4-5 medium was aspirated and cells washed twice with BBM-T (serum free). GRF and analogs were incubated for 4 hours at 37°C in one mL of BBM/well at concentrations of 3.1-400 fmole/mL. Treatment media was removed, diluted 1-20, and assayed by radioimmunoassay utilizing reagents supplied by Dr. A. Parlow (Research and Education Institute, Torrance, CA).

RESULTS AND DISCUSSION

The medium scale solid phase peptide synthesis (SPPS) of GRF(1-44)-NH$_2$ (Figure 2) was performed starting with 50 g of Boc-Leu-benzhydrylamine (BHA)-resin. A total of 43 cycles of SPPS were carried out yielding 203.5 g of GRF(1-44)-BHA-resin. The

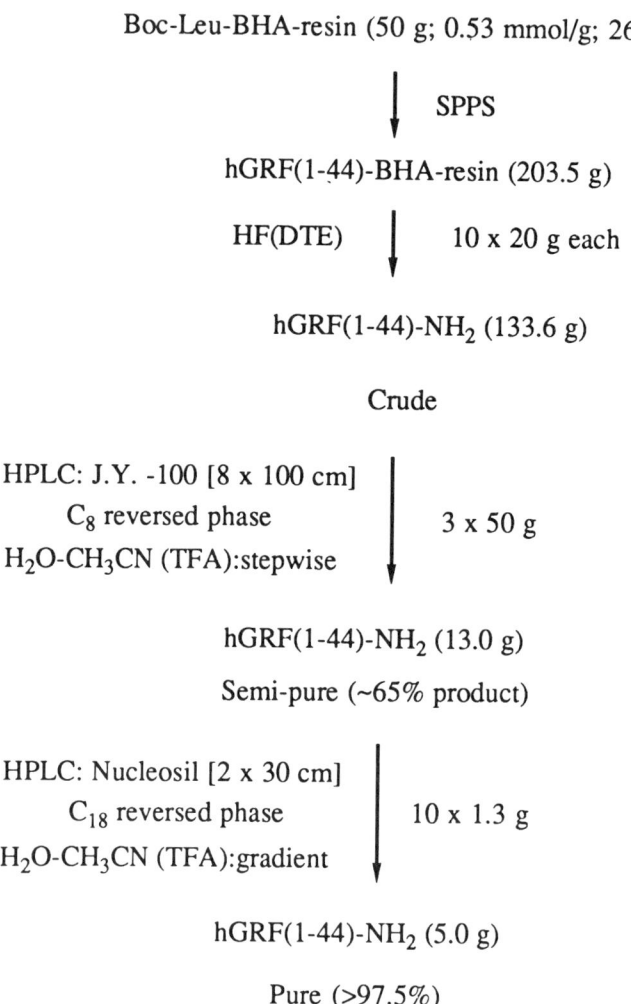

FIGURE 2. Gram-Scale Preparation of hGRF(1-44)-NH$_2$.

peptide-resin was subjected to cleavage using anhydrous HF containing 10% dithioethane (20 g peptide resin/cleavage) giving a total of 134 g of crude GRF(1-44)-NH_2. The crude product was partially purified on the Jobin-Yvon Chromatospac Prep 100 HPLC system to yield semi-pure product (13 g) (estimated purity ~65%). Further purification on a Nucleosil (2 x 30 cm) C-18 column as outlined in Figure 2 yielded a total of 5 g of pure GRF(1-44)-NH_2. The product was shown to be homogeneous by analytical hplc and gave the expected amino acid composition after acid hydrolysis. Fast Atom Bombardment (FAB) mass spectroscopy confirmed a mass $[M + H]^+$ of 5040.5 ± 1.0 (calc. 5040.8). Further characterization by tryptic peptide mapping (Figure 3)

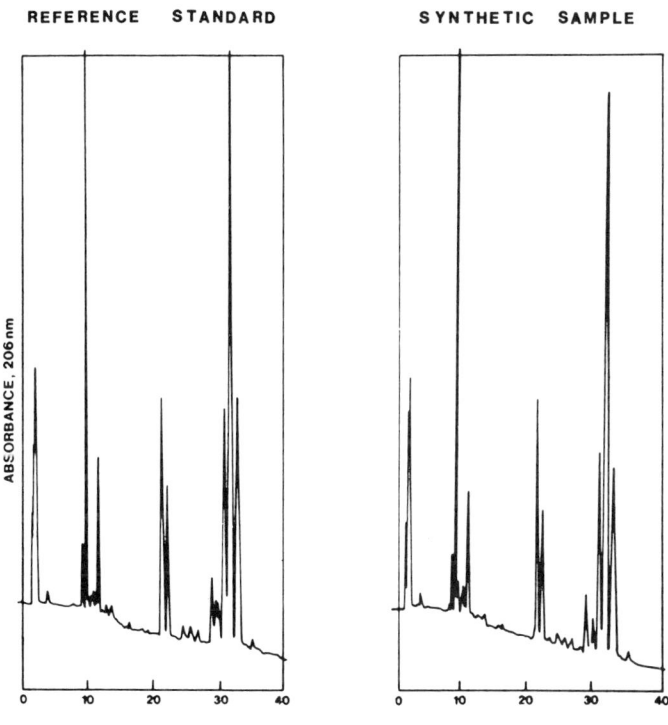

FIGURE 3. Tryptic Digest of hGRF(1-44)-NH_2. Tryptic digestion of GRF at pH 8 at 25° for 20 hours. Analytical hplc on µBondapak C-18 (0.39 x 30 cm). Eluant: H_2O (0.025% TFA)-CH_3CN (0.025% TFA) [0 to 25% in 30 min]; detection: 206 nm; flow rate: 2 mL/min.
Left: 30 µg reference standard
Right: 30 µg purified synthetic product

resulted in an analytical hplc "fingerprint" chromatogram that was identical to that of a reference standard. Other methods of characterization of clinical lots of GRF(1-44)-NH_2 include thin layer high voltage electrophoresis, specific rotation and *in vitro* biological activity (rat pituitary cell culture). Similar scale-up results have been reported for GRF(1-44)-NH_2 as well as GRF(1-29)-NH_2 (19). The initial clinical studies with GRF(1-44)-NH_2 (1 µg/kg every 3 hours) in growth hormone deficient children have demonstrated an acceleration in the velocity of lower leg growth (6). Acceleration of growth has also been reported using GRF(1-40)-OH administered by infusion pump (7). Clinical studies in growth hormone deficient children with our synthetic GRF(1-44)-NH_2 are planned.

As shown in Table 1, replacement of Gly^{15}, with helix forming hydrophobic amino acids, Val, Leu, Aib and Ala resulted in analogs with increasing biological activity (*in vitro*). This observed structure-activity relationship has been attributed to increased α-helicity and maximization of amphiphilic character (12). This conclusion was reinforced by our observation that replacement of Gly^{15} by Sar, a helix-breaking residue, resulted in a profound loss of biological potency.

TABLE 1
REPLACEMENT ANALOGS OF GRF(1-29)-NH_2

GRF ANALOG	RELATIVE POTENCY IN VITRO
GRF(1-44)-NH_2	1.0
GRF(1-29)-NH_2	0.8
[Val^{15}]-GRF(1-29)-NH_2	1.03
[Leu^{15}]-GRF(1-29)-NH_2	3.5
[Aib^{15}]-GRF(1-29)-NH_2	3.7
[Sar^{15}]-GRF(1-29)-NH_2	0.04
[Ala^{15}]-GRF(1-29)-NH_2	4.0
[D-Ala^2]-GRF(1-29)-NH_2	1.6
[N-Me-D-Ala^2]-GRF(1-29)-NH_2	1.2
[$DesNH_2 Tyr^1$]-GRF(1-29)-NH_2	2.5[15]
[$DesNH_2 Tyr^1$,D-Ala^2,Ala^{15}]-GRF(1-29)-NH_2	5.14
Cyclo^{3,12}-[Ala^{15}]-GRF(1-29)-NH_2	0.07
Cyclo^{8,12}-[Ala^{15}]-GRF(1-29)-NH_2	0.79

Analogs of GRF(1-29)-NH$_2$ with replacement of Tyr^1 by desNH$_2$Tyr^1 or incorporation of D-Ala^2 also resulted in increased biological activity. Multiple replacement analogs were prepared and [desNH$_2$Tyr^1,D-Ala^2,Ala^{15}]-GRF(1-29)-NH$_2$ was found to be >5 times more potent than the parent compound *in vitro* (Table 1).

In vivo (porcine) studies with [desNH$_2$Tyr^1,D-Ala^2,Ala^{15}]-GRF(1-29)-NH$_2$ revealed an even greater enhancement (10-fold) in potency which is attributable to the increased stability of the NH$_2$-terminus to enzymatic degradation. This conclusion was reached in a recent *in vivo* plasma study which demonstrated that the primary breakdown in GRF is initiated by a diaminopeptidase which recognizes the Tyr^1-Ala^2- sequence (9) and metabolizes GRF(1-44)-NH$_2$ to GRF(3-44)-NH$_2$. Since desNH$_2$Tyr^1,D-Ala^2, is a poor substrate for this enzyme the t$_{1/2}$ of [desNH$_2$Tyr^1,D-Ala^2,Ala^{15}]-GRF(1-29)-NH$_2$ is substantially greater than the parent compound (20). In a swine performance enhancement study (21) treatment with [desNH$_2$Tyr^1,D-Ala^2,Ala^{15}]-GRF(1-29)-NH$_2$ at a dose of 2.5 μg/kg/day (4 times/day for 35 days) resulted in increased average daily weight gain, improved feed utilization and decreased backfat. A similar performance enhancement was observed in a lactating dairy cow study (22).

Conformational studies of GRF(l-29)-NH$_2$ using 2-dimensional nmr methods and molecular dynamics calculations from Nuclear Overhauser Effects converted into distant constraints (23) were in general agreement with Clore et al. (14). Conformational analysis (2-dimensional nmr) of the more potent analog, [Ala^{15}]-GRF(l-29)-NH$_2$, in 75% methanol and molecular dynamics calculations confirmed the predicted enhancement of secondary structure. A nearly ideal α-helical geometry between residues 4-29 was determined from these studies (23). These observations led to the synthesis of a novel cyclic GRF analog, Cyclo(Asp^8-Lys^{12})[Ala^{15}]-GRF(1-29)-NH$_2$, which was designed to retain the α-helical conformational characteristics of the linear peptide. Cyclo(Asp^8-Lys^{12})-[Ala^{15}]-GRF(l-29)-NH$_2$ was prepared by a solid phase procedure (Figure 4) using a novel benzotriazol-l-yl-oxy-tris(dimethylamino)- phosphonium hexafluoro phosphate (BOP reagent) cyclization reaction (24). Although the cyclic analog was less potent than the linear compound *in vitro* (Table 1), it retained significant bioactivity and induced marked GH release in pigs *in vivo* (25). Two-dimensional nmr and molecular dynamics calculations in 75% methanol confirmed our conclusion that Cyclo(Asp^8-Lys^{12})[Ala^{15}]-GRF (1-29)-NH$_2$ is in a preferred bioactive conformation with a nearly ideal α-helical conformation with enhanced rigidity between residues 2-29 (23).

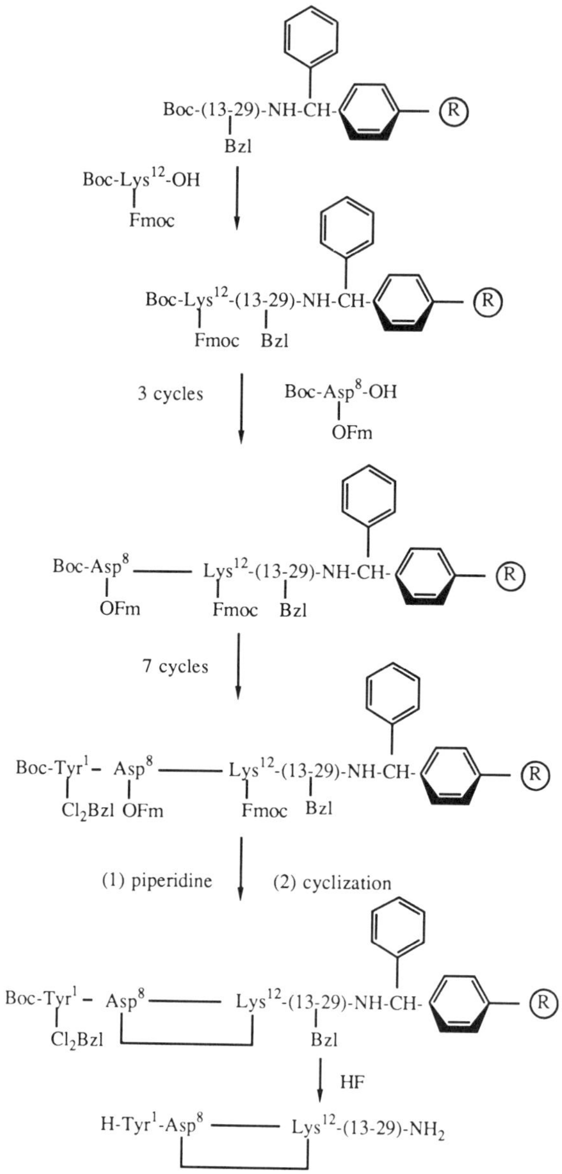

FIGURE 4. (8 →12) CYCLIZATION OF [Ala^15]-GRF(1-29)-NH_2

ACKNOWLEDGMENTS

The authors wish to thank Dr. F. Scheidl and his staff for amino acid analysis, Dr. W. Benz for the mass spectrometry and Dr. Y.-C. Pan for sequence analysis. We also thank Dr. V. Toome for the Circular Dichroism studies and Drs. V. Madison and D. Fry for the molecular dynamics studies. We also acknowledge the assistance of Mr. P. Stricker for the *in vitro* bioassay.

REFERENCES

1. Rivier J, Spiess J, Thorner M, Vale W (1982). Characterization of a growth hormone-releasing factor from a human pancreatic islet tumor. Nature 300:276.
2. Guillemin R, Brazeau P, Böhlen P, Esch F, Ling N, Wehrenberg WB (1984). Growth hormone-releasing factor from a human pancreatic tumor that caused acromegaly. Science 218:583.
3. Gubler U, Monahan JJ, Lomedico PT, Bhatt RS, Collier KJ, Hoffman BJ, Böhlen P, Esch F, Ling N, Zeytin F, Brazeau P, Poonian MS, Gage LP (1983). Cloning and sequence analysis of cDNA for the precursor of human growth hormone-releasing factor, somatocrinin. Proc Natl Acad Sci USA 80:4311.
4. Mayo KE, Vale W, Rivier J, Rosenfeld MG, Evans RM (1983). Expression-cloning and sequence of a cDNA encoding human growth hormone-releasing factor. Nature 306:86.
5. Thorner MO, Spiess J, Vance ML, Rogol AD, Kaiser DL, Webster JD, Rivier J, Borges J, Bloom SR, Cronin MJ, Evans WE, MacLeod RM, Vale W (1983). Human pancreatic growth hormone-releasing factor selectively stimulates growth-hormone secretion in man. Lancet 1:24.
6. Gelato M, Ross J, Pescovitz O, Cassorla F, Skeeda M, Merriam GR (1984). Acceleration of linear growth after repeated doses of growth hormone-releasing hormone. Pediat Res 18:167A (Abstr 430).
7. Thorner MO, Reschke J, Chitwood J, Rogol AD, Furlanetto R, Rivier J, Vale W, Blizzard RM (1985). Acceleration of growth in two children treated with human growth hormone-releasing factor. New Engl J Med 312:4.
8. Peticlerc D, Pelletier G, Lapierre H, Gaudreau P, Couture Y, Dubreuil P, Morisset J, Brazeau P (1987). Dose response of two synthetic human growth hormone-releasing factors on growth hormone release in heifers and pigs. J Anim Sci 65:996.
9. Frohman LA, Downs TR, Williams TC, Heimer EP, Pan Y-C, Felix AM (1986). Rapid enzymatic degradation of growth hormone releasing hormone by plasma *in vitro* and *in vivo* to a biologically inactive product cleaved at the NH_2 terminus. J Clin Invest 78:906.

10. Galyean R, Vale W, Rivier C, Thorner M, Yamamoto G, Rivier J (1987). Human (h) and rat (r) growth hormone releasing factor (GRF). Design of more potent and longer lasting analogs. 10th Am Pept Symp, St Louis, MO, May 1987.
11. Robberecht P, Waelbroeck M, Coy D, De Neef P, Camus J-C, Christophe J (1986). Comparative structural requirements of thirty GRF analogs for interaction with GRF and VIP receptors and coupling to adenylate cyclase in rat adenopituitary, liver and pancreas. Peptides 7: Suppl. 1, pp 53-59.
12. Felix AM, Heimer EP, Mowles TF, Eisenbeis H, Leung P, Lambros T, Ahmad M, Wang C-T (1986). Synthesis and biological activity of novel growth hormone-releasing factor analogs. In Theodoropolous D (ed): "Peptides 1986, Proceedings of the 19th European Peptide Symposium." Berlin: Walter de Gruyter & Co, p. 481.
13. Coy DH, Murphy WA, Sueiras-Diaz J, Coy EJ, Lance VA (1985). Structure-activity studies of the N-terminal region of growth hormone releasing factor. J Med Chem 28:181.
14. Clore CM, Martin SR, Gronenborn AM (1986). Solution structure of human growth hormone releasing factor. J Mol Biol 191:553.
15. Kaiser ET, Kézdy FJ (1984). Amphiphilic secondary structure: design of peptide hormones. Science 223:249.
16. Merrifield RB (1963). Solid phase peptide synthesis I. The synthesis of a tetrapeptide. J Am Chem Soc 85:2149.
17. Heimer EP, Ahmad M, Lambros TJ, Felix AM (1985). Synthesis, purification and characterization of thymosin α_{11}, a new thymic peptide. Int J Pept Prot Res 25:330.
18. Brazeau P, Ling N, Böhlen P, Esch F, Ying S, Guillemin R (1982). Growth hormone releasing factor, somatocrinin, releases pituitary growth hormone in vitro. Proc Natl Acad Sci USA 79:7909.
19. Goud NA, Garsky V, Huelar E, Shakoori F, Syed M, Verlander MS (1987). Large scale synthesis of GRF(1-29) and (1-44) amides. In Theodoropoulos D (ed): "Peptides 1986, Proceedings of the 19th European Peptide Symposium." Berlin: Walter de Gruyter & Co, p. 275.
20. Frohman LA, Downs TR, Heimer EP, Felix AM (1987). Characterization of the primary human growth hormone releasing hormone (GRF) degrading enzyme in plasma as dipeptidylaminopeptidase (DAP) type IV. Presented at: Central Society for Clinical Research, Chicago, Illinois, Oct. 1987. Manuscript in preparation.
21. Etherton, T, Wiggins JP, Chung CS, Walton PE and Mowles TF. Manuscript in preparation.
22. Petitclerc D, Lapierre M, Pelletier G, Gaudreau P, Mowles T, Felix AM, Brazeau P (1987). Effect of a potent analog of GRF on growth hormone releasing and milk production in dairy cows. 82nd Ann Meeting of Amer Dairy Sci Assoc.

23. Madison V, Berkovitch-Yellin Z, Fry D, Greeley D, Toome V (1988). Conformations of three peptides deduced from experiments and molecular energetics. In Tam J, Kaiser ET (eds) "Synthetic Peptides: Approaches to Biological Problems. UCLA Symposia on Molecular and Cellular Biology, New Series, Vol 86.." New York: Alan R. Liss, Inc. (in press).
24. Felix AM, Wang C-T, Heimer EP, Fournier A (1988). Applications of BOP reagent in solid phase synthesis. II. Solid phase side-chain to side-chain cyclization using BOP reagent. Int J. Pept Prot Res 31:231.
25. Campbell RM, Su C-M, Maines SL, Stricker PR, Jensen LR, Heimer EP, Felix AM, Mowles TF (1988). Biological activities of novel cyclic growth hormone-releasing factor analogs. Ann Meeting Amer Soc Anim Sci (in press).

SYNTHESIS AND BIOLOGICAL PROPERTIES OF A HIGH SPECIFIC ACTIVITY RADIOIODINATED, PHOTOLABILE CYCLOSPORINE[1,2]

Roger Tung, Brian Dunlap, Johannes D. Aebi,
William Mellon, Arnold E. Ruoho++,
N. Dhanasekaran++, and Daniel H. Rich*

School of Pharmacy and Department of Pharmacology++
University of Wisconsin-Madison
Madison, WI 53706

ABSTRACT The synthesis of an analog of cyclosporine A (CsA, $\underline{1}$), wherein the [D-Ala]^8 has been replaced with a D-Lys to give [D-Lys^8]-CsA $\underline{2}$ is described. Derivatization of the ε-amine of the D-Lys^8 residue with carrier free ^{125}I labeled AIPPS (N-(3-(p-azido-m-iodophenyl)-propionyl)-succinimide) has been accomplished to yield a high specific activity (2010 Ci/mmol) derivative $\underline{3}$ which strongly inhibits Con-A stimulated thymocyte proliferation (IC$_{50}$ ≅ 50 ng/mL). Upon photolysis, compound $\underline{3}$ has been found to label isolated, purified cyclophilin (Cyp), a cytoplasmic protein shown to specifically bind CsA, and a 19000 MW component of thymocyte cytoplasm.

---

[1]Abstracted in part form the Ph.D. Dissertation of Roger D. Tung (1987), University of Wisconsin-Madison.
[2]Nomenclature and symbols of amino acids and peptides follow the recommendations of the IUPAC-IUB Joint Commission of Biological Nomenclature (Pure Appl. Chem. 1984, 56, 595-624) except that N-methyl amino acids are abbreviated as Me-AA, where AA stands for the usual abbreviation of an amino acid.

*Address correspondence to this author at 425 N. Charter St.

Determination of immunosuppressive activity and competition experiments utilizing photolytic binding of $\underline{3}$ to Cyp in presence of MeAla^6-Cs, C-10-Ene^1-Cs and CsA reveal the Cyp binding of MeAla^6-Cs is considerably greater than C-10-Ene^1-Cs while the immunosuppressive activity of C-10-Ene^1-Cs was greater than that of MeAla^6-Cs. These results are difficult to reconcile with a simple model of Cyp as the primary target for the immunosuppressive activity of CsA and indicate the possibility that other, as yet undetected, binding components exist for mediating the immunosuppressive activity of CsA. We have used a similar competition protocol with affinity label $\underline{3}$ to tentatively identify Cyp as a 19000 MW component labeled in murine thymocyte cytoplasm, which establishes the utility of this protocol for obtaining useful information on putative CsA binding components without purifying such components from crude extracts.

INTRODUCTION

Cyclosporin A (CsA, $\underline{1}$; Figure 1) is an orally active, cyclic undecapeptide which displays potent immunosuppressive and antiparasitic activities.[1] Due to its widespread use in the prevention of organ transplant rejection as well as its considerable therapeutic potential in other areas, CsA is the subject of close study by a number of groups. Nevertheless, many basic questions regarding the etiology of its biological activities remain unanswered. The study of CsA is considerably hampered by the lack of active analogs which might be used as probes of the various functions of CsA. Due to the presence of a unique, highly functionalized amino acid (MeBmt, $\underline{4}$) and extensive N-methylation of the peptide ring, the cyclosporine system has been difficult and time consuming to synthesize.[2] In addition, it is extraordinarily sensitive toward modification of the side chain structures; the addition or deletion of as few as one or two carbons to several of the amino acyl side-chains can produce essentially inactive analogs.

We have recently developed new, simplified chemical methods for synthesizing the major fragments of the cyclosporine ring (the MeBmt $\underline{4}$[3,4]; the 8-11 tetrapeptide[5]; and the 2-7 hexapeptide portions.[6] With these key fragments now readily available to us, we have begun to develop compounds which might be of use in studying the biology of cyclosporines.

The identity of the biologically relevant receptor(s) for CsA is not yet established and multiple receptors are possible.[2,7-11] To attempt to identify CsA receptors, we wanted to synthesize a radioactive, photolabile analog of CsA which might allow photolytic labeling and isolation of CsA receptor(s). To be successful, such an analog must possess both a relatively low dissociation constant to ensure specificity for the receptor(s), and high specific radioactivity to maximize labeling of receptor present in low numbers. In this article we report the synthesis of a new CsA derivative

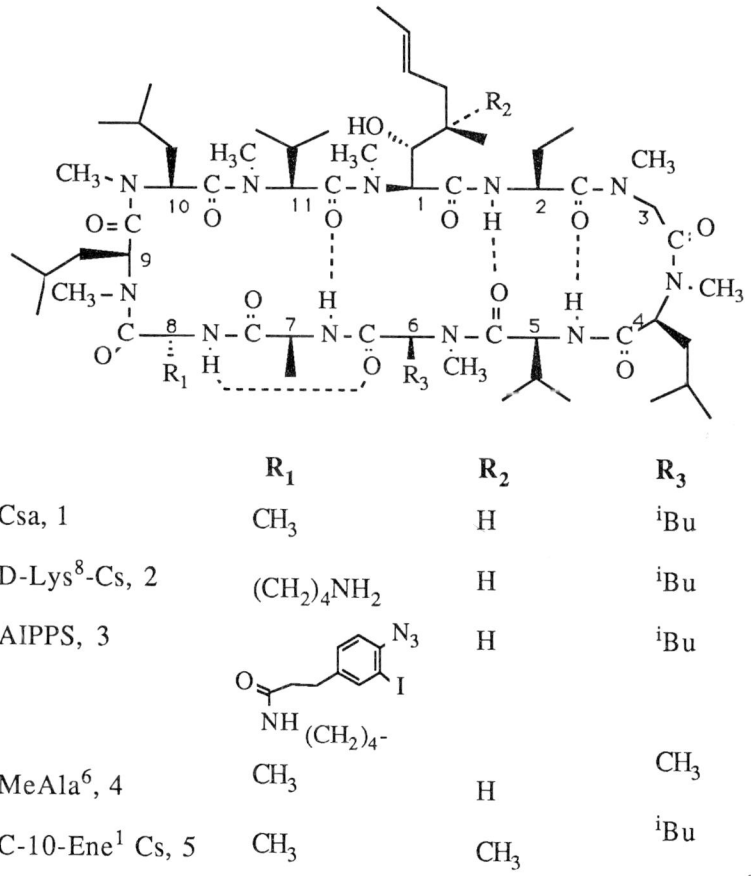

FIGURE 1. Structure of cyclosporin A (1), D-lys8-cyclosporin (2), D-Lys8-AIPPS-cyclosporin (3), MeAla6-Cs (4), and C-10-Ene1-Cs (5).

which meets these requirements and demonstrate its utility as a highly specific probe.

MATERIALS AND METHODS

Cyclosporine A and C-10-Ene[1]-Cs were synthesized by modifications of reported methods.[12,13] AIPPS was synthesized, iodinated with carrier free ^{125}I, and photolyzed by using the method reported by Lowhdes et al.[14] Cyclophilin and MeAla[6]-Cs were obtained as gifts from Dr. J. Boger and Dr. Phil Durette, Merck Sharp & Dohme, Rahway, NJ.

Results (Chemistry).

Due to the number of functional groups involved in the assembly of the cyclosporine ring, considerable attention was necessarily directed towards compatibility of protecting groups. In particular, it was necessary to use a blocking group for the ε-amine of the lysine which would be stable to the basic conditions used for deblocking the C and N terminii of the linear undecapeptide. The presence of the double bond in MeBmt also precluded hydrogenolytically labile protection. Use of the acid labile Boc group for protecting the ε-amino group of lysine required extra synthetic steps relative to a completely orthogonal blocking group due to its removal simultaneously as the tert-butyl ester protecting group is cleaved from residue 11 (Figure 2), and thus its necessary reintroduction. However, these procedures were accomplished in high overall yield (90%) without the need for purification of intermediates, and thus little extra labor was incurred through this route.

The strategy used for assembly of the cyclosporine ring generally followed that previously described by Wenger for the synthesis of cyclosporine itself[15,16] except as modified by us (Figure 2). Notable exceptions were our use of C-terminal tert butyl protection for the 8-11 tetrapeptide[5] which allowed simultaneous deprotection of the N and C terminii in a manner which we have found to be advantageous in other systems.[17]

In general, all steps proceeded well (Figure 3) except for the fragment coupling of the 8-11 tetrapeptide acid with the 1-7 heptapeptide amine. In that

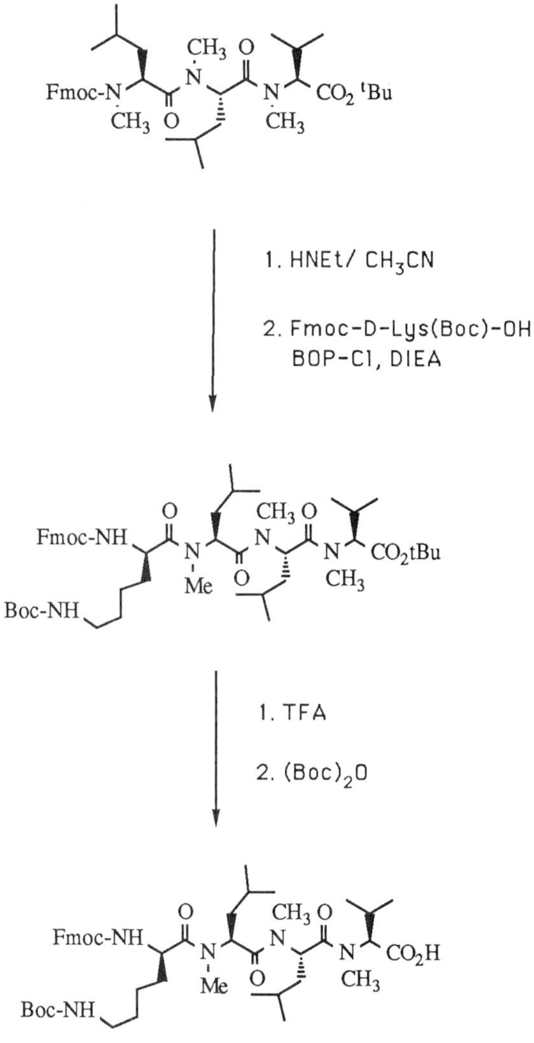

FIGURE 2. Synthesis of D-Lys(Boc)[8]CsA 8-11 Tetrapeptide Derivatives.

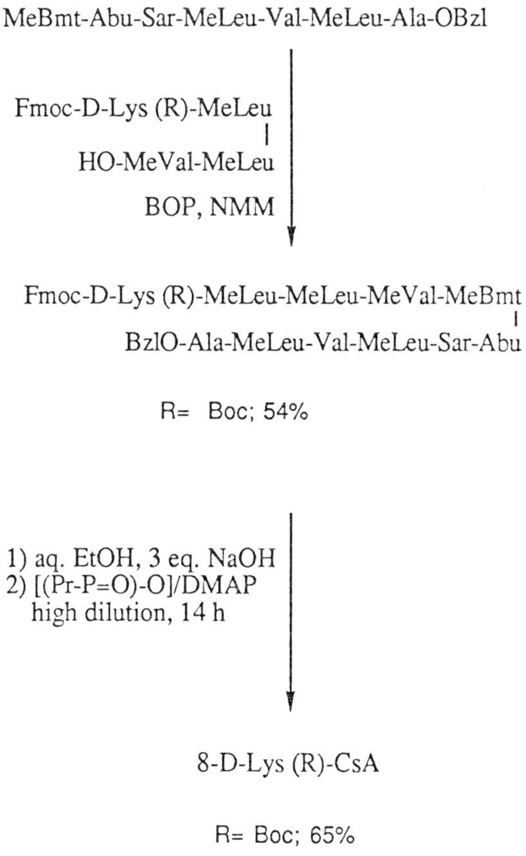

MeBmt-Abu-Sar-MeLeu-Val-MeLeu-Ala-OBzl

Fmoc-D-Lys (R)-MeLeu
　　　|
HO-MeVal-MeLeu
BOP, NMM

Fmoc-D-Lys (R)-MeLeu-MeLeu-MeVal-MeBmt
　　　　　　　　　　　　　　　　　　|
　　　　BzlO-Ala-MeLeu-Val-MeLeu-Sar-Abu

R= Boc; 54%

1) aq. EtOH, 3 eq. NaOH
2) [(Pr-P=O)-O]/DMAP
   high dilution, 14 h

8-D-Lys (R)-CsA

R= Boc; 65%

FIGURE 3. Deprotection and Cyclization of
Protected Undecapeptide to Cyclic D-Lys^8-CsA Analog.

step, a second product was formed in nearly half the amount of the desired compound, with consequent lowering of the reaction yield. Interestingly, separation of the by-product from the desired product was not possible with the chromatography system used by Wenger.[15] Similar results have been obtained in our laboratory in the case of the parent CsA as well as with other analogs; this step remains the greatest problem in the synthesis of the cyclosporines. Nevertheless, the ease with which this analog was constructed bodes well for future synthetic studies on these compounds.

Removal of the Lys-Nε blocking group with strong acid proceeded well, giving a high yield of the free amine derivative $\underline{2}$. Simultaneous introduction of the radiolabeled and photoactivatable moiety was accomplished by reaction of the amine with carrier-free [^{125}I]-AIPPS, in turn synthesized from Na^{125}I with a specific activity of 2010 Ci/mmol to yield $\underline{3}$. This reaction, which proceeded in moderate yield, was also carried out with non-radioactive [^{127}I]-AIPPS to provide cold $\underline{3}$ which was characterized by 500 MHz NMR and by high resolution FAB mass spectrometry.

Biological Results.

The concentration of non-radioactive $\underline{3}$ needed to inhibit Con-A stimulated mitogenesis of 50% [IC$_{50}$] was 50 ng/mL, compared to 5.5 ng/mL for CsA and 350 ng/ml for the parent structure $\underline{2}$. Preliminary measurements of the K$_d$ values for the binding of $\underline{3}$ to whole thymocytes indicate a value of 80 ng/mL (data not shown).

Initial photolabeling experiments were carried out to determine if $\underline{3}$ could be bound to the known CsA binding protein cyclophilin[10,18] and if any specifically photolabeled proteins could be identified in whole thymocytes and cytoplasmic extracts. Thus, we photolysed 10 μg of purified cyclophilin with 2 μCi of radiolabeled $\underline{3}$, and analyzed the products by electrophoresis on a 12% SDS-polyacrylamide gel. The results, Figure 4, lanes 1,4, show the presence of a single radioactive band, whose Mr is slightly higher than the stained band. This is to be expected since the addition of $\underline{3}$ to cyclophilin would increase its MW by 1200 daltons. The cytoplasmic extract from 2x10^7 thymocytes was similarly photolysed with 4 μCi of $\underline{3}$ in the presence (+, lane 3) and absence (-, lane 2) of 10 μM CsA. The products were electro-

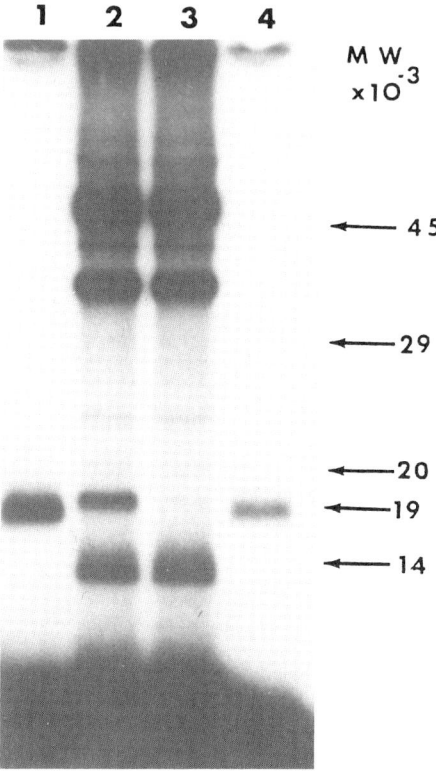

FIGURE 4. Autoradiographic results from purified cyclophilin (lanes 1,4) or thymocyte cytoplasm photolysed in presence of 3 with (lane 3) or without (lane 2) 10 µM CsA and electrophoresed on 12% acrylamide SDS gel. Arrows to right indicate position of stained MW standard proteins on gel.

phoresed in lanes adjacent to the separately prepared labeled cyclophilin, as shown in Figure 4. Comparison of lanes 2 and 3 indicate a protein of Mr 19,000 which is labeled in the absence but not in the presence of CsA.

Whole thymocytes (2×10^7 cells) were incubated with

2 µCi 3 in media with or without 10 µM CsA for 20 min, washed, photolyzed, non-ionic detergent extracted and after removal of insoluble material (e.g., nuclei) the extracts electrophoresed on a 10% polyacrylamide-SDS gel. Comparison of labeled material from cells photolyzed in presence (Fig. 5, lane 2) with that from cells photolyzed in absence (Fig. 5, lane 1) of CsA, indicate several possible specific Cs binding components, but the

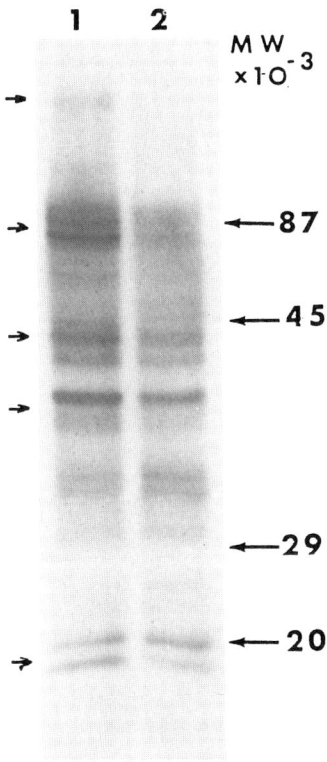

FIGURE 5. Autoradiographic results from whole thymocytes photolysed in presence of 3 with (lane 2) or without (lane 1) 10 µM CsA, and detergent extracts electrophoresed on 10% acrylamide SDS gel. Arrows to right of gel indicate position of stained MW standard proteins on gel. Arrows to left of gel indicate possible specific CsA binding components.

relatively high non-specific labeling precludes any unequivocable conclusions. (Experiments are underway using purified membranes to determine if any of these components are membrane associated.)

We next compared the dose dependency of inhibition of binding of $\underline{3}$ to Cyp by two other analogs of Cs, MeAla^6-Cs ($\underline{4}$),[22] and C-10-Ene^1-Cs ($\underline{5}$).[13] Photolysis of purified Cyp with $\underline{3}$ was carried out in the presence of 3-30 μM inhibitor and was analyzed by autoradiography after electrophoresis on 12% SDS-PAGE. The results (Fig. 6) indicate a dose-dependent inhibition of photo-

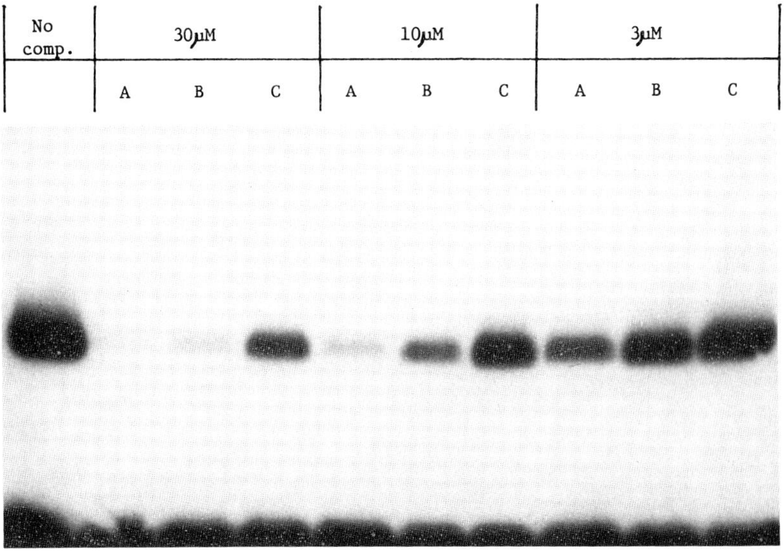

FIGURE 6. Autoradiographic results from purified cyclophilin photolysed with $\underline{3}$ in presence of indicated concentrations of: (A) Cyclosporin A, (B) MeAla^{6}Cs, and (C) C-10-Ene^1-Cs, and electrophoresed on 12% Polyacrylamide/SDS gel.

lytic binding by these two compounds (and CsA). The important point to note is that in comparison to CsA, MeAla^6-Cs $\underline{4}$ is approximately 50% as effective in preventing binding, while C-10-Ene^1-Cs $\underline{5}$ is less than 10% as effective. The immunosuppressive ability (determined by inhibition of Con A/thymocyte mitogenesis) of these analogs relative to CsA shows that $\underline{4}$ is only 1-2% as effective, while $\underline{5}$ is 30-50% as effective.

A protocol identical to the above was used with thymocyte cytoplasmic extracts (2×10^7 cells/point), photolyzing $\underline{3}$ in the presence of 3-30 μM MeAla^6-Cs,

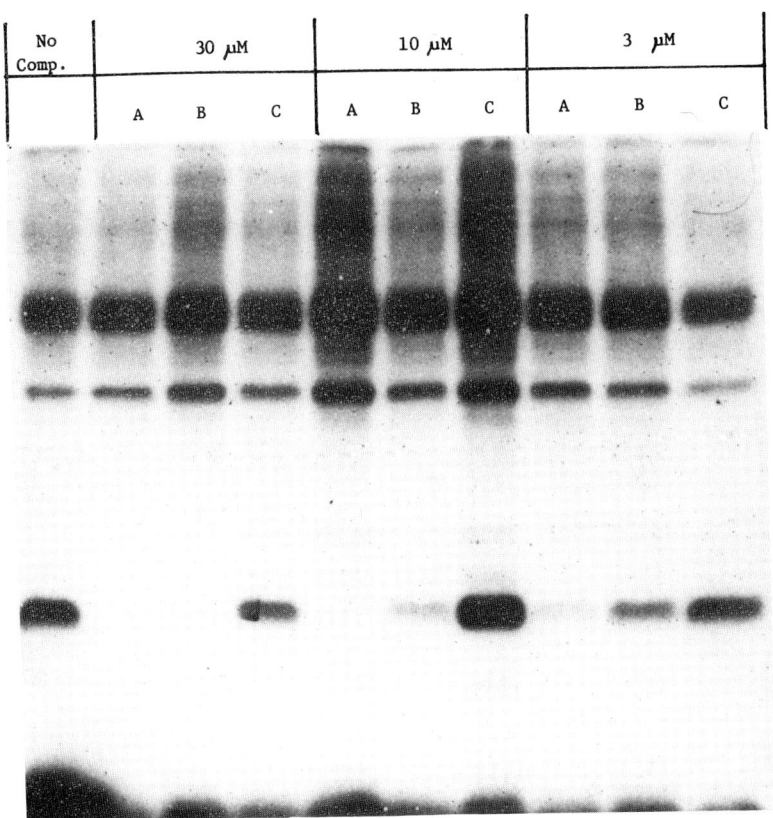

FIGURE 7. Autoradiographic results from thymocyte cytoplasm photolysed with $\underline{3}$ in presence of indicated concentrations of (A) Cyclosporin A, (B) MeAla^6-Cs, (C) C-10-Ene^1-Cs, and electrophoresed on 12% Polyacrylamide/SDS gel. Arrow to right of figure indicates 19000 MW component.

C-10-Ene^1-Cs, or CsA. The autoradiographic results shown in Fig. 7 indicate the pattern of relative blocking of $\underline{3}$'s photolytic attachment to the 19000 MW cytoplasmic component is identical to that in Fig. 6 for Cyp, which leads us to conclude the compound labeled by $\underline{3}$ in thymocyte cytoplasm is cyclophilin.

DISCUSSION

In this work we demonstrated the applicability of our earlier synthetic studies to the construction of new, biologically active cyclosporine derivatives. This is the first account of a CsA analog which possesses a moiety which can be derivatized with a range of substituents while retaining high biological activity. In particular, we have successfully derivatized this compound with a carrier-free photoaffinity ligand, thus providing a biologically active photoaffinity analog of CsA with high specific activity (2000 Ci/mM). While several studies have been reported in which a tritiated photolabile derivative of CsA was used,[19,20,11] the characterization of the derivative was insufficiently precise to allow comparison with our label $\underline{3}$, except to note the use of a different photoaffinity ligand, and the specific activity 623 mCi/mM, which is 3300x lower than $\underline{3}$. The only biological results reported[11] using immune cells (human PBL's) indicate labeling of several low MW compounds in lymphocyte cytoplasm, but no information as to the specificity of labeling.

We have been able to show that our photolabile derivative specifically labels isolated cyclophilin, and that cyclophilin labeling is prevented by addition of CsA. Incubation and photolysis of $\underline{3}$ in thymocyte cytoplasm results in formation of a band on SDS-PAGE which is also protected by CsA, and which co-migrates with authentic $\underline{3}$ labeled cyclophilin.

We[12,13] and others[21,22] have synthesized a number of Cs analogs whose immunosuppressive activity correlates with their Cyp binding activity, leading to speculation that Cyp is the "target" of CsA's immunosuppressive activity. Two recently synthesized analogs, C-10-Ene-Cs[13] and MeAla^6-Cs[22] have proven to be exceptions to this correlation as reported by Durette et al.[22,23] and as shown in this paper. In a simple model of Cyp as the primary target of CsA's immunosuppressive activity, we would expect the relative binding activity of an analog to reflect its relative immunosuppressive activity. Our results and those of Durette et al.[22,23], indicating C-10-Ene^1-Cs is immunosuppressive but binds poorly to Cyp, while MeAla^6-Cs is poorly immunosuppressive but binds well to Cyp, are difficult to reconcile with this simple model, which indicate either a complex mechanism of Cyp mediated immunosuppression or that other, as yet undetected, binding components exist which are the primary targets for mediating CsA's immunosuppressive activity. The identification of the 19000 MW labeled cytoplasmic component of Cyp, using $\underline{3}$ and the above analogs, demonstrates the usefulness of this protocol in obtaining important information on putative Cs binding components without the necessity of isolating such components.

The relative insensitivity of the CsA derivatives that have been functionalized at the D-Lys^8 position to loss of biological activity is consistent with earlier suppositions[2] that the opposite edge of CsA, (including the side-chains of residues 11, 1, 2, and 3) is the most important with respect to receptor recognition. Thus, substituents on the Lys Nε-position may be quite remote from the actual receptor. This may be a reason for the relatively low observed levels of incorporation of the radiolabel upon photolysis of analog $\underline{3}$. Another possible reason is the lipophilic nature of the presumed binding pockets of the cyclophilin. Incorporation of aromatic nitrenes (the initial intermediate formed on photolysis of aromatic azides) is rarely by direct C-H insertion, and more generally, proceeds by attack of a nucleophile on the late formed cyclohepatatetraeneimine intermediates.[24,25] Thus, in the absence of proximal nucleophiles on the receptor, most of the label might be lost to the surrounding structures.

Analog $\underline{3}$ is a highly useful compound due to its high levels of specific radioactivity and its good biological activity. It is clear that there is a considerable potential for further studies involving changes in the length and position of the side-chain bearing photolabile moiety, as well as in the type of photolabile group employed.

ACKNOWLEDGMENTS

This work was supported in part by research grants from

the National Institutes of Health (AR32007) and Merck Sharp and Dohme. We thank J. Boger and P. Durette for samples of cyclophilin and MeAla^6-Cs, and for communicating their cyclophilin binding data for 4 and 5 prior to publication.

REFERENCES

1. Borel JF (1982) In White DG (ed): "Cyclosporine A," Amsterdam: pp 5-17.
2. Wenger RM (1986). Prog Allergy 38:46-64.
3. Aebi JD, Dhaon MK, Rich DH (1987). J Org Chem 52:2881-2886.
4. Tung RD, Rich DH (1987). Tetrahedron 28:1139-1142.
5. Tung RD, Dhaon MK, Rich DH (1986). J Org Chem 51:3350-3354.
6. Tung RD, Sun C-Q, Deyo D, Rich DH. In Marshal GR (ed): "Peptides, Structure and Function: Proc. of the 10th American Peptide Symposium, In Press.
7. Colombani PM, Robb A, Hess AD (1985). Science 228:337-339.
8. Hess AD, Colombani PM (1986). Transplant Proc, 18 Suppl 5:219-237.
9. Hait WN, Harding MW, Handschumacher RE (1986). Science 233:987-988
10. Handschumacher RE, Harding MW, Rice J, Drugge RJ, Speicher DW (1984). Science 226:544-546.
11. Hess AD, Tuszynski T, Engel, P, Colombani, PM, Farrington J, Wenger R, Ryffel B (1986). Transplantation Proceedings 18:861-865.
12. Rich DH, Dhaon MK, Dunlap B, Miller SPF (1986). J Med Chem 29:978-984.

13. Aebi JD, Rich DH (1987). Unpublished observations.
14. Lowndes J, Hokin-Neaverson M, Ruoho AE (1987). Analytical Biochemistry, In Press.
15a. Wenger RM (1983). Helv Chim Acta 66:2672-2702.
15b. Wenger RM (1983). Helv Chim Acta 66:2309-2321.
16. Wenger RM (1984). Helv Chim Acta 67:502-525.
17. Shute RE, Rich DH (1987). Tetrahedron Lett 28:3419-3423.
18. Harding MW, Handschumacher RE, Speicher DW (1986). J Biol Chem 261:8547-8555.
19. Ziegler K, Frimmer M (1986). Biochem Biophys Acta 855:136-142.
20. Frimmer M, Ziegler K (1986). Biochem Biophys Acta 855:143-146.
21. Quesniaux VFJ, Schreier MH, Wenger RM, Hiestand PC, Harding MW, Van Regenmortel MHV (1987). Eur J Immunol 17:1359-1354.
22. Durette PL, Boger J, Dumont F, Firestone R, Frankshun RA, Koprab SL, Lin CS, Melino MR, Pessolano AA, Pisano J, Schmidt JA, Sigal NH, Staruch MJ, Witzel BE (1987). Transplant Proc, In Press.
23. Durette PL et al. unpublished data.
24. Chowdhry V (1979). Ann Rev Biochem 293-325.
25. Fedan JS, Hogaboom GK, O'Donnell JB (1984). Biochem Pharmacol 33:1167-1180.

Index

α/β dimer of human choriogonadotropin, 223–237; *see also* Human choriogonadotropin
Adjuvants, hepatitis B virus vaccine and, 149–150
Affinity chromatography, 57
Alamethicin, 78
Alum, hepatitis B virus vaccine and, 149–150
Amino acid equivalence, 20–21, 27
Amino acid sequences
    feline leukemia virus and, 163
    foot-and-mouth disease virus and, 137
    of Gardner-Arnstein feline leukemia virus, 164, 165
    herpes simplex virus α protein DNA binding and, 249–250
    malarial parasite protein molecule and, 185
Amino-aromatic interactions, 94
Amphibian peptide, bombesin, 295–296, 304–305
Amphipathicity, peptide, 203–204
Amphiphilic helical peptides, 71–85
    characterization and synthesis of, 71–85
    conformational stability in, 73–74
    helix types in, 75–76
    hydration of helix backbones in, 78–84
    parallel packing and, 76–77
    study of
        introduction to, 72
        methods in, 72–73
Amphiphilic mimic, 80–81
Amphiphilic packing mode, 81–82
ANF. *See* Atrial natriuretic factor
Antagonists, gastrin releasing peptide, 295–307
Antamanide, 73
Antibody binding, feline leukemia virus variants and, 159, 166–167
Antibody binding epitope, feline leukemia virus and, 159–170
Antibody binding epitope, malarial parasite protein and, 181–196
Antibody specificity, antigenic determinants and, 19–21
Antigen-antibody interactions, 19–20
Antigenic determinants, 19–30
    amino acid equivalence in, 20–21, 27
    antibody specificity and, 19–21
    cognitive features of continuous, 19–30
    epitope length and, 28
    propensity factors and, 26–29
    replaceability patterns and, 22–25
    residues comprising protein epitopes and, 25–26
    study of
        introduction in, 20–21
        methods and materials in, 21–22
Antigenic domains
    hepatitis B virus envelope and, 211–222
    herpes simplex virus major protein α4 and, 239–254
    human choriogonadotropin topographic analysis and, 223–237
    T cell activation by peptides and, 199–210
Antigenic variation, foot-and-mouth disease virus and, 136
Antigens, processing and presentation of protein. *See* Protein antigen processing and presentation
Anti-peptide antibodies
    crossreactivity of, 171–180
    fibronectin-peptide and, 47–48
Anti-peptide response, foot-and-mouth disease virus and, 135
Anti-peptide sera, polyclonal, foot-and-mouth disease virus and, 171–180
Anti-protein reactivity, malarial parasite protein and, 187–189
Aotus monkey model, malarial parasite protein and, 191

Index

Applications. *See* Synthetic peptides, methods and applications of studies of
α protein, herpes simplex virus and, 239–254; *see also* Herpes simplex virus
Aromatic-aromatic interactions, 93
Artemia, 52–61
Asperlicin, 268
Atrial natriuretic factor, 267–279
  cyclic feature of, 274
  stereo view of bioactive model of ring portion of, 277
Autoradiographic images of [^{32}P]-labeled DNA protein complexes, 245, 246, 247, 248
Autoradiographic results, purified cyclophilin and, 328–332

Backbone of helix, solvation of, 79–84
B-cell epitopes, pre-S sequences of hepatitis B virus envelope protein and, 143–149
Bending of helix, 78–79
Benzodiazepine antagonist of cholecystokinin, 268
Benzotriazol-1-yl-oxy-tris(dimethylamino)-phosphonium hexafluoro phosphate, 315
Beta-galactosidase, foot-and-mouth disease vaccine and, 127, 131, 138
Beta-ribbon helix, 75–76
BHA resin, 65–67
Binding inhibition studies, 298–302
Bioactive conformation model, 267–279
Bioactive conformations of peptide hormones, 255–279
  atrial natriuretic factor and, 267–279
  mimics of secondary structural elements and, 257–266
BioGel P-4 chromatography, 57, 58, 59
Biosynthesis, glycosylation of peptide acceptors by larval brine shrimp microsomes and, 51–61
BNPS-skatole. *See* 2-(2-Nitrophenylsulfenyl)-3-methyl-3-bromoindolenine
Bombesin, 295–296, 304–305
BOP reagent. *See* Benzotriazol-1-yl-oxy-tris(dimethylamino)-phosphonium hexafluoro phosphate
Bovine serum albumin
  gastrin releasing peptide antagonist studies and, 298
  NP-12 conjugated to, 13
Brine shrimp microsomes, 51–61

Calf kidney cells, 128
Carrier-free synthetic peptide antigens, 181–196
cDNA clone, malarial parasite protein molecules and, 181, 182
Chagas' disease, interference with *Trypanosoma cruzi* and, 44
Charge-charge interactions, 89
Charge-dipole interactions, 90–91
Chitin, 51–53
Chitin synthetase, 51
Cholecystokinin, benzodiazepine antagonist of, 268
Choriogonadotropin, human. *See* Human choriogonadotropin
Circular dichroism, 110
Circular dichroism spectra, 103, 104
Class I histocompatibility antigens, 32–34
Class II histocompatibility antigens, 32–34
Class II histocompatibility antigen–T cell receptor complex, 33–34
Cleavage, insulin growth factor I and, 283–294
Competitive inhibition assays, human choriogonadotropin epitope and, 227–229
Conformational studies, growth hormone releasing factor and, 315
Conformation of peptides, 109–123
  in membranes, 97–98
Constrained molecular dynamics, 109–123
Cross reactivity
  amino acid substitutions in foot-and-mouth disease vaccine and, 127
  hepatitis B virus envelope and, 211, 213–221
  subpopulations of anti-peptide antibodies and, 171–180
Crystal structures, 71, 72, 73, 76
  hypothetical multiple channel for ion transport in, 83, 84
CS protein of *Plasmodium falciparum*, 11–13
C-terminal octapeptide of gastrin releasing peptide, 300–305

C terminus of herpes simplex virus α4 protein, 242
Cyanogen bromide, 283, 284, 285
Cyclic encephalins
  biological activity of, 261
  peptide mimics and, 257–266
Cyclization of peptide-resin, 309–319
Cyclophilin, 321–335
Cyclosporine, 321–335
  cyclosporine ring assembly and, 324–327
  deprotection and cyclization of protected undecapeptide and, 326
  immunosuppressive activity of, 322, 332–333
  photolabeling studies and, 327–332
  radioiodinated, 321, 327–332
  structure of, 323
  synthesis of analogs of, 332–333
Cyclosporine ring, 322–327

Dextran 2000, 149
Dioxane, 73
Dipole-dipole interactions, 91–92
Dispersion interactions, 92
Dithiothreitol, malarial parasite protein and, 183
DMSO, selective protein cleavage and, 284, 286, 288
DNA binding protein, herpes simplex virus and, 239–254, see also Herpes simplex virus
Dolichol, 51–52, 60
Domain structure, herpes simplex virus and, 239–254; see also Herpes simplex virus

EDTA
  gastrin releasing peptide antagonist studies and, 298
  malarial parasite protein and, 183
Electrophoresis, autoradiographic images of [^{32}P]-labeled DNA protein complexes and, 245, 246, 247, 248
Electrostatic interactions
  free energy of stabilization and, 94–95
  noncovalent, categories of, 87–88
ELISA. See Enzyme-linked immunoabsorbent assay
Encephalin
  biological activity of, 261
  peptide mimics and, 257–266
Encephalin analogs, 257–266
  biological activity of, 261
Envelope protein, hepatitis B virus. See Hepatitis B virus envelope, protein of, pre-S sequences of
Enzyme-linked immunoabsorbent assay
  efficacy of multiple antigen peptide system and, 13, 14, 15
  hepatitis B virus envelope and, 211, 214, 217, 219
  herpes simplex virus α4 protein and, 243
  human choriogonadotropin epitope identification and, 228–229
  multiple antigen peptide system and, 6
Epitopes, 19–30
  frequency data for residues comprising protein, 25–26
  herpes simplex virus α protein N terminal, 240, 248–249
  identification of, human choriogonadotropin and, 227–229
  length of, 28–29
  mapping of
    herpes simplex virus and, 240–242
    pre-S sequences of hepatitis B virus envelope protein and, 143, 144–146
  replaceability matrix for essential residues in, 24
  T-cell, histocompatibility antigens and, 31–32
Escherichia coli
  foot-and-mouth disease vaccine and, 131–132
  insulin growth factor I synthesis and, 286
Ethylene bridge, peptic mimics and, 262–263
Exogenous peptide acceptors, larval brine shrimp microsomes in glycosylation of, 51–61

Feline leukemia virus, 159–170
  neutralization-resistant variants and, 166–167
  neutralizing epitopes on, 162–165
    characterization of, 165–166
  peptide selection for vaccine for, 161–162

Index

Feline retrovirus groups, 160; *see also* Feline leukemia virus
FeLV. *See* Feline leukemia virus
Fibronectin, *Trypanosoma cruzi* interference and, 43–50
Fibronectin receptor, trypomastigote surface antigen and, 44
Follitropin, 223
Foot-and-mouth disease vaccines, 127–142
    conformation of peptide sequence on virus particle in, 130–131
    immunogenic sites on VP1 in, 130
    overcoming antigenic variation in, 136
    peptides as practical option for, 136–139
    presentation of peptide and, 131–132
    structure of virus and, 129
    T cells in immune response to peptide in, 132–135
    viral immunogenic activity and, 129
Foot-and-mouth disease virus, 171–180
Formic acid, insulin growth factor I synthesis and, 285
Fragment condensation, luteinizing hormone–releasing hormone preparation and, 63–67
Free β subunit of human choriogonadotropin, 223–237
    antigenic determinants associated to. *See* Human choriogonadotropin
FTLV, 160

β-Galactosidase, foot-and-mouth disease vaccine and, 127, 131, 138
Gamma turn, 257–258
    comparison of gamma turn mimics with, 261–264
Gamma turn mimics, 257, 258–259, 260, 261–264
Gardner-Arnstein feline leukemia virus, 159, 161–162, 163
    amino acid sequence of, 164, 165
Gastrin releasing deletion peptide, potency of, 301
Gastrin releasing peptide antagonist, 295–307
    binding inhibition studies and, 298–302
    mitogenesis studies and, 298, 302
Gastrin releasing peptide receptors, 295
Gel permeation chromatography, 58

Glasgow-1 feline leukemia virus, 159, 163, 165
Glucagon, atrial natriuretic factor and, 268
Glutaraldehyde
    hepatitis B virus vaccine and, 149–150
    malarial parasite protein and, 183, 186
Glycopeptides, gel permeation chromatography and, 60
Glycoprotein hormones, quaternary structure of, 223, 224; *see also* Human choriogonadotropin
Glycosylation
    of exogenous peptide acceptors, 51–61
    malarial parasite protein molecules and, 182
    pre-S sequences of hepatitis B virus envelope protein and, 143–144, 150–151
    properties of, 54–56
GRF. *See* Growth hormone releasing factor
Growth factor and peptide hormones, 281–335
    gastrin releasing peptide antagonist and, 295–307
    growth hormone–releasing factor and, 309–319
    high specific activity radioiodinated, 321–335
    insulin-like, 283–294
    synthesis of, 283–294
Growth hormone releasing factor, 109–123, 309–319
    conformational studies of, 315
    high performance liquid chromatography and, 311, 313–314
    solid-phase peptide synthesis of, 311, 312
    structure-function relationships and, 310–311
Growth hormone releasing factor analogs, 119, 120
Guanidine hydrochloride, malarial parasite protein and, 183

Hamster kidney cell line BHK21, 128
HBIg. *See* Hepatitis B immune globulin
hCG. *See* Human choriogonadotropin
Head-to-tail hydrogen bonds, 74, 75
Heat shock proteins, 181–196
Helical dipole, peptide analogs and, 204–207

Helices
    backbone of, solvation of, 79–84
    bending of, 78–79
    content of, 118
    hepatitis B virus envelope and, 211, 215, 216
    parallel packing of, 76–77
    types of, 75–76
α-Helix backbone, insertion of water molecule into, 71–85; *see also* Amphiphilic helical peptides
Helper T cell receptors, foot-and-mouth disease virus and, 138
Hepadnaviruses, hepatitis B virus envelope and, 212, 219
Hepatitis B core particle, foot-and-mouth disease virus and, 138
Hepatitis B core protein, foot-and-mouth disease vaccine and, 127
Hepatitis B envelope proteins
    fine analysis of anti-HBs antibodies with ELISA and, 217
    group *a* antigen of hepatitis B surface antigen and, 217–219
    predicted secondary structure of, 214–216
Hepatitis B immune globulin, 212, 213–214, 219
Hepatitis B surface antigen, 211–222
    group *a* antigen of, 217–219
    model of structure of, 211, 214–221
Hepatitis B vaccines, 143–158, 213–214; *see also* Hepatitis B virus envelope, protein of, pre-S sequences of
Hepatitis B virus, 143–158, 211–222; *see also* Hepatitis B virus envelope, protein of, pre-S sequences of
Hepatitis B virus envelope, 211–222
    protein of, pre-S sequences of, 143–158
        determinants essential for antigenicity and immunogenicity in, 144–146
        glycosylation and, 150–151
        hybrid immunogens of recombinant S-protein and, 152–154
        hybrid peptides derived from nonadjacent regions in, 146–149
        interdomain interactions and, 151–152
        safety of hepatitis B vaccines and, 154–156

        synthetic peptides linked to nonimmunogenic carriers and, 149–150
Herpes simplex virus, 239–254
    α4 protein and DNA binding domains of, 250–252
    DNA binding of
        amino acid sequences of peptides 19 and 26–28 in, 249–250
        C-terminus of α4 and, 242
        peptides 19 and 27 in α9, 245–246
        peptides 19 and 27 specificity in, 246–248
        synthetic peptides affecting α4, 242–245
    mapping of epitopes by monoclonal antibody recognition and, 240–242
hGRF. *See* Human growth hormone releasing factor
High performance liquid chromatography, growth hormone releasing factor and, 311, 313–314
Histocompatibility antigens, 32–34
    mechanisms of presentation of, 36–38
    T-cell epitopes and, 31–32
Histocompatibility molecules, T-cell receptors and, 36–38
HLA-DR2, peptide analogs and, 200, 207
^{3}H-peptide acceptors
    Artemia microsome glycosylation and, 54
    preparation of, 53
HPLC chromatography, 67
Human choriogonadotropin, 223–237
    and antigenic determinant
        specifically associated to, 230–232
        specifically associated to hCG dimer, 229
    monoclonal antibody characteristics and, 227
    synthetic peptides and peptide-carrier conjugates of, 225–229
Human growth hormone releasing factor, 309–319
Human immunodeficiency virus, lentivirus in cats and, 160
Hybrid immunogens of recombinant S-protein, 152–154
Hybrid peptides, pre-S sequences of hepatitis B virus envelope protein and, 146–149

Hydrogen bonds, 89–90
Hydrogen bromide, selective protein cleavage and, 284, 288
Hydropathic analysis, herpes simplex virus α4 protein and, 244
Hydrophobic moments, peptide, 204, 205, 206

Ia antigens, foot-and-mouth disease virus and, 138
Ia molecules, 32–34
Ia-peptide antigen-T cell receptor complex, 33–34
^{125}I-labeled fibronectin, 46
^{125}I-labeled monoclonal antibody, human choriogonadotropin and, 228–229
Immunization procedure, multiple antigen peptide system and, 6
Immunoassays, human choriogonadotropin epitope and, 227–229
Immunogens, synthetic, *Trypanosoma cruzi* and, 43–50
Immunosuppressive activity of cyclosporine, 322, 332–333
Insulin, atrial natriuretic factor and, 268
Insulin growth factor I, 283–294
Interdomain interactions in HBV envelope proteins, 151–152
Ion pairs, 88
Ion transport, hypothetical multiple channel for, 83, 84
IRMA. *See* Solid phase immunodiometric assay
Iron superoxide dismutase, weakly polar interactions and, 93
Isopropanol, 73

Keyhole limpet hemocyanin
  feline leukemia virus and, 159
  malarial parasite protein and, 183

Lambda B1 feline leukemia virus, 163
Larval brine shrimp microsomes, glycosylation of peptide acceptors by, 51–61
Lentivirus in cats, 160
Leprosy, peptide analogs and, 200
LH-RH. *See* Luteinizing hormone-releasing hormone

London dispersion interactions, 92
Luteinizing hormone-releasing hormone, synthesis of, 63–67
Lutropin, 223, 224

Major histocompatibility complex, peptide-binding receptors and, 32–35
Malaria, 181–196; *see also* Malarial parasite protein
Malarial parasite, multiple antigen peptide system and, 9
Malarial parasite protein, 181–196
  carboxyl terminal portion of, 192, 193
MAP. *See* Multiple antigen peptide system
MAP-NP-12 as model, 11–13
MBS. *See* M-maleimidobenzoylsulfosuccinimide-ester
Melittin, 78
Methanol, aqueous
  growth hormone releasing factor conformers in, 118–123
  vasoactive intestinal peptide conformers in, 114–118
Methionine, as cleavage target for insulin growth factor synthesis, 283–292
Methionine sulfoxide, reduction of, 288–292
Methods and applications. *See* Synthetic peptides, methods and applications of studies of
Methoxamine contraction of aorta tissue, 267, 272, 273
MHC. *See* Major histocompatibility complex
Mimics, peptide. *See* Peptide mimics
M-IRMAs. *See* Monoclonal immunoradiometric assays
Mitogenesis studies, gastrin releasing peptide antagonist and, 298, 302
M-maleimidobenzoylsulfosuccinimide-ester, hepatitis B virus vaccine and, 149–150
Molecular dynamics, constrained, 109–123
Molecular energetics, 109–123
Monkey model, malarial parasite protein and, 182, 187, 189–190, 191, 192, 195
Monoclonal antibodies
  feline leukemia virus and, 159, 167
  foot-and-mouth disease virus and, 172–179
  herpes simplex virus and, 239–254; *see also* Herpes simplex virus

human choriogonadotropin antigenic determinants and, 224–237; *see also* Human choriogonadotropin
multiple antigen peptide system for, 11
Monoclonal immunoradiometric assays, human choriogonadotropin epitope and, 228–229
Multiple antigen peptide system, 3–18
assays in, 6–7
comparison of, and monomer in ELISA, 13–15
conjugation and, 16
immunization procedure in, 6
increased sensitivity and, 16
introduction to, 4–5
methods in, 5–7
for monoclonal antibodies, 11
octameric branching, immunologic responses of, 12
for polyclonal antibodies, 10
protein carrier and, 16
prototype of, 3
results in, 7–16
schematic of, 6
in solid-phase immunoassays, 11–13
synthesis of, 5–6, 7–8
technical comments on, 8–10
synthetic scheme for synthesis of, by stepwise solid phase method, 9
Mycobacterium leprae, peptide analogs and, 200

N-acetylated C-terminal octapeptide of gastrin releasing peptide, 300–305
N-chlorosuccinimide, tryptophan cleavage and, 284–285
Neutralization-resistant variants, feline leukemia virus variants and, 159, 166–167
Neutralizing antibody response
feline leukemia virus and, 159–170
foot-and-mouth disease virus and, 134, 135, 177
Neutralizing epitopes on feline leukemia virus, 159, 162–166
2-(2-Nitrophenylsulfenyl)-3-methyl-3-bromoindolenine, 284, 286–288
NOE. *See* Nuclear Overhauser effects
NP-12, suitability of, as model, 11–13

N terminal domain of herpes simplex virus $\alpha 4$ protein, 240, 248–249
Nuclear magnetic resonance, 110
Nuclear Overhauser effects, 110, 110–123
for GRF and VIP analogs, 117
growth hormone releasing factor and, 315

Oil-containing adjuvants, hepatitis B virus vaccine and, 149
Oligosaccharide-lipid pool, labeling of, 58
Oligosaccharides, 51–53, 56–60
Oligosaccharyl transferase activity, divalent cation effect on, 56
Oligosaccharyl transferase assay, 53
Ova peptide, Ia-molecule interaction with, 34
Owl monkey model, malarial parasite protein and, 182, 183, 189–190, 192
Oxygen-aromatic interactions, 93

Paper chromatography, glycosylation reaction and, 53–56
Parallel packing of helices, 76–77
Peptide amphipathicity, 203–204
Peptide analogs, 199–210
comparison of, with alpha helical hydrophobic moments and helical potential, 204–207
construction and testing of, 200–203
hepatitis B virus envelope and, 211, 213
substitutions of individual residues in, 201
T cell activation and, 199–210
epitopes containing residues in pattern in, 207
2F10 control response in, 202
helical potential and, 204–207
HLA-DR2 restricted epitopes and, 207
peptide amphipathicity and, 203–204
Peptide bond cleavage, selective, 283–292
Peptide-carrier conjugates, human choriogonadotropin and, 225–229
Peptide conformation, 109–123
Peptide deletion analogs, T cell activation and, 201–203
Peptide hormones
bioactive conformations of, 255–279
atrial natriuretic factor and, 267–279

mimics of secondary structural elements and, 257-266
growth factors and
cyclosporine in, 321-335
gastrin releasing peptide antagonist and, 295-307
growth hormone-releasing factor and, 309-319
high specific activity radioiodinated, photolabile cyclosporine in, 321-335
minimal ligand analysis and, 295-307
recombinant DNA techniques and, 283-294
selective chemical cleavage at tryptophan and, 283-294
synthesis of insulin-like, 283-294
Peptide hydrophobic moments, 204, 205, 206
Peptide mimetics, atrial natriuretic factor and, 267-279
Peptide mimicking processed protein fragments, synthesis of, 36-38
Peptide mimics, 257-266
Peptide presentation, foot-and-mouth disease virus particle and, 131-132
Peptide and protein structure prediction
conformations of peptides and, 109-123
experiments and molecular energetics and, 109-123
modes of hydration in, 71-85
molecular energetics and, 109-123
protein/lipid interactions and, 97-107
protein polar interactions and, 87-96
water molecule in α-helix backbone in, 71-85
weakly polar interactions in proteins and, 87-96
x-ray structure analysis in, 71-85
Peptides, gastrin releasing, 295-307
Peptide secondary structure
hepatitis B envelope proteins and, 214-216
predicted, 199-210
Peptide selection, feline leukemia virus vaccine and, 161-162
Peptide sequence, 110
on foot-and-mouth disease virus particle, 130-131

Phosphate groups, charge-dipole interactions and, 91
Photochemical cleavage of protected peptides on solid phase supports, 63-67
Photolabile cyclosporine, synthesis and biological properties of, 321-335; see also Cyclosporine
Picornaviridae, foot-and-mouth disease and, 129
Pig kidney cells, foot-and-mouth disease vaccine and, 128
Plasmodium falciparum, 181-196
efficacy of multiple antigen peptide system and, 9
Polar channel formation, 82-84
Polar interactions
charge-dipole, 90-91
dipole-dipole interactions and, 91-92
Polar interactions, weakly. See Weakly polar interactions
Polyclonal antibodies, multiple antigen peptide system for, 10
Polyclonal anti-peptide sera, foot-and-mouth disease virus and, 171-180
Pre-S antibodies of hepatitis B virus, passive administration of, 221
Pre-S proteins, hepatitis B virus envelope and, 212-221
Pre-S sequences
of hepatitis B virus envelope protein. See Hepatitis B virus envelope protein, pre-S sequences of
recombinant S-protein with attached synthetic peptides from, 152-154
Propensity factors, antigenic determinants and, 26-29
Pro residues
bending of helix by, 78
curving of helix by, 71
Protective antibodies, pre-S sequences of hepatitis B virus envelope protein and, 143-144
Protein antigen processing and presentation, 31-42
histocompatibility molecules and, 36-38
introduction to, 31-34
location of T-cell antigenic sites and, 35-36

synthesis of peptide mimicking processed protein fragments and, 36–38
synthetic peptides for studying, 38–39
T-cell receptors and, 36–38
Protein cleavage, selective, 283–292
Protein fragments, peptide mimicking processed, 36–38
Protein/lipid interactions, 97–107
composition of transmembrane regions of receptors and, 98–101
induced conformation characteristic of peptide rather than lipid type, 105
membrane binding and penetration of peptides in, 102–105
Proteins
mimics of secondary structural elements of, 257–266
packing of atoms in, 87–88
T-cell antigenic sites in, location of, 35–39

Quadrupolar interactions, 92–93

Rabbit antibodies, hepatitis B envelope proteins and, 213–221
Rabbit, atrial natriuretic factor and aorta of, 267–277
Rabbit model, malarial parasite protein and, 182, 183–184, 186–187, 190–191, 195
Radioimmunoassays, human choriogonadotropin epitope identification and, 228–229
Radioiodinated cyclosporine, 321, 327–332
Radiolabelled chicken ovalbumin peptide, Ia-molecule interaction with, 34
Radioligand studies, gastrin releasing peptide antagonist and, 298–302
Receptor proteins, 97–107
Receptors, gastrin releasing peptide, 295
Recombinant DNA, pre-S sequences of hepatitis B virus envelope protein and, 143, 152–154
Recombinant S-protein with attached synthetic peptides from pre-S sequence, 152–154
Reduction of methionine sulfoxide, 288–292
Replaceability matrix for essential residues in epitopes, 24
Replaceability patterns

antigenic determinants and, 22–25
frequency of occurrence of residues with different, 23
16-Residue peptide, 72, 74, 75
Residues comprising protein epitopes, frequency data for, 25–26
Resistant variant, feline leukemia virus and, 159, 166–167
Retroviruses, feline leukemia virus and, 160; see also Feline leukemia virus
Reverse turn mimics, 257–258
Ricard subtype B feline leukemia virus, 163

Safety of vaccine, 143, 154–156
Sarma-C feline leukemia virus, 159, 163, 165
Shrimp microsomes, glycosylation and, 51–61
Signal transduction, 97–107
Snyder-Theilen feline leukemia virus, 159, 163
Sodium channel model, 112–113
Solid-phase immunoassays, multiple antigen peptide system in, 11–13
Solid-phase immunodiometric assay, multiple antigen peptide system and, 7
Solid-phase peptide synthesis, growth hormone releasing factor and, 311, 312
Solid-phase supports, photochemical cleavage of protected peptides on, 63–67
Solvation of helix backbone, 79–84
Somatomedin C, 285; see also Insulin growth factor I
Somatostatin, atrial natriuretic factor and, 267–268
SPDP. See N-Succinimidyl-3-(2-pyridyldithio) propionate
S proteins, hepatitis B virus envelope and, 212–221
Streptomyces griseus, chitinase from, 53–54
Substance P, 102–104
N-Succinimidyl-3-(2-pyridyldithio) propionate, 149
Sulfhydryl-reactive groups, hepatitis B virus vaccine and, 149
Surface antigen, Trypomastigote-specific, 43
Swiss 3T3 fibroblasts, thymidine uptake assays in, 295, 298, 302
Synthetic immunogens, *Trypanosoma cruzi* and, 43–50

Index

Synthetic peptide antigens, malarial parasite protein and, 181–196
Synthetic peptide-based vaccines, 125–196
   feline leukemia virus and, 159–170
   foot-and-mouth disease, 127–142
      virus subpopulations in polyclonal anti-peptide sera and, 171–180
   malarial parasite protein and, 181–196
   pre-S sequence of hepatitis B virus envelope protein and, 143–158
Synthetic peptides
   herpes simplex virus α4 protein and, 242–245
   human choriogonadotropin and, 225–229
   methods and applications of studies of, 3–67
      continuous antigenic determinants in, 19–30
      design for synthetic peptide vaccine and immunoassay in, 3–18
      glycosylation of exogenous peptide acceptors in, 51–61
      host cell penetration by *Trypanosoma cruzi* in, 43–50
      larval brine shrimp microsomes in, 51–61
      LH-RH preparation by fragment condensation in, 63–67
      mechanisms of protein antigen processing and presentation in, 19–30
      multiple antigen peptide system and, 3–18
      photochemical approaches to LH-RH preparation by fragment condensation in, 63–67
      strategies for preparing, residues interacting with histocompatibility molecule and, 37
   T-cell receptors and, 36–38
   Trypanosoma cruzi host cell penetration and, 43–50
Synthetic peptide vaccine and immunoassay design, 3–18; *see also* Multiple antigen peptide system
   introduction to, 4–5
   methods in, 5–7
   results in, 7–16

T1c-cell receptors, 33
   peptide strategy for, 37–38
T1H-cell receptors, 33
   peptide strategy for, 37–38
3T3 fibroblasts, infectivity of, 46
T-cell antigenic sites in proteins, location of, 35–39
T-cell epitopes
   histocompatibility antigens and, 31–32
   peptide analogs and, 199–210; *see also* Peptide analogs
   pre-S sequences of hepatitis B virus envelope protein and, 143–149
T-cell receptors, native and denatured proteins and, 36–38
T cells
   foot-and-mouth disease vaccine and, 131, 132–135
   foot-and-mouth disease virus and, 138
Thymocytes, cyclosporine and proliferation of, 321, 332–333
Thyrotropin, 223, 224
Tifluadom, 268
Tongue epithelium fragments, foot-and-mouth disease vaccine and, 128
Transferase, oligosaccharyl, 51–61
Transforming growth factor type α, efficacy of multiple antigen peptide system and, 13–15
Transmembrane segments of membrane proteins, 97–107
Tris hydrochloride, malarial parasite protein and, 183
Trypanosoma cruzi, host cell penetration by, 43–50
Trypomastigote-specific surface antigen, 43
Trypsin, foot-and-mouth disease vaccine and, 129
Tryptophan, selective cleavage at, insulin growth factor I and, 283–294

Vaccines
   feline leukemia virus, 159–170; *see also* Feline leukemia virus
   foot-and-mouth disease. *See* Foot-and-mouth disease vaccines
   hepatitis B, 143–158; *see also* Hepatitis B virus envelope, protein of, pre-S sequences of

malarial parasite protein and, 181–196
safety of, pre-S sequences of hepatitis B virus envelope protein and, 143, 154–156
Vasoactive intestinal peptide, 109–123
ordered structural segments for, 116
Virus-neutralizing antibodies, pre-S sequences of hepatitis B virus envelope protein and, 143–144
VP1, VP3, foot-and-mouth disease vaccine and, 129, 130, 172–173
amino acid sequences and, 137
VP4, foot-and-mouth disease vaccine and, 129

Water molecule, insertion of, into α-helix backbone, 71–85; *see also* Amphiphilic helical peptides
Weakly polar interactions in proteins, 87–96
charge-charge interactions in, 89
charge-dipole and, 90–91
dipole-dipole interactions and, 91–92
dispersion interactions and, 92
hydrogen bonds and, 89–90
ion pairs and, 88
London interactions and, 92
Wheatgerm agglutinin-agarose
affinity chromatography on, 54
glycosylation reaction and, 57
Woodchuck hepatitis virus, 219

X-ray crystallography, peptide mimics and, 257–266
X-ray structure analysis of water molecule insertion in α-helix backbone, 71–85; *see also* Amphiphilic helical peptides

Zervamicin IIA, 72, 73, 74

QP 552 .P4 S96 1988

Synthetic peptides

APR 2 4 1989